Recent Advances in Analytical Techniques

(Volume 3)

Edited by

Atta-ur-Rahman, *FRS*

Honorary Life Fellow, Kings College, University of Cambridge, Cambridge, UK

&

Sibel A. Ozkan

Faculty of Pharmacy, Ankara University, Turkey

General:

1. Any dispute or claim arising out of or in connection with this License Agreement or the Work (including non-contractual disputes or claims) will be governed by and construed in accordance with the laws of the U.A.E. as applied in the Emirate of Dubai. Each party agrees that the courts of the Emirate of Dubai shall have exclusive jurisdiction to settle any dispute or claim arising out of or in connection with this License Agreement or the Work (including non-contractual disputes or claims).
2. Your rights under this License Agreement will automatically terminate without notice and without the need for a court order if at any point you breach any terms of this License Agreement. In no event will any delay or failure by Bentham Science Publishers in enforcing your compliance with this License Agreement constitute a waiver of any of its rights.
3. You acknowledge that you have read this License Agreement, and agree to be bound by its terms and conditions. To the extent that any other terms and conditions presented on any website of Bentham Science Publishers conflict with, or are inconsistent with, the terms and conditions set out in this License Agreement, you acknowledge that the terms and conditions set out in this License Agreement shall prevail.

Bentham Science Publishers Ltd.
Executive Suite Y - 2
PO Box 7917, Saif Zone
Sharjah, U.A.E.
Email: subscriptions@benthamscience.org

BENTHAM SCIENCE

CONTENTS

PREFACE

Recent Advances in Analytical Techniques, Vol. 3 presents comprehensive chapters on recent trends related to ionic liquids, chiral analysis, multi-parameter flow cytometry, electrospun nanofibers, neutron activation analysis, enantioselective chromatography, Ru(II)-polypyridyl complexes.

We hope that the readers will greatly enjoy reading these excellent articles from eminent scientists in the field. We would like to thanks all the authors that contributed to this e-book for their excellent contributions.

We would like to thank the Bentham staff including Ms. Mariam Mehdi (Assistant Manager Publications), Mr. Shehzad Naqvi (Editorial Manager Publications), and Mr. Mahmood Alam (Director Publications) at Bentham Science Publishers for their hard efforts and efficient interactions with the authors in the publication process.

Atta-ur-Rahman, *FRS*
Honorary Life Fellow
Kings College
University of Cambridge
UK

&

Sibel A. Ozkan
Faculty of Pharmacy
Department of Analytical Chemistry
Ankara University
06560 Yenimahalle/Ankara
Turkey

List of Contributors

Alberto Reis	Laboratório Nacional de Energia e Geologia, Unidade de Bioenergia, Lisboa, Portugal
Ali Fouad	Chirality Program, Faculty of Science and Technology, University of Canberra, ACT, Australia
Anupam Ghorai	Department of Chemistry, Guru Ghasidas Vishwavidyalaya, Bilaspur (C.G), India
Amit Kumar Manna	Department of Chemistry, Guru Ghasidas Vishwavidyalaya, Bilaspur (C.G), India
Ashraf Ghanem	Chirality Program, Faculty of Science and Technology, University of Canberra, ACT, Australia
Betul Unal	Faculty of Science, Biochemistry Department, Ege University, Bornova-Izmir, Turkey
Casimiro S. Munita	Nuclear and Energy Research Institute, IPEN-CNEN/SP, , SãoPaulo, SP, Brazil
Chexu Wang	Chirality Program, Faculty of Science and Technology, University of Canberra, ACT, Australia
Dilek Odaci Demirkol	Faculty of Science, Biochemistry Department, Ege University, Bornova-Izmir, Turkey
Fatma Ozturk Kirbay	Faculty of Science, Biochemistry Department, Ege University, Bornova-Izmir, Turkey
Goutam Kumar Patra	Department of Chemistry, Guru Ghasidas Vishwavidyalaya, Bilaspur (C.G), India
Irem Yezer	Faculty of Science, Biochemistry Department, Ege University, Bornova-Izmir, Turkey
I. Nowak	Faculty of Chemistry, Adam Mickiewicz University, Poznań, Poland
I. Rykowska	Faculty of Chemistry, Adam Mickiewicz University, Poznań, Poland
Lyanne Valdez	Forensic Science Program, Department of Chemistry and Physical Sciences, Pace University, New York, NY, USA
Michael D. Glascock	Research Reactor Facility, University of Missouri, Columbia, MO, USA
Michelle Dumit	Forensic Science Program, Department of Chemistry and Physical Sciences, Pace University, New York, NY, USA
Roberto Hazenfratz	Nuclear and Energy Research Institute, IPEN-CNEN/SP, , SãoPaulo, SP, Brazil
Suna Timur	Faculty of Science, Biochemistry Department, Ege University, Bornova-Izmir, Turkey
Teresa Lopes da Silva	Laboratório Nacional de Energia e Geologia, Unidade de Bioenergia, Lisboa, Portugal
W. Wasiak	Faculty of Chemistry, Adam Mickiewicz University, Poznań, Poland
William Maher	Chirality Program, Faculty of Science and Technology, University of Canberra, ACT, Australia

Zhaohua Dai   Forensic Science Program, Department of Chemistry and Physical Sciences, Pace University, New York, NY, USA

CHAPTER 1

Chiral Analysis of Methamphetamine and Related Controlled Substances in Forensic Science

Zhaohua Dai[*], **Lyanne Valdez** and **Michelle Dumit**

Forensic Science Program, Department of Chemistry and Physical Sciences, Pace University, 1 Pace Plaza, New York, NY 10038, USA

Abstract: The chiral analysis of some seized drugs is important for the administration of justice. The chapter reviews the current analytical methods that have been developed and employed in academic, industrial and forensic laboratories, for the enantiomeric identification and quantitation of some controlled substances, with the focus on methamphetamine and related compounds. Although some spectroscopic methods are discussed, separation techniques, including gas chromatography, high performance liquid chromatography, supercritical fluid chromatography and capillary electrophoresis, especially those coupled with mass spectroscopy, are examined in detail. Some important terms and mechanisms associated with chiral analysis are discussed.

Keywords: Chiral, Controlled substance, enantiomer, diastereomer, Gas chromatography, High performance liquid chromatography, Supercritical fluid chromatograph, Capillary electrophoresis, Optical rotation, IR, NMR.

INTRODUCTION

Drug analysis is an important field in forensic science [1 - 5]. Such analysis follows a typical forensic approach: narrowing down possibilities until arriving at a positive identification [6]. Initially, observations are done to describe the contents upon accepting the physical evidence, which is followed by a color/presumptive test performed on a small portion of the evidence to determine the category of the substance. Finally, the evidence is subject to instrumental analysis for a definitive identification. When the evidence is not pure, gas chromatography (GC), high performance liquid chromatography (HPLC) [7], and/or capillary electrophoresis (CE) [8], often coupled with mass spectrometry

---

[*] **Corresponding author Zhaohua Dai:** Forensic Science Program, Department of Chemistry and Physical Sciences, Pace University, 1 Pace Plaza, New York, NY 10038, USA; Tel: (212)346-1760; Fax: (212)346-1256; E-mail: zdai@pace.edu

Atta-ur-Rahman & Sibel A. Ozkan (Eds.)

[9], are often used to give insight into the concentration of each component [10 - 17].

Color tests are manually performed and color changes are visually examined, which takes at least a couple of minutes for each sample. Test reagents often contain corrosive or carcinogenic chemicals. Also, varying the incubation time can result in misleading results. Current color tests for basic drugs are not very selective and are susceptible to interference. They are for screening, not confirmatory analysis. Furthermore, these assays are surely **NOT** able to distinguish between **enantiomers of chiral drugs that are each other's non-superimposable mirror images** such as **methamphetamine and related substances**, which is critical in forensic drug analysis [18]. The two individual components of an enantiomer-pair have identical chemical and physical properties, differing only in the way they react with other chiral compounds and the direction in which they rotate plane-polarized light, and therefore, cannot be differentiated by conventional spectroscopic and chromatographic methodologies.

Chiral analysis is of great practical importance [19]. All living things on earth are composed of chiral components: amino acids, carbohydrates, proteins, enzymes, nucleosides, DNA, and a number of alkaloids and hormones, *etc.* The majority of drugs derived from natural products are chiral with specific absolute-configurations, *i.e.* enantiomerically pure. In other words, they are single enantiomers [20, 21]. Nowadays, many synthetic pharmaceuticals and illicit drugs are chiral, too. Biological processes involve recognition between chiral molecules in hand-shake like events and react differently toward each component of a pair of mirror-image chiral compounds. Often, though not always, only one of the enantiomers of a pharmaceutical drug has therapeutic value, while the other is useless or toxic. Many illicit drugs are effective and addictive only if they contain a stereogenic center or stereogenic centers of specific configuration, while their corresponding enantiomers are not. Dextromethorphan, which plane-polarized light with a wavelength of 589 nm (sodium D line) rotates such polarized light to the right and it is the dextrorotatory enantiomer whose name can be prefixed with dextro, *d*- or (+)-, is an uncontrolled substance found in Over-the-Counter cough and cold pharmaceuticals [and more recently in Ecstasy mimics]. However, levomethorphan, which rotates plane-polarized light sodium D line to the left and it is the levorotatory enantiomer whose name can be prefixed with levo, *l*- or (-)-, is a Schedule II narcotic analgesic [22]. Cocaine derived from the extraction of coca leaves is the *l*-enantiomer, while synthetic cocaine usually contains both *d*- and *l*-cocaine, and their diastereomers. The L- isomers of some barbiturates are depressants, while their D- isomers are convulsants. (The usage of D/L should be avoided for compounds other than carbohydrates and amino acids [23] because it is confusing. D and L do not necessarily correspond to *d* and *l*, respectively).

Currently, chiral drug analysis in the forensic science field focuses on methamphetamine (Fig. 1) and related compounds because of legal and intelligence considerations. So will this chapter. Although both enantiomers of almost every chiral controlled substance are under strict and similar regulation, the US Sentencing Commission guidelines do distinguish between the enantiomers of methamphetamine, a DEA Schedule II drug. One gram of *l*- or *d*-methamphetamine is equivalent to forty grams or ten kilograms of marijuana, respectively, for sentencing purpose [22]. Pharmaceuticals containing low levels of *l*-methamphetamine, such as Vicks® Vapor Inhaler, are not controlled, while sentences are heavier for offenses involving contrabands containing more than 80% *d*-methamphetamine hydrochloride [24], which is known as "ice". Amphetamine (Fig. 1), a US DEA Schedule II drug, is the parent compound of a type of Schedule I and II psychoactive compounds such as MDMA (Ecstasy) and methamphetamine (the *N*-methylated amphetamine, Fig. 4). The racemic (*d*, *l*)–amphetamine sulfate salt was used as a medicine. Adderall and its generics used to treat attention disorders contain both *d*-amphetamine and *l*-amphetamine salts (3:1) [25]. When abused, the *d* enantiomer is predominant [26, 27]. With the increased emphasis on routine drug testing, there is a need for the stereoselective determination of these drugs to address current and possibly future legal complications. Changing public opinion and legal landscape, which resulted in the recent legalization of recreational marijuana in a number of US states and the corresponding changes in federal and local enforcement practices, may bring different treatment by law regarding the real addictive forms of other drugs and their corresponding enantiomers.

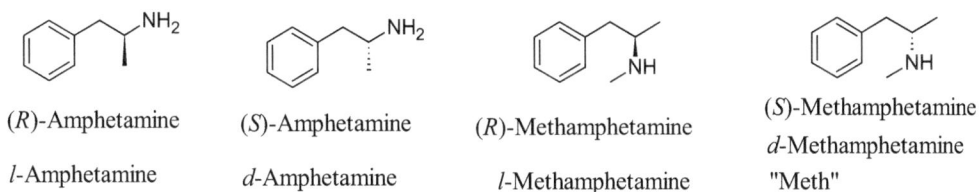

(*R*)-Amphetamine   (*S*)-Amphetamine   (*R*)-Methamphetamine   (*S*)-Methamphetamine

l-Amphetamine   *d*-Amphetamine   *l*-Methamphetamine   *d*-Methamphetamine

"Meth"

Fig. (1). Structures of stereoisomers of amphetamine and methamphetamine
(*R*)-amphetamine: *l*; (*S*)-amphetamine: *d*; (*R*)-methamphetamine: *l*; (*S*)-methamphetamine: *d*

Enantiomeric analysis can provide valuable information on illicit drug production [27]. Only one of the enantiomers of ephedrine or pseudoephedrine (*l*-ephedrine and *d*-pseudoephedrine, respectively, usually extracted from the plant *Ephedra sinica, "Mahuang"*) can be easily converted to enantiopure *d*-methamphetamine (Fig. 2) in "meth" labs [28], as covered in "Breaking Bad". Racemic *d, l*-methamphetamine, is most commonly produced using 1-phenyl-2-propanone (a.k.a. phenylacetone or P2P) as the starting material (Fig. 3). Since 2006 when the purchase of *l*-ephedrine and *d*-pseudoephedrine (the active ingredient in

Sudafed®) started to be limited by the Combat Methamphetamine Epidemic Act [29], the chiral composition of seized "meth" samples in the U.S. has shifted toward skewed *d:l* ratios [30], which points to the use of the P2P route followed by enrichment through optical resolution [27]. The enantiomeric excess (*ee*) of these compounds can be used to track them to their sources (chemical forensics) by matching a batch from a particular lab with exhibits picked up off a pusher.

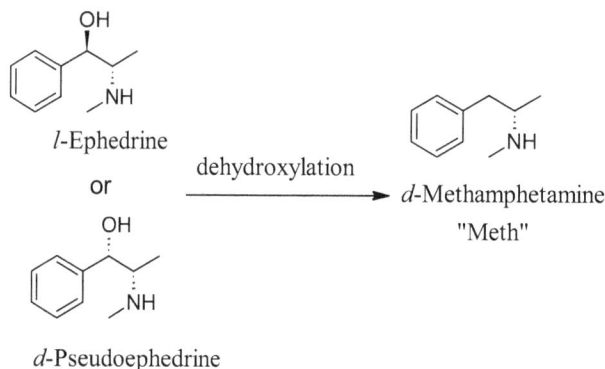

Fig. (2).   Enantiopure *d*-methamphetamine can be prepared using natural products *l*-ephedrine and *d*-pseudoephedrine.

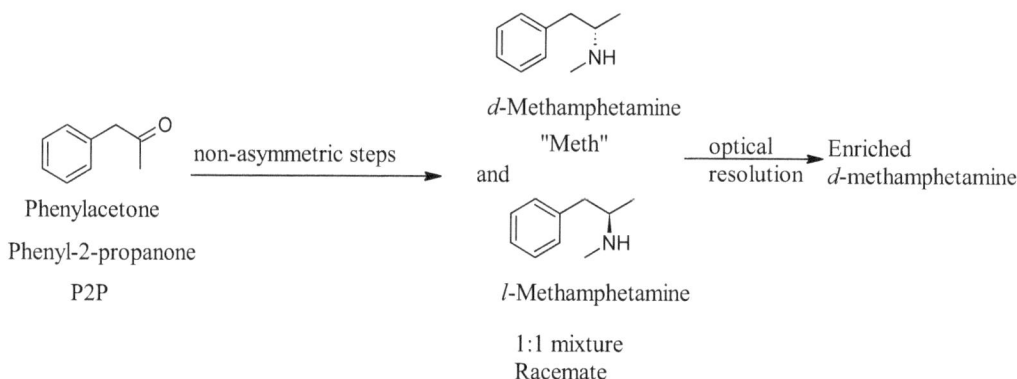

Fig. (3). P2P can be converted through non-stereoselective synthesis to racemic methamphetamine, which can be enriched to skewed ratios through optical resolution.

RELEVANT STEREOCHEMISTRY TERMS

Before we discuss the chiral analysis of selected controlled substance, it is useful to refresh our minds with some other keys terms used in stereochemistry. *l*-Methamphetamine is also called (*R*)-methamphetamine, while *d*-methamphetamine is also called (*S*)-methamphetamine. (*R*)- or (*S*)- represents the **absolute configurations** at their stereocenter(s), which has four different substituents [31]. The absolute configuration of each chiral center [usually carbon] is assigned based on the spatial arrangement (the direction of the priority) of the four different groups attached to it. Priority is assigned following the Cahn-Ingold-Prelog (CIP) rules [32], with the highest priority numbered 1. An atom with a higher atomic number is given a higher priority. For two atoms of the same element, the isotope with a higher atomic mass has a higher priority. Priority is determined at the first point of difference. First, draw a picture so that the atom with lowest priority (Group #4), for example H, seems pointing away from you and ignore the lowest-ranked group to draw plane projection (Fig. **4**). Then start tracing from the highest priority (smallest number) to the lowest one (biggest number). The *R* configuration is given if the priority of the groups moves clockwise. Otherwise the *S* configuration is given [32].

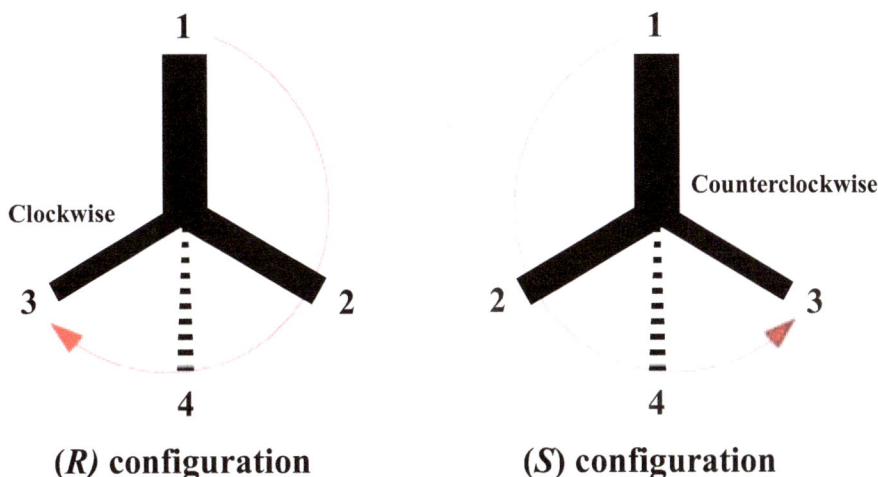

Fig. (4). Assignment of the *(R)* and *(S)* configurations. Only the dash representing the group with the lowest priority points away from the reader. The solids lines are plane projections of the three groups with higher priorities, which correspond to their thickness with a thicker line representing higher priority.

Although the *R* designation originated from the German word rectus (right-handed), an (*R*)-enantiomer does not necessarily rotate plane-polarized light (589 nm) to the right and therefore is not necessarily a *d*- or (+)- enantiomer. For the same reason, an (*S*)-enantiomer is not necessarily an *l*- or (-)- enantiomer. This is evident for the enantiomers of methamphetamine.

For molecules with two or more stereocenters, two kinds of stereoisomers exist: enantiomers and diastereomers. Enantiomersare molecules that have the same formula and same connectivity but are non-superimposable mirror images of each other. Stereoisomers that are not enantiomers are diastereomers. A molecule with two chiral carbon atoms has up to four stereoisomers (*R*, *S*)-, (*S*, *R*)-, (*R*, *R*)-, and

(*S*, *S*)- isomers. The (*R*, *S*)- and (*S*, *R*)- isomers are a pair of enantiomers. So are the (*R*, *R*)- and (*S*, *S*)- isomers. All others are pairs of diastereomers. If both stereocenters have exactly the same four substituents and there is a plane of symmetry in the (*R*, *S*)- or (*S*, *R*)- structures, they represent the same *meso* isomer, which is **not** chiral. Unlike enantiomers, the properties of diastereomers can differ and their differentiation is easier. An example of a molecule with 2 chiral centers is 2-methylamino-1-phenyl-1-propanol, a.k.a. ephedrine/pseudoephedrine. Among its four stereoisomers (Fig. **5**), there are two pairs of enantiomers: (1*R*, 2*S*)-(-)-ephedrine and (1*S*, 2*R*)-(+)-ephedrine, and (1*S*, 2*S*)-(+)-pseudoephedrine and (1*R*, 2*R*)-(-)- pseudoephedrine, respectively. There are 4 pairs of diastereomers: (1*R*, 2*S*)-(-)-ephedrine and (1*R*, 2*R*)-(-)-pseudoephedrine, (1*R*, 2*S*)-(-)-ephedrine and (1*S*, 2*S*)-(+)-pseudoephedrine, (1*S*, 2*R*)-(+)-ephedrine and (1*R*, 2*R*)-(-)-pseudo-ephedrine, and (1*S*, 2*R*)-(+)-ephedrine and (1*S*, 2*S*)-(+)-pseudoephedrine, respectively.

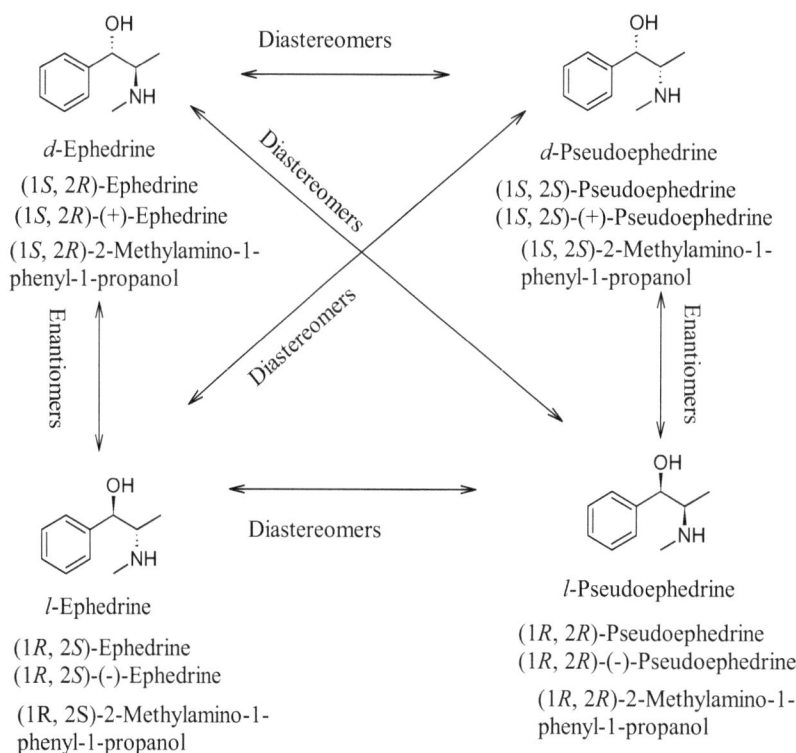

Fig. (5). Relationships between the stereoisomers of ephedrine and pseudoephedrine.

Each stereocenter's absolute configuration can be determined by structural analysis, while the optical rotation of each chiral stereoisomer has to be measured. The prefix (*R*)-(+)-, (*R*)-(-)-, (*S*)-(+)- or (*S*)-(-)- can be added to the name of a chiral compound with a single chiral center, depending on its absolute configuration and optical rotation. Similar combinations can be added as prefixes to compounds with two or more chiral centers. For example, the name (1*S*, 2*R*)-(+)-ephedrine tells that this compound is *d*-ephedrine since its optical rotation sign is + and it has two chiral centers: the lower numbered one being *S* and the higher numbered one being *R*. The enantiomer of *d*-ephedrine is *l*-ephedrine, a.k.a. (-)-ephedrine, or (1*R*, 2*S*)-ephedrine, or (1*R*, 2*S*)-(-)-ephedrine. Please note that there is **no** such a compound as (1*S*, 2*R*)-(-)-ephedrine. For a specific enantiomer, there is only one combination of absolute configuration and optical rotation. A homogeneous sample containing exactly the same quantities of two enantiomers does not rotate plane polarized light. It is called a racemate or a racemic solution, whose name is prefixed with (±)- or (*d*, *l*)-.

It is useful to keep in mind the following relationship:

(*R*)-amphetamine: *l*; (*S*)-amphetamine: *d*; (*R*)-methamphetamine: *l*; (*S*)-methamphetamine: *d*; (1*R*, 2*S*)–ephedrine: *l*; (1*S*, 2*R*)–ephedrine: *d*; (1*R*, 2*R*)–psudoephedrine: *l*; (1*S*, 2*S*)–psudoephedrine: *d*.

Chiral Analysis

In forensic science, the scientific analysis has to comply with a couple of standards to be admissible in court. Mainly such analysis should be relevant to the specific task and should be reliable in the eyes of the judge. Consequently, this chapter covers mature and reliable methods are that used most often in government forensic science laboratories, not necessarily the latest in academic and industrial research and development. We are discussing spectroscopic and separation methods in the following sections, with more attention on separation methods because most forensic samples are mixtures.

SPECTROSCOPIC ANALYSIS

Pure chiral compounds can be easily analyzed by chiroptical spectroscopic measurements (see the definition of selected one in Table 1) using polarized light. Plane polarized light consists of left- and right-circularly polarized components, which interact differently with a chiral medium. They travel with different velocities through the medium, giving rise to a circular birefringence or anisotropic refraction (n_L - $n_R \neq 0$), observable as a rotation of the plane of polarization. They can be differentially absorbed by achiral medium if the wavelength is right, resulting in a circular dichroic effect or anisotropic absorption (A_L - $A_R \neq 0$, or $\Delta\varepsilon \neq 0$). The rotatory power of a substance varies with the wavelength, which can be recorded as optical rotatory dispersion (ORD). Nowadays, ORD has largely been replaced by electronic circular dichroism (CD)

[33]. Vibrational circular dichroism (VCD) and Raman optical activity (ROA) measurements recently gain some attention in chiral analysis [34].

Table 1. Terms and definitions in Chiroptical Spectroscopy.

Term	Symbol	Definition
Optical rotation	α	The rotation of plane-polarized light as it travels through certain non-racemic materials.
Specific rotation	$[\alpha]$	The observed angle of optical rotation α when plane-polarized light of wavelength λ is passed through a sample with a path length (l) of 1.0 dm and a sample concentration (c) of 1.0 g mL^{-1} at temperature T.
Circular dichroism	CD	A form of spectroscopy based on the differential absorption ($\Delta\varepsilon$) of left- and right-handed circularly polarized light in non-racemic molecules.
Vibrational Circular Dichroism	VCD	Circular dichroism in the infrared and near infrared ranges.
Raman optical activity	ROA	A form of spectroscopy based on the intensity differences of the Raman intensities scattered by the molecules excited by incident right- and left-circularly polarized light.

Optical Rotation

When it comes to controlled substance analysis, optical rotation(α) is still sometimes used although it is a long existing chiroptical method. A pair of enantiomers rotate plane-polarized light in opposite directions when all other conditions are the same. One rotates plane-polarized light with a wavelength of 589 nm (Sodium D line) to the left and it is the levorotatory enantiomer whose name can be prefixed with levo, *l*- or (-)-. The other rotates such polarized light to the right and it is the dextrorotatory enantiomer whose name can be prefixed with dextro, *d*- or (+)-. The values of the specific rotation [α], the observed angle of optical rotation α when plane-polarized light of wavelength λ is passed through a sample with a path length (l) of 1.0 dm and a sample concentration (c) of 1.0 g mL^{-1} at temperature T, of a pair of enantiomers are of the same amplitude but opposite signs (enantio = opposite). As a result, enantiomers are also called optical isomers.

The optical rotation of achiral compounds is zero (0). So is that of racemates, although they consist of components that are optically active. To obtain accurate optical rotation values, however, the concentrations of the samplesare generally high in the g/dL range. The specific rotation values of selected substances are listed in Table 2.

Table 2. Specific Rotation of Some Compounds [35].

Compounds	$[\alpha]_D$ ($_D$ represents sodium D line)
Testosterone	+109°
(*S*)-Amphetamine	+37.6°
(*R*)-Methamphetamine hydrochloride	−18.05°
(1*R*, 2*S*)-Ephedrine hydrochloride	−34°
(1*S*, 2*S*)-Pseudoephedrine	+52°
Diacetylmorphine (HEROIN)	-166°
Morphine (natural product)	-132°
Cocaine (natural product)	-16°
Dextromethorphan	+39.6°
Sucrose	+66.47°
Fructose	-94.37°
Acetic Anhydride	0
P2P	0

For a mixture of two enantiomers, the optical rotation depends on their concentration difference. To obtain the concentration of each enantiomer, both optical rotation and UV-Vis measurements are required, and both the absolute value of specific rotation [α] and extinction coefficient (ε) should be known. The concentration effects of the two enantiomers are subtractive on optical rotation and additive on UV-Vis absorbance since the two enantiomers should give the same UV-Vis signal. The following equations can be established:

$A = \varepsilon b(c_1 + c_2)$ (UV-Vis spectroscopy)

$\alpha = [\alpha]l\,(c_1 - c_2)$ (Polarimetry)

where c_1 and c_2 are the concentrations of the enantiomers, respectively, and b and l are the optical pathlengths in UV-Vis and optical rotation measurements, respectively. UV-Vis measurements are typically done at lower concentrations than that of optical rotation. Therefore, the dilution factors must be accounted for in the first equation when determining the concentrations of the enantiomers.

Circular Dichroism

Circular dichroism has also been used to analyze controlled substances, amphetamine [36] and ephedrine/pseudoephedrine [37] (Fig. **6**) for example, and in many cases it requires less concentrated samples than in optical rotation measurements. Recently, CD has been employed as a selective detector in chiral HPLC analysis.

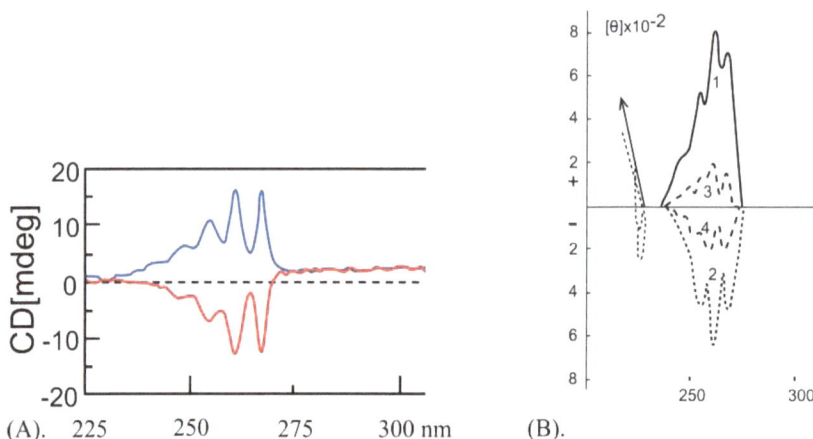

Fig. (6).   Circular Dichroism spectra of (A)*d*-amphetamine(blue) and *l*-amphetamine(red) [36];(B)ephedrines(1: *l*, 2: *d*) and pseudoephedrines[3: *l*, 4: *d*] [37]. Reproduce with permission from reference [37]. Copyright © 1969, NRC Research Press.

IR AND NMR

Enantiomers produce exactly the same infrared (IR) and nuclear magnetic resonance (NMR) spectra since the energy sources in these methods are not polarized. However, the physical properties of diastereomers can differ and consequently their IR and NMR spectra are different. If a pair of enantiomers are derivatized by an enantiopure chiral reagent, or are mixed with an enantiopure chiral shifting agent, diastereomeric products or complexes can be formed, which can produce different IR and NMR spectra. When mixed with *d*-mandelic acid, *d*-, *l*- and *d, l*-amphetamines gave quite different IR spectra in the 600-800 cm^{-1} region (Fig. **7**) [38].

For some substances, the IR of a pure enantiomer and a racemate can be the same. For example, (+)-, (-)- and (±)-cocaine hydrochloride have exactly identical IR spectrum because (+)- and (-)-cocaine hydrochloride do not interact with each other and they do not form a true racemate in solid state when mixed in 1:1 ratio [39]. Their carbonyl stretching peaks are all at 1730 and 1711 cm^{-1}. However, if the two enantiomers can pack in a way that they affect each other, *i.e.* they serve as each other's chiral shifting agent that they form a true racemate in solid state, the racemate and the pure enantiomer can produce different signals, as is shown in the IR spectra of free bases of (±) and (-)-cocaine: The carbonyl stretching peaks are at 1734 and 1706 cm^{-1} in the *l*- enantiomers, while they are at 1750 and 1705 cm^{-1} [39]. There are other differences as well.

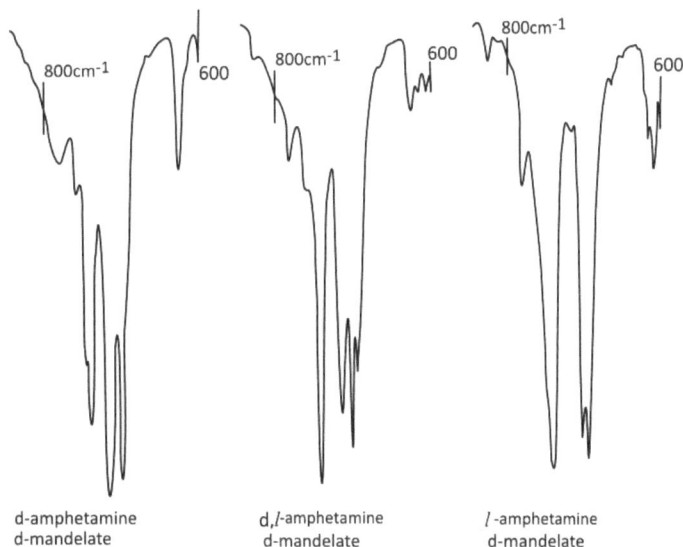

Fig. (7). The IR spectra in the 600-800 cm^{-1} region of amphetamine *d*-mandelates [38]. Reprinted with permission from reference [38]. Copyright © 1970, American Chemical Society.

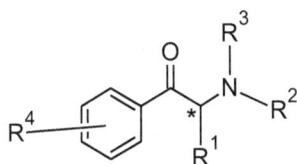

Cathinones

Cathinone: R^1 = CH$_3$; R^2, R^3, R^4 = H
Methcathinone: R^1, R^2 = CH$_3$; R^3, R^4 = H

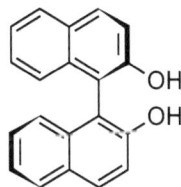

(R)-(+)-1,1-bi-2-naphthol

Fig. (8). Structures of cathinones and *(R)*-(+)-1, 1'-bi-2-naphthol.

Similarly, enantiomers can be differentiated by NMR with the help of chiral shifting agents. With the help of the chiral shifting agent *(R)*-(+)-1, 1'-bi-2-naphthol (Fig. **8**), the enantiomers of methamphetamine, ephedrine, pseudo-ephedrine and methcathinone (Fig. **8**) in a mixture were determined by NMR, without involving any separation methods [40]. The chemical shifts of *N*-methyl singlets in the 1.9-2.4 ppm increase in the following order:*(R)*-methcathinone< *(R)*-methamphetamine < *(S)*-methamphetamine < *(S)*-methcathinone < (1*R*, 2*S*)-ephedrine < (1*R*, 2*R*)-(-)-pseudoephedrine < (1*S*, 2*R*)-(+)- ephedrine < (1*S*, 2*S*)-(+)-pseudoephedrine. The chemical shifts of the *C*-methyl doublets in the 0.7-1.2 ppm increase in the following order: (1*S*, 2*S*)-(+)-pseudoephedrine < (1*R*, 2*R*)-(-)-pseudoephedrine < *(R)*-methamphetamine, *(S)*-methamphetamine, (1*R*, 2*S*)-(-)-

ephedrine, (1*S*, 2*R*)-(+)-ephedrine] < (*S*)-methcathinone < (*R*)-methcathinone, although those of (*R*)-methamphetamine, (*S*)-methamphetamine, (1*R*, 2*S*)-(-)-ephedrine, and (1*S*, 2*R*)-(+)-ephedrine overlaparound 0.8 ppm.

Separations

When the evidence is not pure or when the purity of the evidence is unknown, separation methods [41] are oftenused in the confirmation step of forensic drug analysis. Such methods include gas chromatography (GC) [21], high performance liquid chromatography (HPLC) [7, 21], and/or capillary electrophoresis (CE) [42 - 45], with mass spectrometers, more recently tandem mass spectrometers, as the preferred detectors.

Chromatography is a method used for the physical separation of the components ina mixture by partitioning them between a stationary phase and a mobile phase. Different components are partitioned at different ratios due to their different affinities to the stationary phase (and in some cases, the mobile phase). In gas chromatography, the mobile phase is known as a carrier gas, usually hydrogen, helium, or nitrogen, since it does not participate in the chemistry of separations and is only responsible for carrying the sample through the stationary phase, which is usually a nonvolatile compound immobilized to a solid support or column inner wall. High affinity to the stationary phase results in a longer retention time, while low affinity to the stationary phase results in a shorter retention time. Similar molecules, such as diastereomers, have similar retention times, while enantiomers have exactly the same retention times on achiral columns (stationary phases). Separation of the components is plotted as signal strengths (y-axis) against the different retention times (x-axis), displaying as chromatograms. The area of each peak on the chromatogram can be used for quantitative analysis.

Two types of techniques are employed in chiral separations [20, 46]. Indirect ones involve separation using achiral stationary and mobile phases after samples are reacted with an enantiopure chiral derivatization agent (CDA). Direct ones utilize chiral selectors, which can be either chiral stationary phases (CSPs) or chiral mobile phase additives (CMPAs). In such techniques, a pair of enantiomers are either turned into a pair a diastereomers covalently (chiral derivatization) or non-covalently form transient diastereomeric complexes with CSPs or CMPAs.

Direct chiral separations are based on chiral recognition: ability of CSP, or CMPAs to interact differently with each enantiomer to form transient-dias-tereomeric complexes. It requires a minimum of three interactions (*"3-Point" Interaction Rule*) through:

- H-bonding
- π-π interactions
- Dipole stacking
- Inclusion complexing
- Steric bulk hindrance

At least one such interaction must be stereochemically dependent so that the CSP or CMPA interact smore strongly with one enantiomer than with the other to effect separation.

The 3-point rule also explains the often-different effects of a pair of enantiomer in the body since they interact with enzymes, proteins [47], and other chiral receptors (similar to CSPs and CMPAs in chromatography), which is illustrated in Fig. **9**. The groups in the high-affinity or effective enantiomer must interact with the corresponding regions of the binding site of its receptor/chiral selector to have a perfect alignment (Fig. **9**, left). In contrast, the other enantiomer cannot bind in the same most effective way with this chiral receptor/selector no matter what (Fig. **9**, middle and right).

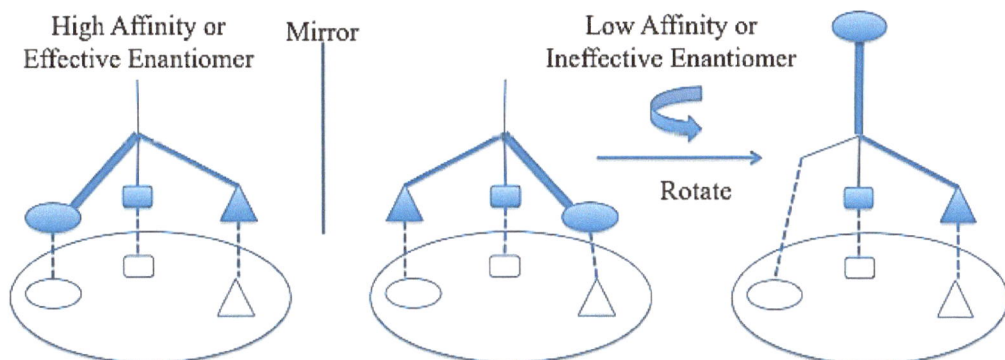

Chiral Stationary Phase, Chiral Mobile Phase Additive or Chiral Receptor

Fig. (9). Achiral receptor or chiral stationary phaseinteracts differently with a pair of enantiomers through the "3-point" rule. The interaction with one enantiomer is stronger (left) than the other (middle or right).

For example, cyclodextrin is a chiral molecule that serves as a chiral stationary phase in GC and HPLC, and a chiral mobile phase additive in HPLC and CE. Cyclodextrin, a.k.a. cycloamylose, is a cyclic chain of D-glucose joined together by α-1, 4-glycosidic linkages and is found to have a truncated cone shape. Alpha, beta (Fig. **10**), and gamma cyclodextrins consist of six, seven, and eight D-glucose units, respectively. They are synthesized from the enzymatic hydrolysis of starch. The hydroxyl groups of a cyclodextrin molecule are located on the outer surface; the primary hydroxyl groups are located towards the narrow portion, while the secondary hydroxyl groups are located towards the wider portion of the molecule (Fig. **10**) [48]. In chiral recognition, its 3-point interactions with chiral compounds include:

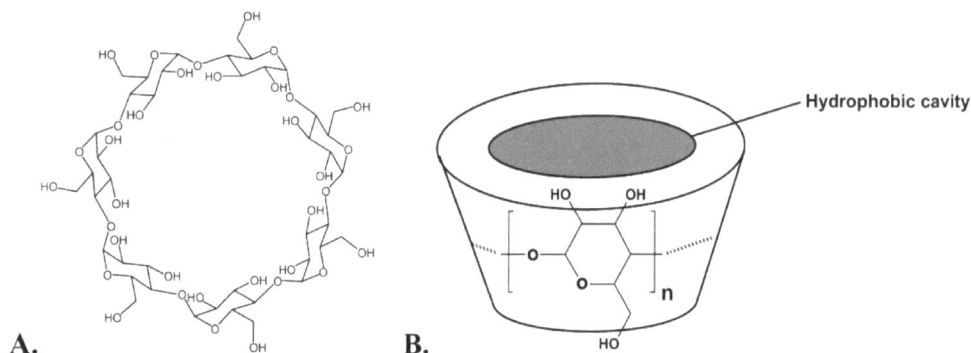

Fig. (10).   (A)Structure, and (B) Three-dimensional representation of β-cyclodextrin [48].Reprinted with permission from reference [48]. Copyright © 2009, American Chemical Society.

• H-bonding between the hydroxyls and polar groups of analytes
• Inclusion of hydrophobic portion of analyte into non-polar cavity
• Steric hindrance, *etc.*

One enantiomer will be able to better fit in the cavity than the other.

Other types of CSPs are used in chiral HPLC analysis and will be discussed in more detail later.

In the indirect technique, chiral analytes bearing certain functional groups, such as amine (primary and secondary, but not tertiary), hydroxyl, carboxyl, carbonyl and thiol, are reacted with a CDA before they are separated using readily available inexpensive conventional achiral stationary and mobile phases. Although the CDA needs to be readily available with extremely high enantiopurity, various CDAs and reactions are available to optimize the formation of diastereomeric derivatives that can be more easily separated (and detected). The enantiomeric impurity can be made to elute before the main peak by choosing the advantageous enantiomeric form of the CDA so that quantitation can be improved.

The majority of published chiral drug analysis work has been done using GC following the indirect technique: chiral derivatization before separation on achiral stationary phases. Although chiral mobile phase additives, which are usually not desirable for mass spectrometry detection, are of course never used in GC and rarely used in HPLC (because of higher cost, lower efficiency and possible introduction of foreign species in preparative HPLC), they are used quite often in CE. Direct chiral separations using CSPs are more widely used in HPLC than in GC. Reaction with a CDA to form diastereomers is not required in the direct technique, which is an advantage. However, derivatization with anachiral agent is still often employed to effect appropriate molecular interactions with the chiral discriminator for better separation and/or to impart certain properties to the analytes for easier vaporization (in GC), better shaped peaks, shorter analysis time or better detection. Since many controlled substances have no derivatizable functional groups, direct chiral separations using CSPs are the most used for such compounds.

We will mainly discuss the chiral separations of amphetamine, methamphetamine and related compounds, focusing on the indirect chiral GC (chiral derivatization and achiral stationary phases), direct chiral HPLC and supercritical fluid chromatography (SFC, chiral stationary phases) and, CE (chiral additives), although chiral derivatization followed by separation on chiral stationary phases will also be discussed in some cases.

Gas Chromatography

GCMS is the workhorse in forensic drug analysis because it combines the separation efficiency of GC with the sensitivity and specificity of mass spectrometry (especially tandem mass spectrometry (MS^n), which is a Category A method for drug analysis because its identification power according to the Scientific Working Group for the Analysis of Seized Drugs (SWGDRUG) [24, 49]. In GC-MS, a sample is injected into the injection port, where it is vaporized and swept into the column by the carrier gas. As the sample passes through the column, different components are separated before they go to a mass spectrometer, which often employ electron impact (EI) ionization where the sample is bombarded by high-energy electrons to be ionized and fragmented. The molecular ion and fragments are sorted by their mass-to-charge (m/z) ratio, which can be used to elucidate the structure of the original molecule. The height/area of the peaks can be used to find out the concentration of analytes. Most published chiral drug analysis work employs chiral derivatization and separation by GC on achiral stationary phases [50 - 52]. Analytes that are not enantiomers, including diastereomers, are separated on achiral columns based on their differences in boiling points and/or polarity. Although there are direct analyses on chiral GC columns, derivatization with an achiral agent is often still needed, offering little advantage in sample preparation. Less chiral stationary phases are available for GC than for HPLC.

Indirect Method: Derivatization

A commercially available reagent *S*-(-)-*N*-(trifluoroacetyl) prolyl chloride, a.k.a. trifluoroacetyl-L-prolyl chloride (L-TPC, Fig. **11**) has been widely used to derivatize chiral drugs for gas chromatography analysis [51].

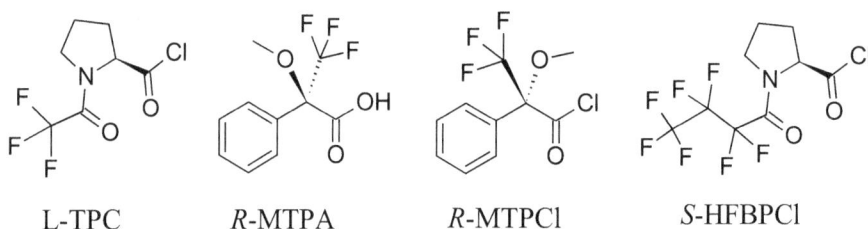

L-TPC   *R*-MTPA   *R*-MTPCl   *S*-HFBPCl

Fig. (11). Some chiral derivatization agents in chiral GC drug analysis.

In an early study, amphetamine samples were derivatized with L-TPC and then analyzed by a capillary gas chromatograph/mass spectrometer(GCMS) system [53].

There is a small amount of D-TPC in commercial L-TPC. Therefore, such derivatization can theoretically produce 4 products: *l*-amphetamine-L-TPC (*RS*), *d*-amphetamine-L-TPC (*SS*) (Fig. **12**), *l*-amphetamine-D-TPC (*RR*), and *d*-amphetamine-D-TPC (*SR*) since *d*-amphetamine and L-TPC are of the *S* configuration while *l*-amphetamine and D-TPC are of the *R* configuration. However, only two peaks showed up on the chromatogram when the column was achiral (SP-2100), which was expected since *SR* and *RS* co-elute and so do *RR* and *SS* because they are pairs of enantiomers. Therefore, to find out the percentages of *d*- and *l*-amphetamines in a mixture, which corresponds to the percentages of *SS* and *RS*, respectively, correction for the co-elution of their corresponding enantiomers (*RR* or *SR*, respectively), and *vice versa*, is generally needed.

After correction for enantiomerical TPC impurity, calculated results obtained from the achiral column (SP-2100) were found to be statistically the same as those obtained from a chiral column (Chirasil-Val). Peak areas correction may become a serious problem if the exact enantiomeric purity of the derivatization reagent is not available. Another drawback is that the manual derivatization method seems quite tedious even for standards.

An automated sample preparation system was applied to the chiral GC analysis of amphetamineand methamphetamineon an achiral column (HP Ultra 1 column coated with 100% dimethylpolysiloxane) after derivatization with L-TPC [9].

This automated procedure was compared with two existing methods which employ manual on-column or off-line derivatization. The derivatization procedure did not affect chromatographic performance, giving the same base-line resolutions for enantiomers and the same elution order of the L-TPC derivatives these drugs: (*R*)-amphetamine (8.35 min)<(*S*)-amphetamine (8.70 min)<(*R*)-methamphetamine (10.81 min)< (*S*)-methamphetamine (11.17 min). However, sensitivity was better for both automated and manual off-line derivatization. The automated procedure gave the best quantitative reproducibility based on peak areas because the automated "just-in-time" derivatization can minimize the time-dependent decomposition of the derivatives before GC analysis.

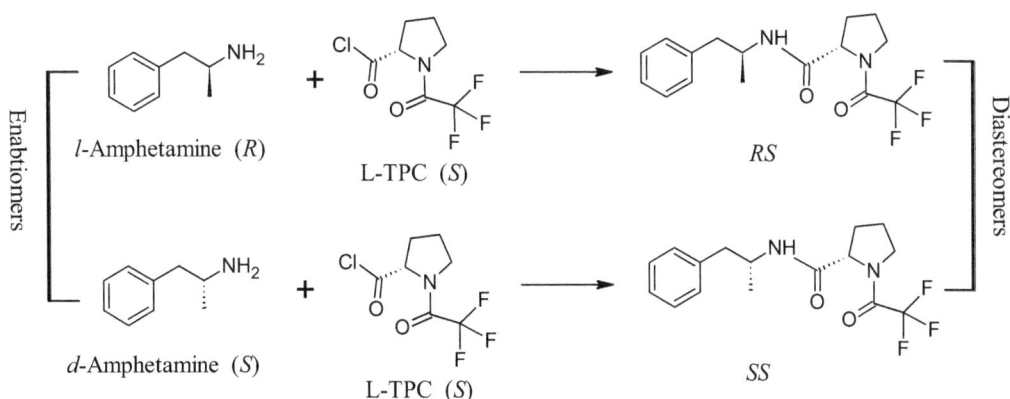

Fig. (12). Derivatization of enantiomers *d*- and *l*-amphetamines with L-TPC gives a pair of diastereomeric products.

To accurately determine a drug's enantiomeric content, the raw peak areas were corrected using the following equation:

$$c = \frac{50(a - p)}{50 - p}$$

where *c* and *a* are the corrected percentage ans draw percentage obtained from the peak area of an enantiomer, respectively, and *p* is the percentage of D-TPC impurity.

L-TPC derivatization was used for chiral determination of amphetamines and related designer drugs in 12 hair samples from suspected amphetamines abusers [54]. The substances were extracted from hair with a saturated carbonate buffer under sonication for 30 min. The solution was extracted with hexane. L-TPC was then added to the collected organic layer, which was subsequently analyzed by GC/MS in SIM mode. *R* and *S* isomers were separated and detection limits were

0.1 ng/mg for amphetamine, methamphetamine and MDA, and 0.2 ng/mg for MDMA and MDEA. Both isomers of amphetamine and/or MDMA were found to be in these hair samples.

In Sweden, drug abusers diagnosed with Adult ADHD have been successfully treated with Metamina®(*d*-amphetamine). Hair was dissolved in sodium hydroxide, followed by extraction with iso-octane and chiral derivatization with L-TPC. GCMS analysis of hair samples uncovered existence of *l*-amphetamine, which suggested intake of additional racemic illicit amphetamine and which was not shown by enantiomer ratios and concentrations of amphetamine in blood [55].

Developing analytical methods for the determination of cathinones, which have structures (Fig. **8**) and stimulating effects similar to amphetamines, is of great interest. Under optimized conditions, 14 amphetamine/cathinone-related compounds were derivatized with L-TPC and resolved into their enantiomers by GCMS with an HP5-MS column [56]. Unlike amphetamines, racemic methcathinone derivatives showed different peak areas for each of the produced diastereomeric isomers. Derivatization with either of the pure methcathinone enantiomers led to both diastereomers, which pointed to the presence of racemic species during derivatization through at least one of the 3 effects: racemization of L-TPC, different reaction rates of the two enantiomers and different yields of diastereomers due to keto-enol tautomerization of the analytes. If TPC were the problem in this case, which was possible mainly due to the keto-enol tautomerization of the α-proton on the chiral carbon and the carbonyl group under certain conditions (Fig. **19**), different peak areas would have been observed for amphetamines. After derivatization with another CDA, (*R*)-(+)-α-methoxy-α-(trifluoromethyl)phenylacetic acid (*R*-MTPA, Fig. **11**), which does not possess any α-protons on the chiral carbon atom, three model chiral substances showed different peak areas as well. Therefore, racemization during derivatization was not due to the CDA, being *R*-MTPA or L-TPC, in this case, but was due to the keto-enol tautomerization of the methcathinone (Fig. **13**).

Since *R*-MTPA does not possess any α-protons on its chiral carbon and is known to be more stable than L-TPC, it is gaining popularity in chiral forensic drug analysis [40]. The enantiomers of the related substances methamphetamine, ephedrine, pseudoephedrine and methcathinone were determined by GCMS on an achiral DB-5 (15 m x 0.25 mm, 0.25 μm film thickness) column after derivatization with *R*-MTPA (actually the anhydride of *R*-MTPA, which was obtained by mixing *R*-MTPA with 1 equivalent of DCC), giving a resolution of at least 1.6 between all derivatives. All the *R*-MTPA derivatives eluted between 8-14 minutes with the following order: (*S*)-methamphetamine <(*R*)-methamphetamine <(*S*)-methcathinone <(*R*)-methcathinone <(1*R*, 2*S*)-ephedrine <(1*S*, 2*R*)-ephedrine

<(1*S*, 2*S*)-pseudoephedrine <(1*R*, 2*R*)-pseudoephedrine [40]. The carrier gas was helium with a linear velocity of 60 cm/s. *R*-MTPA derivatives of khat-related compounds [57] were reported to give better resolution than their corresponding TPC derivatives.

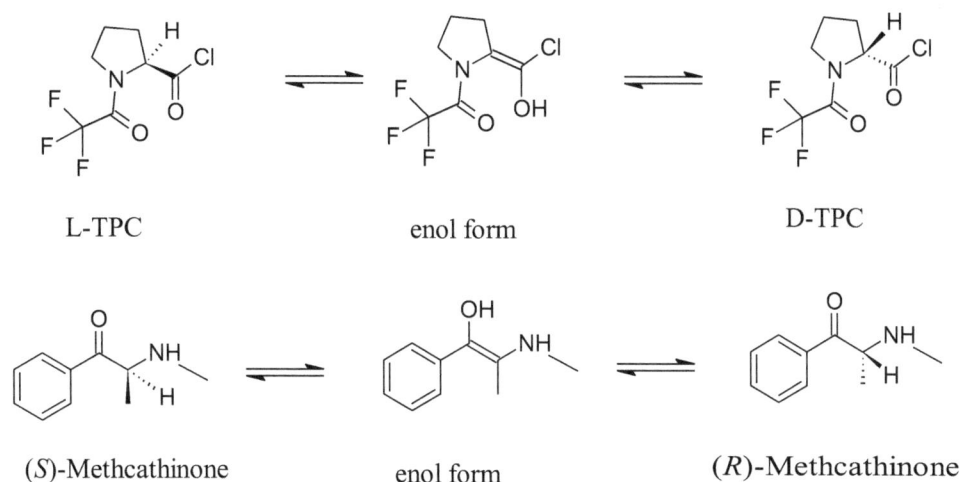

L-TPC   enol form   D-TPC

(*S*)-Methcathinone   enol form   (*R*)-Methcathinone

Fig. (13). The racemization of TPC and methcathinone through keto-enol tautomerization [56].

Another CDA without any α-H, *R*-(-)-α-methoxy-α-trifluoromethylphenylacetic chloride(*R*-MTPCl, Fig. **11**), which is the acid chloride of (*S*)-(-)-α-methoxy- α-(trifluoromethyl) phenylacetic acid (*S*-MTPA) has been used to derivatize amphetamine and methamphetamine and a wide range of phenethylamines directly in the non-protic elution solvent that was used to elute the compounds from a polymer-based solid-phase extraction column [58], which eliminates an elution solvent dry-down step that may adversely affect extraction recovery and derivatization before GC analysis. This CDA was used to derivatize both classic amphetamines (amphetamine and methamphetamine) and related designer drugs (MDA, MDMA and MDEA) after liquid-liquid extraction of whole blood samples (0.5 g) originating from traffic and criminal cases and post mortem cases before enanitomeric determination by GC on a HP-5MS(5% phenyl methyl siloxane) column and detected by SIM-MS [59]. Adding 0.02% triethylamine resulted in more complete derivatization, which made it possible to detect the amphetamines in adequately low concentrations. The linear range was from 0.004 to 3 μg/g per enantiomer. Please note that the same single enantiomer derivatized with *R*-MTPA and *R*-MTPCl, respectively, will produce a pair of diastereomers.

(*S*)-(-)-*N*-(heptafluorobutyryl)prolyl chloride (*S*-HFBPCl, Fig. **11**), an analogue of L-TPC with 6 F atoms in one molecule, was produced through a modified procedure to afford high optical purity and long shelf-life (at least 10 months) at - 20 °C [60]. It was used to produce derivatives of amphetamine and methamphetamine that were readily ionized in the negative-ion chemical ionization mode thanks to the electronegativity of the heptafluorobutyryl group in a GCMS assay of their enantiomers in small volumes of plasma or serum, which was applied to toxicology cases (Fig. 21) [60]. No or very little racemization of the CDA occurred during synthesis or derivatization. GC separation was achieved on a HP-5MS column (30 m × 0.25 mm; 0.25 μm film thickness). Mass spectrometer transfer line was heated to 280 °C and negative-ion-chemical ionization (NICI) was employed using methane (2 mL/min) as the reagent gas. The derivatives of selected compounds were well resolved between 9.5 and 12 minutes with the following elution order: (*R*)-amphetamine < (*S*)-amphetamine < (*R*)-methamphetamine < (*S*)-methamphetamine.

Direct Method: Chiral GC Column

Indirect chiral GC-MS has a few disadvantages when compared to other analytical techniques. There is one extra step required in the sample preparation involving derivatization of the analytes, commonly using L-TPC. The purity of such derivatization reagents is often not 100%, leading to result bias if not corrected. Using chiral columns may address such problems.

In an interesting early study, *d*- and *l*-amphetamine mixtures were derivatized with L-TPC and then analyzed by a chiral phase, Chirasil-Val, capillary GCMS [53]. One significant advantage in using the chiral column is that the four possible isomers resulting from the reaction of *d*- and *l*-amphetamines with L- and D-TPC (impurity in commercial L-TPC) are completely resolved with the following elution order in increasing retention times: *d*-amphetamine-D-TPC (*SR*) < *l*-amphetamine-L-TPC (*RS*) < *l*-amphetamine-D-TPC (*RR*) < *d*-amphetamine-L-TPC (*SS*), which facilitates the determination of the percentage of D-TPC by dividing the peak area of *SR* by that of the sum of *SR* and *SS*; or by dividing the peak area of *RR* by that of the sum of *RR* and *RS*. When a low concentration of *l*-amphetamine (relative to *d*-amphetamine) is used, the peak area of *RR* is small; similarly, when a low concentration of *d*-amphetamine is used, the peak area of *SR* is small. With the purity of L-TPC determined, the enantiomeric ratio of *d*- and *l*-amphetamine of each test solution was also determined by an achiral column (SP-2100) as described earlier and the results agree with that obtained by the chiral column, which is directly the ratio of *SS*:*RS*. Reconstructed chromatograms of representative samples analyzed by the two columns are shown in Fig. (**14**). Although impurities are normally encountered in illicit samples (Fig. **14b**), single ion chromatograms (Fig. **14c**) of such samples can be used to clearly display the peaks of the drugs of interest, which was amphetamine in this case.

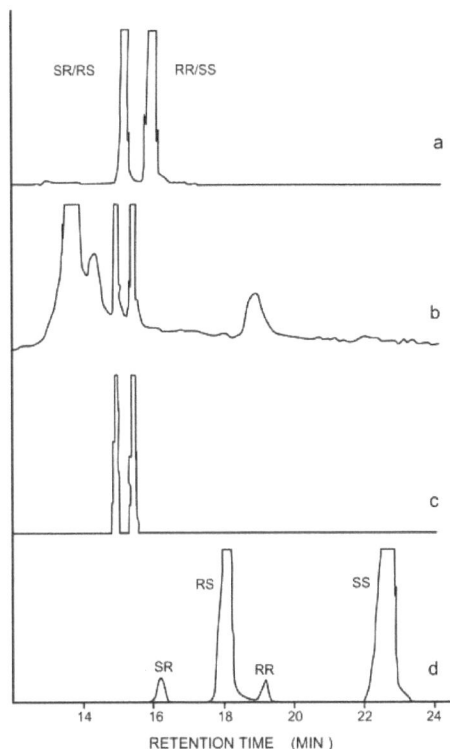

Fig. (14). (a) Total ion chromatogram of a control *d*- and *l*-amphetamine mixture obtained on the SP-2100 column. (b) Total ion chromatogram of illicit amphetamine obtained on the SP-2100 column. (c) Single ion (m/e = 237) chromatogram of illicit amphetamine obtained on the SP-2100 column. (d) Total ion chromatogram of the control *d*- and *l*-amphetamine mixture obtained on the Chirasil-Val column [53]. Reprinted with permission from reference [53]. Copyright © 1981, American Chemical Society.

Gas chromatography (GC) on modified cyclodextrin chiral stationary phases (CSPs) has shown to be highly effective in separating the enantiomers ofamphetamine, methamphetamine, ephedrine, and pseudoephedrine, even if all of them are present in the same sample [61]. Interestingly, reversal of the elution order of enantiomers in chiral GC has been observed in a few cases. Not only CSPs made of different cyclodextrin derivatives or different size cyclodextrins could reverse the enantiometric retention order of many compounds [41], but also different derivatives of the same pair of enantiomers (using achiral derivatization agents), such as trifluoroacetylated amphetaminesand acetylated amphetamines, can have the opposite enantiomeric elution order on the same CSP. For example, on a 30 m Chiraldex G-PN capillary (ID 0.25 mm) column under isothermal conditions (140 °C), (*R*)-trifluoroacetylated amphetamine eluted **before** its (*S*)-enantiomer, while (*R*)-acetylated amphetamine eluted **after** its (*S*)-enantiomer [62].

High-Performance Liquid Chromatography

Indirect Method: Derivatization followed by Separation on C18 Columns

An LC–MS/MS method with precolumn derivatization was developed for chiral analysis methamphetamine and amphetamine in plasma and oral fluid in no more than 10 minutes [63]. After sample collection and preparation, derivatization with 1-fluoro-2, 4-dinitrophenyl-5-L-alanineamide (Marfey's reagent, Fig. **15**) was performed, followed by HPLC analysis under isocratic conditions (methanol: water, 60:40) with a 2.6 μ m C18 column. Analytes were identified by tandem mass spectrometry in ESI **negative** mode. The linear range was from 1 to 500 μg/L for all enantiomers, with RSDs and accuracy of ≤11.3% and 85.3–108%, respectively.

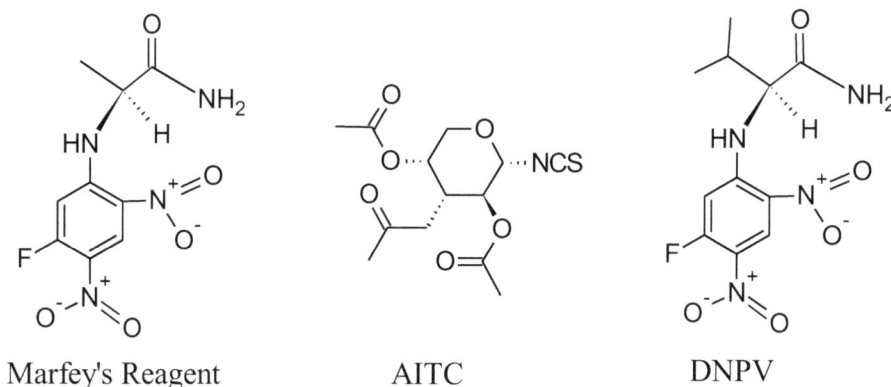

Marfey's Reagent   AITC   DNPV

Fig. (15). Structures of some chiral derivatization agents in enantiomeric HPLC analysis.

In another study, the ultra-high pressure liquid chromatography(UHPLC or UPLC) analysis of enantiomers using a C18 column (1.7 μm particles) was carried out after derivatization using two different agents, 2, 3, 4-tri-O-acety--α-D-arabinopyranosyl isothiocyanate (AITC, Fig. **15**) and Marfey's reagent. Separation of several amphetamine-related compounds contained within the same sample was achieved in 2–5 min with high efficiency and selectivity [64].

An LC-MS/MS methodforthe chiral pharmacokinetic analysis of the metabolites of 3, 4-Methylenedioxymethamphetamine (MDMA, ecstasy, a racemic drug of abuse) in human blood plasma was developed [65]. Blood plasma samples were prepared by simple protein precipitation, followed by derivatization using N_α-(2, 4-dinitro-5-fluorophenyl) L-valinamide (DNPV, Fig. **15**) and separation on standard reversed phase column. The method was selective, sensitive, accurate and precise for all tested analytes such as 4-hydroxy 3-methoxymethamphetamine

(HMMA) 4-sulfate and HMMA 4-glucuronide, except 3, 4-dihydroxy-methamphetamine (DHMA) 3-sulfate.

Direct Separation on Chiral LC Columns

Direct chiral separations using CSPs are more widely used in HPLC than in GC. There are more than a hundred commercially available HPLC CSPs. Cyclodexrin-based phases can be useful, a few better types are now the most widely used: carbohydrate (cellulose, amylose), polyacrylamide, diallyltartardiamide, Pirkle phases, chirobiotic phases (vancomycin, teicoplanin) [66]. However, no CSP can separate all classes of racemates. Although the examination of the interaction mechanism between CSP and chiral analytes can be of some help, column selection is based mostly on empirical data, experience and screening. One often checks literature for structurally-related compounds in screening. However, choosing a chiral column based on analyte structure alone does not guarantee success. Column companies suggest the use of their screening kits using various CSPs and mobile phases. Of course, stationary phase selectivity is the biggest factor. On top of that, a wide range of mobile phases also need to be available in screening. There are systems that can screen multiple columns simultaneously. Since most chiral compounds of forensic concern, especially methamphetamine and related compounds have been in circulation for some time, column screening is not as demanding as in the pharmaceutical industry and will not be discussed in detail here.

Enantiomers of both amphetamines and norephedrineswere clearly separated by HPLC using a chiral crown ether column (CROWNPACK CR(+)) [67]. Even 0.1% *d*-amphetamine in bulk methamphetamine could be determined. When urine from a methamphetamine abuser was examined, *d*-amphetamine was detected as a metabolite while there was no detectable *l*-amphetamine.

A commercially available CHIRALPAK® AS-H column (amylose tris((*S*)-methylbenzylcarbamate coated on 5-μm silica gel) was found to be able to enantiomerically resolve 19 out of 24 cathinone-related compounds [68]. HPLC (normal phase) was run under isocratic conditions with a mixture of hexane, isopropanol, and triethylamine (97:3:0.1).

Two chiral LC-MS methodologies utilizing CBH (cellobiohydrolase) and Chirobiotic V (vancomycin) columns for the quantification of amphetamine, methamphetamine, MDA (methylenedioxyamphetamine), MDMA, propranolol, atenolol, metoprolol, fluoxetine and venlafaxine in river water and sewage effluent were compared [69]. Lower method detection limits (0.3–5.0 ng/L and 1.3–15.1 ng/L for river water and sewage effluent, respectively) for all compounds were achieved using the Chirobiotic V column, except for methamphetamine and MDMA. However, the CBH column gave better resolution (Rs = 2.5 for amphetamine enantiomers *vs* 1.2 with Chirobiotic V), which equals to the difference (d = t_{RB} - t_{RA}, t_{RB} > t_{RA}) in retention times between the two peaks divided by the average of their combined widths (in time units) and which is a measure of how well two peaks are separated with bigger value representing better separation.

Another study compared several chiral stationary phases for the chiral determination of cathinones [70]. Using reversed phase mobile phases, no chiral separation for mephedrone was observed on Chirobiotic V, Cyclobond I 2000 DNP, Whelk-O1 and AmyCoat columns. However, using normal phase solvents, chiral separation was achieved except on the Chiralcel OJ-H column. After optimization, a 250 mm × 4.6 mm I.D. Whelk-O1 gave the best enantioselectivity (1.59) and enantioresolution (5.90) column using anisopropanol(IPA)-hexane-trifluoroacetic acid(TFA)–triethylamine(TEA)(10:90:0.05:0.05) mobile phase. For all the other 6 tested cathinones, enantiomeric separation was achieved on RegisCell with using IPA–hexane–TFA (15:85:0.1), while it was not successful on Whelk-O1.

A simple, rapid, and accurate method was developed for the determination of amphetamine and methamphetamine enantiomers in urine using LCMS/MS on a Chirobiotic™ V2 column (vancomycin, 250 mm × 2.1 mm, 5 μm) with methanol containing 0.1% (v/v) glacial acetic acid and 0.02% (v/v) ammonium hydroxide as the mobile phase [71]. The ESI mass spectrometer was operated in the positive ion multiple-reaction monitoring (MRM) mode. Only the *S*-isomers (*d*-isomers) of methamphetamine and amphetamine were detected in 72 of 86 urine samples from suspected methamphetamine abusers. In the rest 14 samples, both enantiomers of methamphetamine and/or amphetamine were detected and the concentrations of *R*-methamphetamine ranged from below the LOQ (0.05 mg/L) to 13.76 mg/L. Another example used the same column and mobile phase flow rateand successfully separated the enantiomers of these two drugs [72].

Sports Forensics

Chiral analysis has been used in sports to ensure fair play [73]. Methamphetamine, amphetamine, cathinone and methcathinone are able to influence the outcome of horse races and they are all Class I drugs according to the Association of Racing Commissioners International (ARCI) guidelines [74]. Their *S*-enantiomers are more potent CNS stimulants than their corresponding *R*-enantiomers. In animal studies, such effects increase in the order of (*S*)-amphetamine, (*S*)-cathinone, (*S*)-methamphetamine and (*S*)-methcathinone. A method for chiral analysis of these drugs in equine urine and plasma using LC-

MS/MS was developed and applied in the forensics of horse racing [73]. Following a liquid-liquid extraction (under basic pH using MTBE) from equine plasma (1 mL) and urine (0.1 mL), enantiomer differentiation and confirmation were achieved using liquid chromatography with a Chirobiotic V2 column coupled with mass spectrometry detection (Fig. **16**). A mixture of 95:5 (v/v) methanol and aqueous 50 mM ammonium formate/0.01% formic acid was used as the mobile phase with a flow rate of 0.45 mL/min. ESI positive mode multiple-reaction-monitoring (MRM) was used due to its high sensitivity and specificity. Each separation could be done within 6 minutes with low pg/mL LOD.

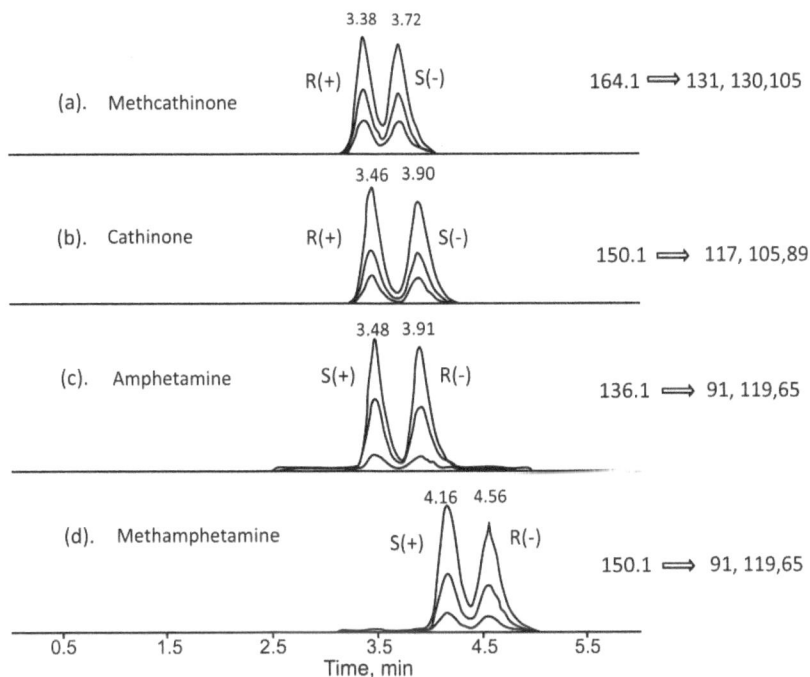

Fig. (16). LCMS separation of the enantiomers of methcathinone, cathinone amphetamine, and methamphetamine extracted from equine plasma spiked at 0.1 ng/mL using a Chirobiotic™ V2 column [73]. Reproduced with permission from reference [73].Copyright © 2015 by original authors and Scientific Research Publishing Inc.

Chiral Drugs in Waste Water

A sewage epidemiology approach to solving problems associated with the estimation of drugs usage in communities via enantiomeric profiling using LCMS analysis was developed [75]. Chiral drugs were extracted from waste water in the U. K.before separation was undertaken on Chiral-CBH column (100 mm x 2 mm, 5 µm) under isocratic conditions. The mobile phase was 1 mM ammonium acetate in 10% IPA aqueous solution (pH = 5.0). Components were monitored and analyzed by an ESI triple quadrupole mass spectrometer in positive MRM mode.

It was found that in raw waste water methamphetamine was racemic or enriched with the (S)-(+)-enantiomer. Amphetamine and MDMA was enriched with their (R)-(-)-enantiomers, pointing to their abuse as racemates. MDA was enriched with (S)-(+)-enantiomer, pointing to possible link with MDMA abuse instead of intentional MDA use. Only (1R, 2S)-(-)-ephedrine and (1S, 2S)-(+)-pseudoephedrine, not their corresponding synthetic enantiomers, were detected. There was more (1S, 2S)-(+)-pseudoephedrine in the winter and more (1R, 2S)-(-)-ephedrine during the spring and summer, while the cumulative concentration of ephedrines decreases between February and August. Non-enantioselective measurement of ephedrine concentrations did not seem to be a reliable indicator of actual potency of ephedrines used. Enantio-profiling might be more useful in determining whether drug residue results from consumption of illicit drug or metabolism of other drugs, verifying the potency of used drugs and monitoring drugs abuse patterns.

A similar study was done on raw and treated wastewater collected from four treatment plants [76]. (1R, 2S)-(-)-ephedrine and (1S, 2S)-(+)-pseudoephedrine were found to be in all samples, while amphetamine, methamphetamine, MDMA and MDEA were detected in some. Their enantiomeric fractions were found to be variable from sample to sample, pointing to possible influence from wastewater treatment processes.

Chiral Mobile Phase Additives

An HPLC method using a chiral mobile phase additive for the chiral separation of cathinones and amphetamines on a LiChrospher 100 RP-18e (End-capped C18, 250 x 4 mm, 5 μm) column under isocratic conditions was developed [77]. The optimized chiral mobile phase was a mixture of methanol and water (2.5:97.5), with 2% sulfated ß-cyclodextrin as the chiral additive. Enantiomers of amphetamine and 5 substituted amphetamines were baseline separated within 23 min. Of 25 cathinones, 3 were completely and 14 partially enantiomerically separated.

In one UPLC study, hydroxypropyl β-CD (HPBCD) was directly added to the mobile phase and the chiral separations of amphetamine-related compounds took less than 5 minutes. However, column lifetime and chromatographic efficiencies were compromised [64]. Peaks appeared broad due to the coexistence of two mechanisms: a nonstereoselective hydrophobic recognition and a chiral recognition. Since chiral recognition in the mobile phase tends to broaden peaks, efficiency was low in comparison with conventional UPLC, although HPBCD is the best chiral in selector for amphetamines and high efficiency separation of enantiomers were achieved in capillary electrophoresis(CE) with sharp peaks [64].

Capillary Electrophoresis

Employing chiral additives in the run buffer, highly efficient chiral separations by capillary electrophoresis (CE) are achievable without prior derivatization. Natural and/or modified cyclodextrins are the most commonly chiral additives. Although CE has limited application in commercial labs, it has been used for chiral drug analysis in the forensic science field.

Pseudoephedrine in tablet and syrup from different manufacturers was investigated using a system to separate its *d-* and *l-* isomers. Chiral separation was successfully achieved using a bare fused-silica capillary and phosphate buffer with carboxymethyl-β-cyclodextrin (CMBCD) as the chiral additive [78]. Fig. (**17**) shows electropherograms of standard *l-* and *d-* pseudoephedrines.

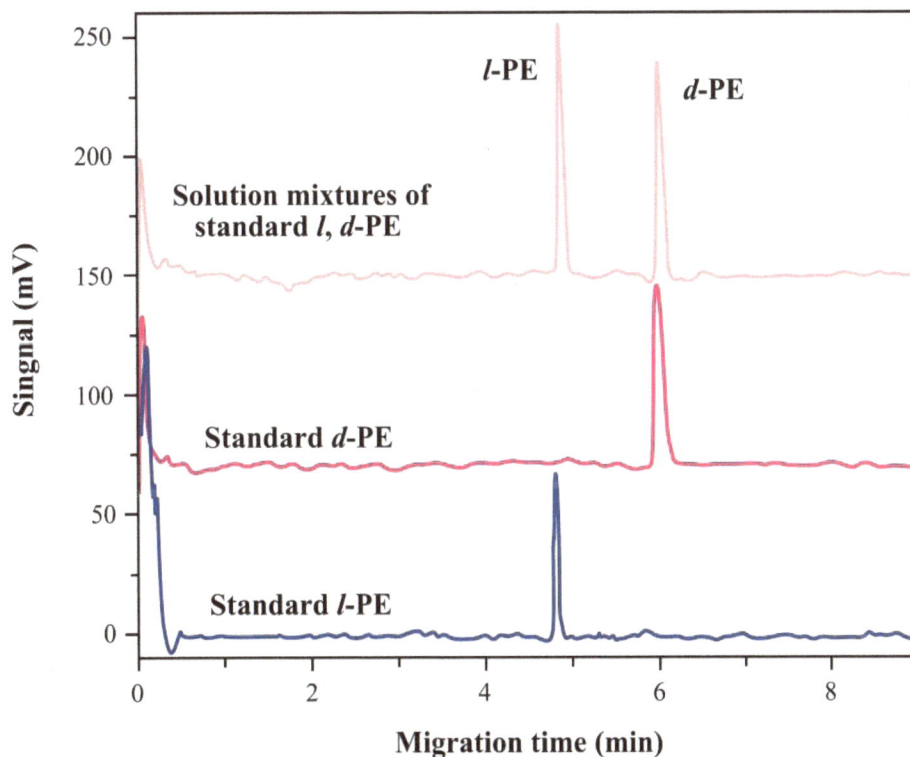

Fig. (17). Electropherograms of *l-* and *d-*pseudoephedrines. CE conditions: capillary (59.00 cm long, 38.30 cm to the detector), phosphate buffer (20.0 mM, pH 2.70) with CMBCD (4.0 mM), electrical field: 420 V/cm, electrokinetic injection (15 kV, 1 s) [78].

Dynamically coated the capillaries give faster and more robust electroosmotic flow compared to uncoated ones, especially at lower pH that's required for the chiral CE analysis of basic drugs [79]. CE analysis using an improved dynamic coating procedure achieved the chiral differentiation and determination of *d*- and *l*- methamphetamines and their precursors and/or byproducts in thousands of seized samples (Fig. **18**) [8]. A fused-silica capillaru (50 μm ID, 64.5 cm total length, 56 cm to the detector) kept at 15°C was treatcd with rapid sequential flushes of 0.1N sodium hydroxide, water, CElixir Reagent A (a buffer containing a polycationinc coating agent), for 1 min each, and finally CElixir B (a buffer containing a polyanionic coating agent, pH=2.5) together with hydroxypropyl--cyclodextrin (HPBCD) for 2 min. Pressure injection (50 mbar, 16 s) of samples was followed by a pressure injection (35 mbar, 1 s) of water to create a water plug, whose stacking effect and the advantage from a relatively large sample concentration helped achieve the sensitive determination of enantio-impurity down to 0.2%. Enantiomers of n-butylamphetamine were used as internal standards. Although the narrow CE peaks and possibly shifts in migration time can make identification using migration time alone problematic, more reproducible relative migration times and relative corrected areas to the standard(s) can be used to produce reliable qualitative and quantitative results. The relative standard deviations of migration time, relative migration time (to the 2nd internal standard peak j), corrected area (area/migration time), and relative corrected area are no bigger than 0.13%, 0.05%, 2.0%, and 0.92%, respectively.

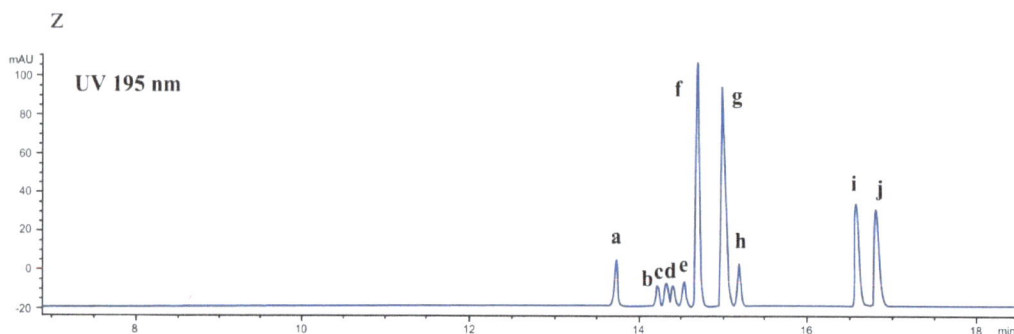

Fig. (18). Dynamically coated CE separation of standard mixture of (a) *l*-pseudoephedrine, (b)*d*-ephedrine, (c)*l*-amphetamine, (d)*l*-ephedrine, (e)*d*-amphetamine, (f)*l*-methamphetamine, (g)*d*-methamphetamine, (h)*d*-pseudoephedrine, (i) n-butylamphetamine (1), and (j) n-butylamphetamine (2) [8].Reproduced with permission from reference [8].

As a tertiary amine, methorphan is not susceptible to derivatization, rendering indirect enantiomeric analysis non-viable. However, using capillary electrophoresis with a dynamically coated capillary andthe chiral additive HPBCD, the chiral differentiation of *d*- and *l*-methorphans (Fig. **19**) was obtained in under 4 minutes with excellent peak shapes [80].

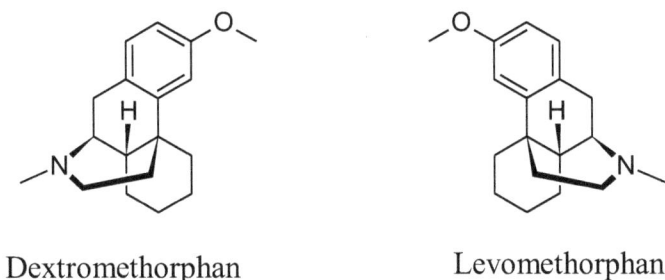

Dextromethorphan   Levomethorphan

Fig. (19). Structures of *d*- and *l*-methorphans.

A capillary dynamically coated with an SDS micelle allowed for the analysis of neutral, acidic, and weakly basic drugs [79]. Using cyclodextrin as an additive can enhance the separation selectivity of achiral analytes aside from enabling chiral separations [79].

CE-MS

Capillary electrophoresis coupled to mass spectrometry(CEMS) was used for the chiral determinations of several types of drugs andadulterants with high sensitivity and selectivity [81]. The running buffer was 1 mol/L formic acid containing 1 mmol/L 2, 3-diacetyl-6-sulfato-β-cyclodextrin and 10 mmol/L 2-hydroxypropyl-β-cyclodextrin. Enantiomeric separation of β-phenylethylamines, methadone and tetramisole was achieved within 30 minutes (20°C, +25 kV) as shown in Fig. (**20**). The concentration of chiral selectors was very low in order to avoid the contamination of the mass spectrometer and ion suppression. Comparing their chiral impurities, different batches of illicit methamphetamine samples could be differentiated.

The advantages of mass spectrometric detection compared to UV- and laser-induced fluorescence (LIF) detections in forensic analysis using CE were reviewed [82].

Chiral CE-MS/MS analysis using a chemically modified capillary containing sulfonated groups was developed for amphetamine, methamphetamine, norephedrine, norpseudoephedrine, ephedrine, pseudoephedrine, dimethyl-amphetamine and methylephedrine [83]. The running buffer was 10 mM formic acid aqueous solution containing 20 mM highly sulfated γ-cyclodextrin (pH 2.5). All of the 8 pairs of enantiomers were well resolved and identified within 60 min in one run. The sulfonated capillary showed larger electroosmotic mobility than an untreated fused-silica capillary over pH 2–9. What's more important, the electroosmotic mobility was relatively constant, the RSDs of the migration times being no more than 0.3% without any standardization. Such fine precision allowed easy identification of the impurities, although some of these analytes were quite similar in terms of migration times and mass spectral patterns. Seized methamphetamine samples were found to contain such impurities as *l*-ephedrine and *d*-pseudoephedrine with 0.2% LOD.

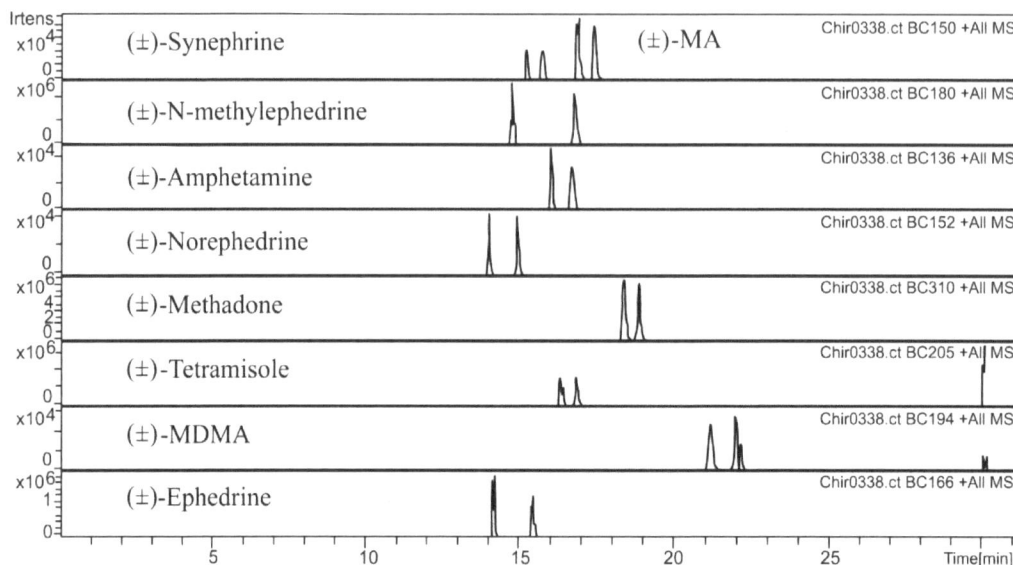

Fig. (20). CE separation of the enantiomers of β-phenylethylamines, methadone and tetramisole [81].

Supercritical Fluid Chromatography (SFC)

Supercritical fluids possess gas-like mass transfer and liquid-like solvation properties, making them uniquely useful in separations. The tunable solvation properties (through density/pressure programming) of the mostly commonly used supercritical carbon dioxidenot only improve the efficiency and speed of supercritical fluid extraction (SFE), but also enable supercritical fluid chromatography (SFC) to solve many problems that bother conventional chromatographic separations [84]. Compared with GC, capillary SFC requires no sample vaporization and can provide high resolution at much lower temperatures, which allows for fast analysis of thermo-labile compounds. Compared with HPLC, SFC is "greener" and faster. CO_2 can be collected as a byproduct or directly from the atmosphere, contributing no new chemicals to the environment. Most columns used in normal phase HPLC can be used in SFC, while the mobile phase flow rate in SFC is much faster than normal phase HPLC because of the low viscosity of supercritical CO_2. The diffusion of solutes in supercritical fluids is about ten times faster than that in liquids, which results in a decrease in resistance to mass transfer in the column and allows for faster separations without compromising efficiency. SFC chromatography can be scaled up to the preparatory level. Product recovery is much easier because much less solvent/additive needs to be removed than in preparative HPLC. Nowadays, SFC is more prevalent than HPLC for normal phase analysis in pharmaceutical research and development.

SFC is becoming more popular for the analysis of forensic evidence. It was utilized for the chiral analysis of methamphetamine, amphetamine, ephedrine, and pseudoephedrine recently [30]. Determination of the chiral composition of illicit methamphetamine, even for samples with skewed ratios as low as 0.1% *d-* or 3% *l-*, was successful with a Trefoil AMY1 (150 x 2.1 mm, 2.5 μm) column using a supercritical CO_2 mobile phase (flow rate 2.5 mL/min), as shown in Fig. (**21**). Ethanol containing 1% cyclohexylamine (CHA) as an additive was added as an organic modifier with a 8-30% gradient in 6 minutes. Using a single quadrupole mass spectrometer detector (SQD) as the detector in positive ESI and selected ion recording (SIR) mode, the resolution was 1.2, detection limit was 0.2 μg/mL, and linearity range was 0.5 to 200 μg/mL. Amphetamine enantiomers were base-line resolved. However, this condition is not ideal for other analytes.

Using MeOH/EtOH (50/50) containing 0.3% CHA additive as the modifier (gradient 5-7.5% in 6 min), all four ephedrine/pseudoephedrine diastereomers were well resolved, while amphetamine enantiomers were baseline resolved (Fig. **22**). For methamphetamine enantiomers, however, the resolution was only 0.9 and detection limit for the enantio-impurity (*l*-methamphetamine) was 3%.

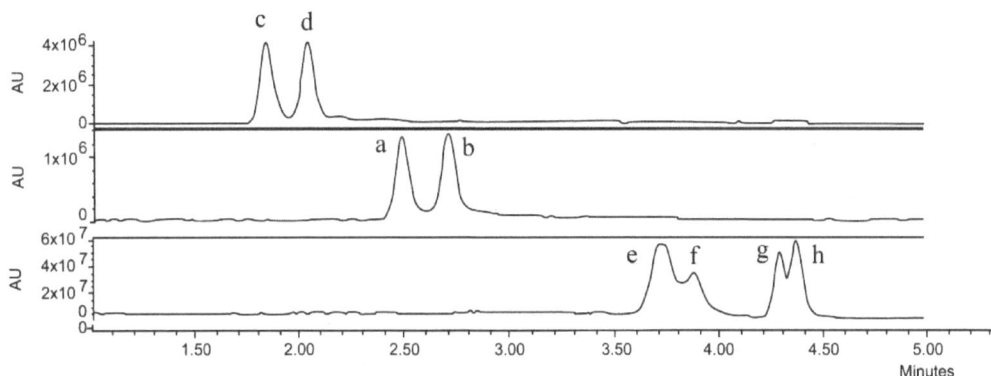

Fig. (21). SFC methamphetamine chiral separation: a, *d*-methamphetamine; b, *l*-methamphetamine; c, *d*-amphetamine; d, *l*-amphetamine; e, *d*-pseudoephedrine; f, *l*-pseudoephedrine; g, *d*-ephedrine; h, *l*-ephedrine.Co-solvent: EtOH and CHA(1%), gradient: 8-30% in 6 min, flow rate: 2.5 mL/min [30]. Reproduced with permission from reference [30].

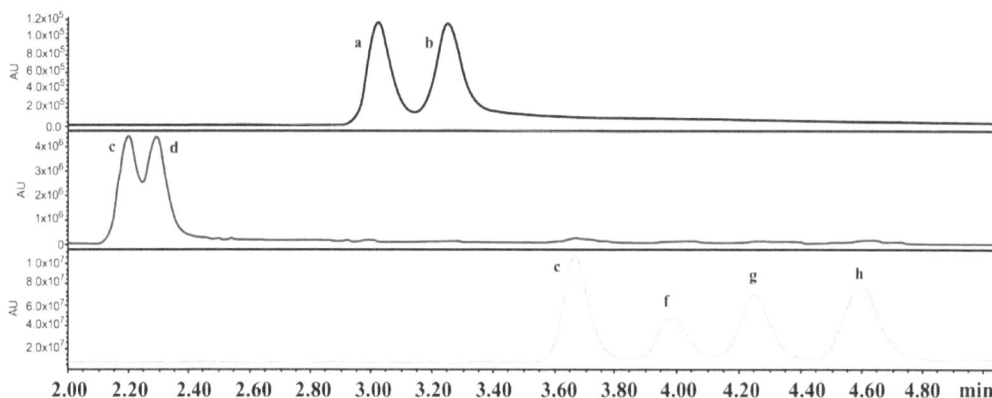

Fig. (22). SFC chromatograms of co-solvent screening. Co-solvent: MeOH/EtOH (50/50) with 0.3% CHA, gradient: 5-7.5% in 6 min, flow rate: 2.5 mL/min [30]. Reproduced with permission from reference [30].

SFC-MS

Mass spectrometers can be used as detectors for SFC. Atmospheric pressure chemical ionization (APCI) and electrospray ionization (ESI) are most prevalent because chromatographic effluents can be directly introduced into mass spectrometers with such sources. Chiral SFC-MS is especially useful in drug discovery. It is quite likely to be routinely used in the forensic science field [85].

CONCLUSION

The chiral analysis of some seized drugs, specially amphetamine, methamphetamine and related compounds, is important for legal and intelligence purposes. Therefore, the developmentof rapid, precise, and reproducible chiral separation techniques is very important. Gas chromatography (GC), especially gas chromatography-mass spectrometry (GC-MS) has been the workhorse in forensic drug analysis. L-TPC has been traditionally and effectively employed to derivatize chiral drugs before the determination of enantiomeric composition by GC. L-TPC derivatization coupled with separation on a chiral column can remove or quantify the interference from D-TPC impurities. Recent advance in chiral GC analysis brought about more enantiopure chiral derivatization agents, which are less susceptible to racemization, to produce more reproducible results. HPLC, which allows for the direct chiral analysis and have been successfully used in the forensic science field, often does not offer high separation efficiency and resolution. In addition, for the chiral GC or HPLC analysis of chiral solutes, expensive chiral columns (usually specific for a class of compounds) and/or tedious derivatization are required. CE has significantly greater resolving power than HPLC. CE, which uses relatively inexpensive chiral additives such as cyclodextrins in the run buffer, allows for the direct chiral determination of seized drugs using conventional capillaries. Dynamically coated capillaries, using an initial coating with a polymeric cation and subsequent coating with a polymeric anion or micelle, offer faster separation times, improved precision, and increased selectivity for the analysis of basic, acidic, and neutral drugs of forensic interest. With dynamic coatings, capillary electrophoresis analysis of a wide variety of drugs and adulterants was be done using the same capillary. Supercritical fluid chromatography has several main advantages over other conventional chromatographic techniques (GC and HPLC). Compared with HPLC, SFC provides rapid separations without the use of organic solvents. In addition, SFC separations can be done faster than HPLC separations. SFC has been recently used for chiral drug analysis and its popularity is set to increase in the forensic science field. With the advent and the advancement of ultrafast chiral separations [86] and miniaturized mass spectrometers [87], it is conceivable that HPLC, CE and SFC fitted with multiple columns, each with its own high-resolution mass-selective detector and its own chiroptical detector (such as CD and VCD) [88, 89] will be commercially available and find their application in high-throughput forensic drug analysis, including chiral analysis. Such instrument would also spur more scientific investigation on new designer drugs and their toxicological effect, helping develop mitigation agents and vaccines.

CONSENT FOR PUBLICATION

Not applicable.

CONFLICT OF INTEREST

The author confirms that he has no conflict of interest to declare for this publication.

ACKNOWLEDGEMENTS

Declared none.

REFERENCES

[1]   Bell, S. *Forensic chemistry*. Pearson Prentice Hall: Upper Saddle River, NJ, **2006**.

[2]   Levine, B. *Principles of forensic toxicology: revised and updated.* 2nd ed. American Association for Clinical Chemistry: Washington, D.C., **2006**.

[3]   Cole, M.D.; Caddy, B. *The analysis of drugs of abuse.* CRC Press: New York, **1994**.

[4]   Smith, F. *Handbook of forensic drug analysis*, 1st Ed. Academic Press: Salt Lake City, **2004**.

[5]   Ballou, S.; Goodpaster, J.; MacCrehan, W.; Reeder, D. Forensic analysis. *Anal. Bioanal. Chem.,* **2003**, *376*(8), 1149-1150.
[http://dx.doi.org/10.1007/s00216-003-2068-x] [PMID: 12851733]

[6]   Smith, J.P.; Sutcliffe, O.B.; Banks, C.E. An overview of recent developments in the analytical

detection of new psychoactive substances (NPSs). *Analyst (Lond.),* **2015**, *140*(15), 4932-4948.
[http://dx.doi.org/10.1039/C5AN00797F] [PMID: 26031385]

[7]   Klein, R.F.; Hays, P.A. Detection and analysis of drugs of forensic interest. *Microgram J.,* **2003**, *1*(1-2), 55-153.

[8]   Lurie, I.S.; Bozenko, J.S., Jr; Li, L.; Miller, E.E.; Greenfield, S.J. CChiral separation of methamphetamine and related compounds using capillary electrophoresis with dynamically coated capillaries. *Microgram J.,* **2011**, *8*(1), 24-28.

[9]   Sievert, H-J. Determination of amphetamine and methamphetamine enantiomers by chiral derivatization and gas chromatography-mass spectrometry as a test case for an automated sample preparation system. *Chirality,* **1994**, *6*, 295-301.
[http://dx.doi.org/10.1002/chir.530060413]

[10]   Bell, S. *Drugs, poison, and chemistry.* Infobase Publishing: New York, **2009**.

[11]   Brettell, T.A.; Butler, J.M.; Almirall, J.R. Forensic science. *Anal. Chem.,* **2009**, *81*(12), 4695-4711.
[http://dx.doi.org/10.1021/ac9008786] [PMID: 19422250]

[12]   Smith, M.L.; Nichols, D.C.; Underwood, P.; Fuller, Z.; Moser, M.A.; Flegel, R.; Gorelick, D.A.; Newmeyer, M.N.; Concheiro, M.; Huestis, M.A. Methamphetamine and amphetamine isomer concentrations in human urine following controlled Vicks VapoInhaler administration. *J. Anal. Toxicol.,* **2014**, *38*(8), 524-527.
[http://dx.doi.org/10.1093/jat/bku077] [PMID: 25217541]

[13]   SWGDRUG. Scientific Working Group for the Analysis of Seized Drugs (SWGDRUG) Recoomendations. Online, **2011**.

[14]   Bono, J. *Report of the Scientific Working Group for the Analysis of Seized Drugs (SWGDRUG) Conference (Montreal)*; **2003**.

[15]   Person, E.C.; Meyer, J.A.; Vyvyan, J.R. Structural determination of the principal byproduct of the lithium-ammonia reduction method of methamphetamine manufacture. *J. Forensic Sci.,* **2005**, *50*(1), 87-95.
[http://dx.doi.org/10.1520/JFS2004204] [PMID: 15831001]

[16]   Mills, T., III; Roberson, J.C.; Matchett, C.C.; Simon, M.J.; Burns, M.D.; Ollis, R.J., Jr *Instrumental data for drug dnalysis,* 3rd ed; CRC Press: New York, **2005**.

[17]   Oulton, S.R. Separation and identification of ephedrine, pseudoephedrine, and methamphetamine mixtures. *Microgram Bulletin,* **1997**, *30*(12), 289-296.

[18]   Burks, R.M.; Pacquette, S.E.; Guericke, M.A.; Wilson, M.V.; Symonsbergen, D.J.; Lucas, K.A.; Holmes, A.E. DETECHIP: a sensor for drugs of abuse. *J. Forensic Sci.,* **2010**, *55*(3), 723-727.
[http://dx.doi.org/10.1111/j.1556-4029.2010.01323.x] [PMID: 20202067]

[19]   Zheng, X.; Yoo, M.J.; Hage, D.S. Analysis of free fractions for chiral drugs using ultrafast extraction and multi-dimensional high-performance affinity chromatography. *Analyst (Lond.),* **2013**, *138*(21), 6262-6265.
[http://dx.doi.org/10.1039/c3an01315d] [PMID: 23979112]

[20]   Nguyen, L.A.; He, H.; Pham-Huy, C. Chiral drugs: an overview. *Int. J. Biomed. Sci.,* **2006**, *2*(2), 85-100.
[PMID: 23674971]

[21]   Liu, J-T.; Liu, R.H. Enantiomeric composition of abused amine drugs: chromatographic methods of analysis and data interpretation. *J. Biochem. Biophys. Methods,* **2002**, *54*(1-3), 115-146.
[http://dx.doi.org/10.1016/S0165-022X(02)00136-7] [PMID: 12543495]

[22]   United States Sentencing Commission Guidelines Manual, **1994**; p 90.

[23]   Slocum, D.W.; Surgarman, D.; Tucker, S.P. The two faces of D and L nomenclature. *J. Chem. Educ.,* **1971**, *48*(9), 597.

[http://dx.doi.org/10.1021/ed048p597]

[24]   *US Federal Sentencing Guidelines Manual.* **2015**, Chapter 2 D.

[25]   Cody, J.T.; Valtier, S.; Nelson, S.L. Amphetamine enantiomer excretion profile following administration of Adderall. *J. Anal. Toxicol.,* **2003**, *27*(7), 485-492.
[http://dx.doi.org/10.1093/jat/27.7.485] [PMID: 14607004]

[26]   UNODC. Methods for the identification and analysis of amphetamine, methamphetamine. United Nations Office on Drugs and Crime: **2006**.

[27]   White, M. *FSS report on Methyamphetamine: chemistry, seizure, statistics, analysis, synthetic routes and histrory of illicit manufacture in the UK and the USA for the ACMD Working Group on Methylamphetamine*; Forensic Science Service, Drugs Intelligence Service: **2004**

[28]   Doerr, J.A.; Kalchik, M.F.; Massetti, J. Mexican National Methamphetamine Laboratories in Central California-Trends and adaptations over time. **1997**.

[29]   Title VII of the PATRIOT Improvement and Reauthorization Act of 2005 [Public Law 109-177]. **2005**.

[30]   Li, L. Enantiomeric determination of methamphetamine, amphetamine, ephedrine, and pseudoephedrine using chiral supercritical fluid chromatography with mass spectrometric detection. *Microgram J.,* **2015**, *12*(1-4), 19-30.

[31]   Moss, G. Basic terminology of stereochemistry. *Pure Appl. Chem.,* **2009**, *68*(12), 2193-2222. [IUPAC Recommendations 1996].

[32]   Cahn, R.S.; Ingold, C.K.; Prelog, V. The specification of asymmetric configuration in organic chemistry. *Experienta,* **1956**, *12*, 81-124.
[http://dx.doi.org/10.1007/BF02157171]

[33]   Nakanishi, K.j.; Berova, N.; Woody, R. Circular dichroism: In: *principles and applications*; Wiley-VCH: New York, **2000**; p. 877.

[34]   Berg, R.W. CShim, I.; White, P. C.; Abdali, S. Raman optical activity and raman Spectra of amphetamine species—quantum chemical model calculations and experiments. *Am. J. Anal. Chem.,* **2012**, *3*, 410-421.
[http://dx.doi.org/10.4236/ajac.2012.36054]

[35]   In *The Merck Index*; The Royal Society of Chemistry, **2017**.

[36]   JASCO. Application book: circular dichroism and polarimeters. Application CD-01-04, Circular dichroism spectrometry for the analysis of amphetamines; pp18.

[37]   Mitscher, L.A.; Kautz, F.; LaPidus, J. Optical rotatory dispersion and circular dichroism of diastereoisomers. I. The ephedrines and chloramphenicols. *Can. J. Chem.,* **1969**, *47*, 1957-1963.
[http://dx.doi.org/10.1139/v69-318]

[38]   Heagy, J.A. Infrared method for distinguishing optical isomers of amphetamine. *Anal. Chem.,* **1970**, *42*(12), 1459-1459.
[http://dx.doi.org/10.1021/ac60294a009] [PMID: 5458229]

[39]   Allen, A.; Cooper, D.; Kiser, W.; Cottreli, R. The Cocaine Diastereoisomers. The Cocaine Diastereoisomer. *J. Forensic Sci.,* **1981**, *26*(1), 12-26.
[http://dx.doi.org/10.1520/JFS11325J]

[40]   LeBelle, M.J.; Savard, C.; Dawson, B.A.; Black, D.B.; Katyal, L.K.; Zrcek, F.; By, A.W. Chiral identification and determination of ephedrine, pseudoephedrine, methamphetamine and methcathinone by gas chromatography and nuclear magnetic resonance. *Forensic Sci. Int.,* **1995**, *71*(3), 215-223.
[http://dx.doi.org/10.1016/0379-0738(94)01669-0] [PMID: 7713461]

[41]   Armstrong, D.W.; Li, W.Y.; Pitha, J. Reversing enantioselectivity in capillary gas chromatography with polar and nonpolar cyclodextrin derivative phases. *Anal. Chem.,* **1990**, *62*(2), 214-217.

[http://dx.doi.org/10.1021/ac00201a023] [PMID: 2310015]

[42]   Lurie, I.S.; Hays, P.A.; Parker, K. Capillary electrophoresis analysis of a wide variety of seized drugs using the same capillary with dynamic coatings. *Electrophoresis,* **2004**, *25*(10-11), 1580-1591. [http://dx.doi.org/10.1002/elps.200405894] [PMID: 15188245]

[43]   Northrop, D.; Person, E. C.; Knops, L. *Capillary electrophoretic analysis of clandestine methamphetamine laboratory evidence,* **2005**.

[44]   Tagliaro, F.; Pascali, J.P.; Lewis, S.W. Capillary electrophoresis in forensic chemistry A2 - Siegel, J. A.*Encyclopedia of Forensic Sciences*; Saukko, P.J.; Houck, M.M., Eds.; Academic Press: Waltham, **2013**, pp. 567-572. [http://dx.doi.org/10.1016/B978-0-12-382165-2.00243-9]

[45]   Tagliaro, F.; Bortolotti, F. Recent advances in the applications of CE to forensic sciences (2001-2004). *Electrophoresis,* **2006**, *27*(1), 231-243. [http://dx.doi.org/10.1002/elps.200500697] [PMID: 16421953]

[46]   Nagori, B. P.; Deora, M. S.; Saraswat, P. Chiral drug analysis and their application. *Int. J. Pharmaceut. Sci. Rev. Res. 6*, 106-113.

[47]   McConathy, J.; Owens, M.J. Stereochemistry in drug action. *Prim. Care Companion J. Clin. Psychiatry,* **2003**, *5*(2), 70-73. [http://dx.doi.org/10.4088/PCC.v05n0202] [PMID: 15156233]

[48]   van de Manakker, F.; Vermonden, T.; van Nostrum, C.F.; Hennink, W.E. Cyclodextrin-based polymeric materials: synthesis, properties, and pharmaceutical/biomedical applications. *Biomacromolecules,* **2009**, *10*(12), 3157-3175. [http://dx.doi.org/10.1021/bm901065f] [PMID: 19921854]

[49]   SWGDRUG. Supplemental Document SD-2 for Prt IVB: Quality assurance/validation of analytical methods, **2006**.

[50]   Hensley, D.; Cody, J.T. Simultaneous determination of amphetamine, methamphetamine, methylenedioxyamphetamine (MDA), methylenedioxymethamphetamine (MDMA), and methylenedioxyethylamphetamine (MDEA) enantiomers by GC-MS. *J. Anal. Toxicol.,* **1999**, *23*(6), 518-523. [http://dx.doi.org/10.1093/jat/23.6.518] [PMID: 10517560]

[51]   Cody, J.T. Determination of methamphetamine enantiomer ratios in urine by gas chromatography-mass spectrometry. *J. Chromatogr. A,* **1992**, *580*(1-2), 77-95. [http://dx.doi.org/10.1016/0378-4347(92)80529-Y] [PMID: 1400833]

[52]   Valdez, L. Chiral analyses of ephedrines. Honors College Thesis, Pace University, New York, **2013**.

[53]   Liu, J.H.; Ku, W.W. Determination of enantiomeric N-trifluoroacetyl-L-prolyl chloride amphetamine derivatives by capillary gas chromatography/mass spectrometry with chiral and achiral stationary phases. *Anal. Chem.,* **1981**, *53*(14), 2180-2184. [http://dx.doi.org/10.1021/ac00237a008]

[54]   Strano-Rossi, S.; Botrè, F.; Bermejo, A.M.; Tabernero, M.J. A rapid method for the extraction, enantiomeric separation and quantification of amphetamines in hair. *Forensic Sci. Int.,* **2009**, *193*(1-3), 95-100. [http://dx.doi.org/10.1016/j.forsciint.2009.09.016] [PMID: 19896309]

[55]   Nyström, I.; Trygg, T.; Woxler, P.; Ahlner, J.; Kronstrand, R. Quantitation of *R*-(-)- and S-(+--amphetamine in hair and blood by gas chromatography-mass spectrometry: an application to compliance monitoring in adult-attention deficit hyperactivity disorder treatment. *J. Anal. Toxicol.,* **2005**, *29*(7), 682-688. [http://dx.doi.org/10.1093/jat/29.7.682] [PMID: 16419401]

[56]   Mohr, S.; Weiß, J.A.; Spreitz, J.; Schmid, M.G. Chiral separation of new cathinone- and amphetamine-related designer drugs by gas chromatography-mass spectrometry using trifluoroacetyl-

l-prolyl chloride as chiral derivatization reagent. *J. Chromatogr. A,* **2012**, *1269*, 352-359.
[http://dx.doi.org/10.1016/j.chroma.2012.09.079] [PMID: 23058937]

[57]   LeBelle, M.J.; Lauriault, G.; Lavoie, A. Gas chromatographic-mass spectrometric identification of chiral derivatives of the alkaloids of KHAT. *Forensic Sci. Int.,* **1993**, *61*(1), 53-64.
[http://dx.doi.org/10.1016/0379-0738(93)90249-A]

[58]   Holler, J.M.; Vorce, S.P.; Bosy, T.Z.; Jacobs, A. Quantitative and isomeric determination of amphetamine and methamphetamine from urine using a nonprotic elution solvent and *R*(-)-α-meth-xy-α-trifluoromethylphenylacetic acid chloride derivatization. *J. Anal. Toxicol.,* **2005**, *29*(7), 652-657.
[http://dx.doi.org/10.1093/jat/29.7.652] [PMID: 16419395]

[59]   Rasmussen, L.B.; Olsen, K.H.; Johansen, S.S. Chiral separation and quantification of R/S-amphetamine, R/S-methamphetamine, R/S-MDA, R/S-MDMA, and R/S-MDEA in whole blood by GC-EI-MS. *J. Chromatogr. B Analyt. Technol. Biomed. Life Sci.,* **2006**, *842*(2), 136-141.
[http://dx.doi.org/10.1016/j.jchromb.2006.05.011] [PMID: 16797258]

[60]   Peters, F.T.; Kraemer, T.; Maurer, H.H. Drug testing in blood: validated negative-ion chemical ionization gas chromatographic-mass spectrometric assay for determination of amphetamine and methamphetamine enantiomers and its application to toxicology cases. *Clin. Chem.,* **2002**, *48*(9), 1472-1485.
[PMID: 12194924]

[61]   Jin, H.L.; Beesley, T.E. Enantiomeric separation of amphetamine and methamphetamine by capillary gas chromatography. *Chromatographia,* **1994**, *38*(9), 595-598.
[http://dx.doi.org/10.1007/BF02277160]

[62]   Armstrong, D.W.; Rundlett, K.L.; Nair, U.B.; Reid, J.L. Enantioresolution of amphetamine, methamphetamine, and Deprenyl (selegiline) by LC, GC, and CE. *Curr. Sep.,* **1996**, *15*(2), 57-61.

[63]   Newmeyer, M.N.; Concheiro, M.; Huestis, M.A. Rapid quantitative chiral amphetamines liquid chromatography-tandem mass spectrometry: method in plasma and oral fluid with a cost-effective chiral derivatizing reagent. *J. Chromatogr. A,* **2014**, *1358*, 68-74.
[http://dx.doi.org/10.1016/j.chroma.2014.06.096] [PMID: 25065924]

[64]   Guillarme, D.; Bonvin, G.; Badoud, F.; Schappler, J.; Rudaz, S.; Veuthey, J-L. Fast chiral separation of drugs using columns packed with sub-2 microm particles and ultra-high pressure. *Chirality,* **2010**, *22*(3), 320-330.
[PMID: 19544347]

[65]   Steuer, A.E.; Schmidhauser, C.; Liechti, M.E.; Kraemer, T. Development and validation of an LC-MS/MS method after chiral derivatization for the simultaneous stereoselective determination of methylenedioxy-methamphetamine (MDMA) and its phase I and II metabolites in human blood plasma. *Drug Test. Anal.,* **2015**, *7*(7), 592-602.
[http://dx.doi.org/10.1002/dta.1740] [PMID: 25371386]

[66]   Andersson, S.; Allenmark, S.G. Preparative chiral chromatographic resolution of enantiomers in drug discovery. *J. Biochem. Biophys. Methods,* **2002**, *54*(1-3), 11-23.
[http://dx.doi.org/10.1016/S0165-022X(02)00126-4] [PMID: 12543489]

[67]   Makino, Y.; Ohta, S.; Hirobe, M. Enantiomeric separation of amphetamine by high-performance liquid chromatography using chiral crown ether-coated reversed-phase packing: application to forensic analysis. *Forensic Sci. Int.,* **1996**, *78*(1), 65-70.
[http://dx.doi.org/10.1016/0379-0738(95)01865-4]

[68]   Mohr, S.; Taschwer, M.; Schmid, M.G. Chiral separation of cathinone derivatives used as recreational drugs by HPLC-UV using a CHIRALPAK® AS-H column as stationary phase. *Chirality,* **2012**, *24*(6), 486-492.
[http://dx.doi.org/10.1002/chir.22048] [PMID: 22544697]

[69]   Bagnall, J.P.; Evans, S.E.; Wort, M.T.; Lubben, A.T.; Kasprzyk-Hordern, B. Using chiral liquid chromatography quadrupole time-of-flight mass spectrometry for the analysis of pharmaceuticals and

illicit drugs in surface and wastewater at the enantiomeric level. *J. Chromatogr. A,* **2012**, *1249*, 115-129.
[http://dx.doi.org/10.1016/j.chroma.2012.06.012] [PMID: 22749459]

[70]   Perera, R.W.; Abraham, I.; Gupta, S.; Kowalska, P.; Lightsey, D.; Marathaki, C.; Singh, N.S.; Lough, W.J. Screening approach, optimisation and scale-up for chiral liquid chromatography of cathinones. *J. Chromatogr. A,* **2012**, *1269*, 189-197.
[http://dx.doi.org/10.1016/j.chroma.2012.11.001] [PMID: 23174478]

[71]   Wang, T.; Shen, B.; Shi, Y.; Xiang, P.; Yu, Z. Chiral separation and determination of R/S-methamphetamine and its metabolite R/S-amphetamine in urine using LC-MS/MS. *Forensic Sci. Int.,* **2015**, *246*, 72-78.
[http://dx.doi.org/10.1016/j.forsciint.2014.11.009] [PMID: 25460108]

[72]   Moore, I.; Sakuma, T.; Herrera, M.J.; Kuntz, D. LC-MS/MS chiral separation of "*d*" and "*l*" amphetamines and methamphetamines. Sciex, AB, Ed. **2013**; Publication number: 7400113-01, Document number: RUOMKT- 02-0403-A.

[73]   Wang, C.C.; Hartmann-Fischbach, P.; Krueger, T.R.; Lester, A.; Simonson, A. Fast and sensitive chiral analysis of amphetamines and cathinones in equine urine and plasma using liquid chromatography tandem mass spectrometry. *Am. J. Anal. Chem.,* **2015**, *6*, 995-1003.
[http://dx.doi.org/10.4236/ajac.2015.613094]

[74]   Uniform classification guidelines for foreign substances and recommended penalties model rule. Association of Racing Commissioners International, I., Ed., **2016**.

[75]   Kasprzyk-Hordern, B.; Baker, D.R. Estimation of community-wide drugs use via stereoselective profiling of sewage. *Sci. Total Environ.,* **2012**, *423*, 142-150.
[http://dx.doi.org/10.1016/j.scitotenv.2012.02.019] [PMID: 22404981]

[76]   Kasprzyk-Hordern, B.; Kondakal, V.V.; Baker, D.R. Enantiomeric analysis of drugs of abuse in wastewater by chiral liquid chromatography coupled with tandem mass spectrometry. *J. Chromatogr. A,* **2010**, *1217*(27), 4575-4586.
[http://dx.doi.org/10.1016/j.chroma.2010.04.073] [PMID: 20537654]

[77]   Taschwer, M.; Seidl, Y.; Mohr, S.; Schmid, M.G. Chiral separation of cathinone and amphetamine derivatives by HPLC/UV using sulfated ß-cyclodextrin as chiral mobile phase additive. *Chirality,* **2014**, *26*(8), 411-418.
[http://dx.doi.org/10.1002/chir.22341] [PMID: 24909415]

[78]   Khlangthamniam, B.; Tophan, P.; Wilairat, P.; Chantiwas, R. In *Invetigation of chiral analysis of pseudoephedrine in decongestant by simple capillary electrophoresis system*, The 2nd ASEAN Plus Three Graduate Research Congress (2ndAGRC), Bankok, **2014**; pp 881-885.

[79]   Lurie, I.S.; Hays, P.A.; Parker, K. Capillary electrophoresis analysis of a wide variety of seized drugs using the same capillary with dynamic coatings. *Electrophoresis,* **2004**, *25*(10-11), 1580-1591.
[http://dx.doi.org/10.1002/elps.200405894] [PMID: 15188245]

[80]   Lurie, I.S.; Cox, K.A. Rapid chiral separation of dextro- and levo- methorphan using capillary electrophoresis with dynamically coated capillaries. *Microgram J.,* **2005**, *3*(3-4), 138-141.

[81]   Dieckmann, S. , **2008**. In *Enantiomeric identification of chiral drugs, adulterants and impurities by capillary electrophoresis-ESI-mass spectrometry (CE-ESIMS)*, Dieser Tagungsband XV. GTFCh-Symposium, 2008.

[82]   Huck, C. W.; Huck-Pezzei, V.; Bakry, R.; Bachmann, S.; Najam-ul-Haq, M.; Rainer, M.; Bonn, G. K. Capillary electrophoresis coupled to mass spectrometry for forensic analysis. *Open Chem. Eng. J.,* **2008**, *1*, 30-43.

[83]   Mikuma, T.; Iwata, Y.T.; Miyaguchi, H.; Kuwayama, K.; Tsujikawa, K.; Kanamori, T.; Inoue, H. The use of a sulfonated capillary on chiral capillary electrophoresis/mass spectrometry of amphetamine-type stimulants for methamphetamine impurity profiling. *Forensic Sci. Int.,* **2015**, *249*, 59-65.

[http://dx.doi.org/10.1016/j.forsciint.2015.01.015] [PMID: 25679984]

[84]   Wang, S-M.; Ling, Y.-C.; Giang, Y-S. Forensic applications of supercritical fluid extraction and chromatography. *Forensic Sci. J.,* **2003**, *2*(1), 5-18.

[85]   Płotka, J.M.; Biziuk, M.; Morrison, C.; Namieśnik, J. Pharmaceutical and forensic drug applications of chiral supercritical fluid chromatography. *Trends Analyt. Chem.,* **2014**, *56*, 74-89.
       [http://dx.doi.org/10.1016/j.trac.2013.12.012]

[86]   Barhate, C.L.; Regalado, E.L.; Contrella, N.D.; Lee, J.; Jo, J.; Makarov, A.A.; Armstrong, D.W.; Welch, C.J. Ultrafast chiral chromatography as the second dimension in two-dimensional liquid chromatography Experiments. *Anal. Chem.,* **2017**, *89*(6), 3545-3553.
       [http://dx.doi.org/10.1021/acs.analchem.6b04834] [PMID: 28192943]

[87]   Xu, W.; Manicke, N.E.; Cooks, G.R.; Ouyang, Z. Miniaturization of mass spectrometry analysis Systems. *JALA Charlottesv Va,* **2010**, *15*(6), 433-439.
       [http://dx.doi.org/10.1016/j.jala.2010.06.004] [PMID: 21278840]

[88]   Salvadori, P.; Bertucci, C.; Rosini, C. Circular dichroism detection in HPLC. *Chirality,* **1991**, *3*, 376-385.
       [http://dx.doi.org/10.1002/chir.530030424]

[89]   Tran, C.D.; Grishko, V.I.; Huang, G. Chiral detection in high-performance liquid chromatography by vibrational circular dichroism. *Anal. Chem.,* **1994**, *66*(17), 2630-2635.
       [http://dx.doi.org/10.1021/ac00089a007] [PMID: 7943734]

CHAPTER 2

Low-Cost Feedstocks for Biofuels and High Value Added Products Production: Using Multi-Parameter Flow Cytometry as a Tool to Enhance the Process Efficiency

Teresa Lopes da Silva[*] and Alberto Reis

Laboratório Nacional de Energia e Geologia, Unidade de Bioenergia, Estrada do Paço do Lumiar, 22, 1649-038 Lisboa, Portugal

Abstract: Currently there is an increase search for microorganisms as a potential source of biofuels and other bioproducts with commercial interest. A way to reduce microbial cultivation costs consists of using low-cost feedstocks, such as lignocellulosic materials, industrial by-products, effluents and wastewaters to cultivate microbes which, in some cases, allow the simultaneous effluent biological treatment and the production of high value-added products. However, such feedstocks may contain, or their pre-treatment may release inhibitory compounds that affect microbial metabolism, thus the process efficiency. Multi-parameter flow cytometry (FC) is an advanced and powerful technique that has been used for bioprocess monitoring, showing many advantages over the conventional microbiological/analytical techniques that are still currently used in most laboratories for process control. This chapter will describe the cell stress response to the most known inhibitors present in low-cost feedstocks, emphasizing the advantages of using the at-line FC technique in these process monitoring.

Keywords: Crude Glycerol, Cell Membrane Damage, Enzymatic Activity, Flow Cytometry, Hydrophobic Wastes, Inhibitors, Intracellular pH, Intracellular Products, Low-Cost Feedstocks, Lignocellulosic Materials, Microorganisms, Reactive Oxygen Species, Wastewaters.

INTRODUCTION

Microalgae, yeasts and bacteria have been cultured on low-cost substrates to produce biofuels and high value added products, with the concomitant effluent treatment [1]. However, this approach raises a few hurdles. Despite low-cost feedstocks contain nutrients that allow microbial growth, they also contain

---

[*] **Corresponding author Teresa Lopes da Silva:** Laboratório Nacional de Energia e Geologia, Unidade de Bioenergia, Estrada do Paço do Lumiar, 22, 1649-038Lisboa, Portugal

inhibitory compounds, or, in other cases, the feedstock pre-treatment step necessary to release monomeric sugars to be converted by microorganisms, may also release inhibitory compounds that may induce cell dead or damage that affect its metabolism, which will reduce the process performance and yield. In order to understand the effect of these inhibitors on microbial cell physiology, it is essential to monitor, near real time, and during the process development, the cell physiological status. However, most of the published works that studied the effect of these inhibitory compounds on the microbial populations usually used conventional microbiological methods to monitor microbial fermentations, such as dry cell weight or serial dilution methods, which present a few problems and limitations, as the results are usually only available a considerable time after the sample is taken, often when the process is over, too late to change the process control strategy. Other methods such as optical density or capacitance monitoring, though faster, only provide average data, not giving any information on cell status. FC gives simultaneous information, near real time, on several cell functions and compartments, being ideal to assess the cellular stress response to environmental adverse conditions and to understand the cell mechanisms that are triggered by these conditions. The understanding of such mechanisms will allow the development of more tolerant microbial strains to these inhibitors and more efficient bioprocess control strategies, as the at-line multi-parameter cytometric information allows changing the control strategy during the time course of the bioprocess, by changing the operational conditions, in order to achieve the highest products yields and optimal process performances. Nevertheless, the use of FC in low cost feedstocks bioconversion processes is still rare. A detailed description of the effects of the most known inhibitors present in low-cost feedstocks on the microorganisms will be presented, highlighting the benefits using FC to monitor such processes.

LOW-COST FEEDSTOCKS INHIBITORS AND THEIR EFFECT ON MICROORGANISMS

The use of low-cost feedstocks for biofuels and high-value added products is crucial for reducing the process cost and, simultaneously, exempting the environment from polluted wastes [1]. A large variety of low cost raw materials have been considered and used as feedstock for the sustainable biofuels microbial production such as lignocellulosic biomass, hydrophobic wastes, wastewaters and crude glycerol (Table 1).

Table 1. Inhibitory compounds present in most used low-cost feestocks for microbial biofuels and bioproducts prodution.

Low-cost raw materials	Source	Feedstock	Substrates	Inhibitors
Lignocellulosic biomass	Agro and food industry, silviculture crops residues	Sugarcane bagasse, corn stove, maize starch, rice hull, cassava starch, rice straw, wheat straw, corncobs, wood residues, saw dusts	Fermentable sugar	Aldehydes (HMF, furfural)
				Organic acids (acetic, formic, levulinic acids)
				Phenolic compounds (catechol, hydroquinone, furfuryl alcohol)
Hydrophobic wastes	Agro-food industry	Volatile fatty acids	Acetic acid, propionic acid, butyric acid	Acetic acid
Wastewaters	Agro and food industry	Distillery wastewater, bioethanol wastewater, fruit processing wastewater, livestock, domestic wastewater	Organic matter, inorganic N, inorganic P	Heavy metals Recalcitrant chemicals Antibiotics
Crude glycerol	Biodiesel industry	Waste oil-based biodiesel industry	Glycerol, free fatty acids	Methanol Ethanol Salts

Lignocellulosic Materials

Lignocellulosic materials are promising feedstock as natural and renewable resource for biofuels and bioproducts production, as considerable amount of such materials as waste byproducts are generated through agricultural practices mainly from various agro-food based industries. Great amounts of lignocellulosic biomass can potentially be biochemically converted into different high value products including biofuels, fine chemicals, and cheap energy sources for microbial fermentation and enzyme production [2]. However, the use of lignocellulosic biomass as feedstock requires a pre-treatment step necessary to generate monomeric sugars from the polysaccharide components of the lignocellulosic biomass for the subsequent fermentation process. The lignocellulosic materials pre-treatment step inevitably generates degradation compounds, *e.g.*, organic acids (acetic, formic, levulinic acids), furan derivates [the aldehydes 5-hydroxymethylfurfural (HMF) and furfural] and phenolic compounds (cathecol, hydroquinone, furfuryl alcohol). These compounds exist in the hydrolysates (obtained after the lignocellulosic materials pre-treatment step) that are used in culture medium for microbial growth, and their effect on microbial cells has been extensively described in second generation bioethanol production processes [3]. To obtain fermentable sugars in high concentrations for fermentations in order to achieve high bioethanol productivities, the raw materials loading ratio during the pre-treatment step must be increased to a considerably high level, which will unavoidably lead to high concentrations of inhibitors in the fermentation broth, which will reduce the microbial metabolism rates and, consequently, the final products yield. Therefore it is important to understand the effect of these toxic compounds on the microorganisms in order to find strategies to overcome the microbial inhibition.

Weak acids (acetic, formic, levulinic acids) have been described as being associated with the alcoholic fermentation inhibition, limiting the microbial growth and ethanol productivity [4 - 8]. Yeast cytosolic pH decrease, leading to an ATP production reduction, resulting in the depolarization of the cytoplasmic membrane potential has also been described [3].

The effect of furan derivates on ethanologenic microorganisms has been extensively described using traditional microbial techniques for bioprocess monitoring (optical density, cell dry weight). Using such techniques, Palmqvist *et al.* 1999 [6], Taherzadeh *et al.* 2000 [9] and Liu *et al.* 2004 [10] have reported that, for high inhibitor concentrations (which depends on the studied strain), the specific growth rate, the cell mass yield, and the volumetric and specific ethanol productivities decreased. Almeida *et al.* (2007) [3] also have reported reduction in *S. cerevisiae* intracellular ATP production and enzymatic inhibition. Using fluorescence microscopy to observe *S. cerevisiae* cells stained with the fluorescent dye 2' 7'-dichlorofluorescein diacetate (DCF), Allen *et al.* (2010) [11] reported reactive oxygen species (ROS) accumulation in yeast cells grown in the presence of 2.5 g/L and 5.0 g/L furfural.

The effect of phenolic compounds, also generated during the pre-treatment step of lignocellulosic materials, on ethanologenic yeasts, has been reported as biomass and ethanol productivity reduction and membrane swelling (indirectly detected by rhodamine-β-chloride fluorescence read in a spectrofluorometer, which in turn is related to the membrane expansion) [3, 12]. Loss of yeast cells membrane integrity and generation of reactive oxygen species (ROS) due to lignocellulosic hydrolysates phenolic compounds presence has also been reported [3].

The negative effect of lignocellulosic inhibitors on lipid and biomass production by oleaginous microorganisms has also been reported [13 - 16].

Hydrophobic Wastes

Hydrophobic wastes, such as volatile fatty acids (VFAs) (acetic acid, propionic and butyric acids) obtained from acidogenic fermentation of sludge, food wastes or biodegradable organic wastes have also been used as carbon source for biofuels production [1, 17 - 19] having been described inhibitory effects of these compounds on microbial growth, particularly acetic acid whose toxic effects have been already described in the previous section.

Wastewaters

Wastewaters containing abundant organic compounds have being considered a source of nutrients that may be used for cost-effective microbial biofuels and bioproducts production. Several industrial effluents from industry (distillery, bioethanol wastewater, fruit processing, domestic wastewater, livestock) can be used in media formulations for microbial growth. The composition of wastewater is a complex mixture of natural organic and inorganic materials as well as man-made compounds. Three quarters of organic carbon in sewage are present as carbohydrates, fats, proteins, amino acids, and volatile acids. The inorganic constituents include large concentrations of sodium, calcium, potassium, magnesium, chlorine, sulphur, phosphate, bicarbonate, ammonium salts and heavy metals [20]. The effect of heavy metals (Hg, As, Cd, Pb, Cr) on microalgae cells has been extensively studied, being reported negative effects such as impairment of photosynthetic mechanism, blockage of cell division, inhibition of enzyme activity in microalgae cells [21]. Metals also influence the morphology of microalgal cells, as cell shape and size, misshape chloroplasts and mitochondria swelling [22]. Wastewaters also contain non-biodegradable antibiotics and recalcitrant chemicals that affect several cell functions and compartments such as efflux pumps and membrane damage [23].

Crude Glycerol

Crude glycerol is a by-product of the biodiesel industry, being considered a waste product because of the cost associated with its disposal [24]. In addition, left unattended crude glycerol from the biodiesel industry is also a threat to environment. Previously only pure glycerol was used as the carbon source, as the impurities present in crude glycerol (methanol, ethanol, salts, metals and soaps) can inhibit the growth of some microorganisms hindering the biological conversion of crude glycerol [25]. However, purification of crude glycerol is a cumbersome, and hence, utilization of crude glycerol as intact source for any industrial product is a value added approach. The conversion of glycerol impurities through biological routes using microorganisms is a viable way to enhance the economy of the process. A few microorganisms have been found to be tolerant to the impurities in crude glycerol, allowing the use of this raw material as a sole carbon and energy source in numerous microbial fermentations for microbial growth and production of various high value products, such as 1,3-propanediol (1,3-PDO), citric acid, 2,3-butanediol, ethanol, succinate, propionic acid, glyceric acid, biosurfactants, hydrogen, among others [26 - 28], being considered a suitable material for biofuels and value-added compounds microbial production [29]. Sarma *et al.* (2013) [30] studied the effect of the glycerol impurities, *i.e.*, methanol, NaCl and soaps on the hydrogen producer *Enterobacter aerogenes* bacterium and concluded that the soaps and methanol have a significant inhibitory effect on the microbial hydrogen production. However, most of the works reporting the use of crude glycerol as substrate in microbial cultivations does not give any information on the cell metabolism or physiological status [31, 32].

FC DETECTION OF CELLULAR STRESS RESPONSE TO THE INHIBITORS IN LOW-COST SUBSTRATES

FC coupled to low-cost feedstocks bioconversion processes to obtain biofuels and bioproducts may provide important *at-line* information about the cell physiological status, as well as adaptation or resistance cell mechanisms, which is important to find strategies to overcome such harmful effects on the process yield. Table **2** displays the works that describe cellular stress response, detected by FC, to inhibitors usually present in low-cost substrates.

Table 2. FC detection of cellular stress response to inhibitors usually present in low-cost substrates.

Substrate	Micoorganism	Bioproducts	Inhibitors	FC detection	Stain	Reference
Glucose and xylose	*Trichosporon fermentans*	Lipids	Acetic acid, Formic acid, Levulinic acid, 4-Hydroxybenzoicacid, Furoic acid, Caproic acid, Gallic acid, Ferulic acid, Syringic acid, Vanillic acid	Membrane integrity	Propidium iodide	[14]
Glucose and galactose	*Saccharomyces cerevisiae*	Bioethanol	Acetic acid	Mitochondrial degradation / Acetic acid-induced cell death	Mitochondrial matrix-targeted green fluorescent protein (GFP) / FITC Annexin V and PI	[46]
Yeast biomass biorefinery effluents and residues	Microbial consortium	Lipids Carotenoids Biogas	Weak acids Phenolic compounds	Membrane integrity	PI	[47-48]
Glucose	*Saccharomyces carlsbergensis*	Bioethanol	Acetic Acid	Membrane integrity Membrane depolarization ROS	PI / DiOC6(3) DHR123	[38]

(Table 2) contd.....

Substrate	Micoorganism	Bioproducts	Inhibitors	FC detection	Stain	Reference
Glucose	*Saccharomyces carlsbergensis*	Bioethanol	Formic Acid	Membrane integrity Membrane depolarization ROS	PI DiOC6(3) DHR123	[39]
Glucose	*Saccharomyces carlsbergensis*	Bioethanol	HMF	Membrane integrity Membrane depolarization ROS Lipids	PI DiOC$_6$(3) DHR 123 Nile Red	[40]
Glucose	*Saccharomyces carlsbergensis*	Bioethanol	Furfural	Membrane integrity Membrane depolarization ROS Lipids Cell size/internal complexity	PI DiOC$_6$(3) DHR 123 Nile Red FSC/SSC	[41]
Carob pulp syrup Sugarcane molasses	*Rhodosporidum toruloides*	Lipids Carotenoids	Phenolic compounds	Carotenoids Membrane Integrity Membrane potential	Autofluorescence PI DiOC$_6$(3)	[45]
Glucose	*Saccharomyces cerevisiae*	Bioethanol	Vanilin Furfuraldehyde Acetic acid	Membrane integrity pHi ROS	PI pHluorin Dihydroethidium (DHE)	[43]
Glucose	*Saccharomyces cerevisiae*	Bioethanol	HMF Furfural	ROS	2' 7'-dichlorodihydrofluorescein diacetate (H2DCF-DA)	[44]
Tris-Acetate-Phosphate liquid medium	*Chlamydomonas reinhardtii*	n.s.	Cadmium	Enzymatic activity ROS	FDA DHR 123	[50]
Municipal wastewater	Cyanobacteria Green microalgae	Lipids	n.s.	Lipid content	BODYPI	[53]

n.s. – not specified

FC Principles

FC is an advanced technique that measures characteristics of cells or particles in a fluid stream as they pass through a light source (laser). The power of a flow cytometer consists of its capacity to measure multiple parameters related not only to cells, but also to chromosomes, proteins or nucleic acids that are attached to a particle (eg, microsphere) as long as the particles or cells are suspended in a fluid [33]. In addition, single cells cross a laser beam at a rate of hundreds or thousands of cells per second, providing statistically accurate data about fluorescence intensity and other cellular parameters. A flow cytometer is made up of three main systems: fluidic, optic, and electronic. The fluidic system transports particles in a stream to the laser beam for interrogation (Fig. **1**). The optic system consists of lasers to illuminate the particles in the sample stream and optical filters to direct the resulting light signals to the appropriate detectors. The electronic system converts the detected light signals into electronic signals that can be processed by the computer (Fig. **2**).

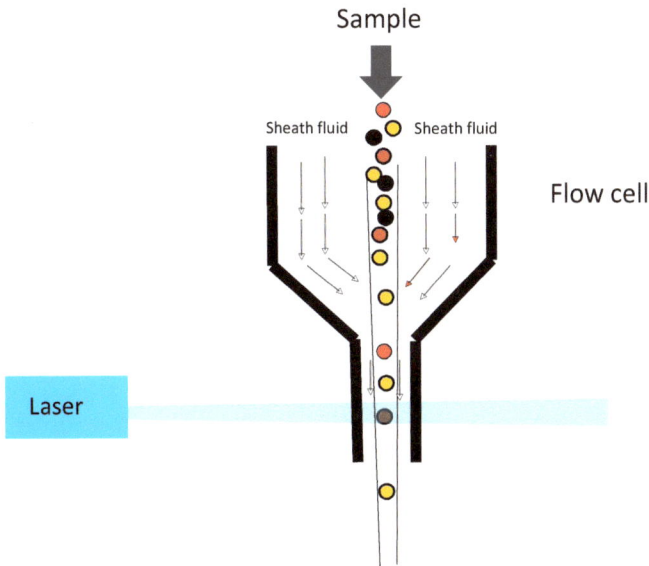

Fig. (1). Hydrodynamic focusing in flow cytometric analysis.

There are two different types of flow cytometers: non-sorting and sorting. Non-sorting type can detect light scattering and fluorescence emission, while the sorting type has the ability to sort particles, named fluorescent activated cell sorters (FACS) and have the capacity to separate or sort fluorescent-labelled cells from a mixed cell population [34].

Cell detection in flow cytometers is based on two types of light scattering: (i) forward scatter (FSC), which is measured in the plane of the beam and gives information on cell size; and (ii) side scatter (SSC), which is measured at 90° to the laser beam plane, and can provide information on cell granularity or internal structure of the cell.

Because scattering signals have a characteristic profile for each particle, they can distinguish the different cell types (and debris) in heterogeneous samples. For instance, because of their different size and shape, microalgae can be distinguished from bacteria based on their scattering signals [35], allowing early bacterial contamination detection in microalgal cultivations. In addition, absolute cell counts can be obtained by adding a microsphere standard to the sample.

Fluorescence emitted by intracellular compounds with specific fluorescence wavelengths, such as nicotinamide adenosine dinucleotide phosphate (NADPH), or pigments such as chlorophyll or phycoerythrin are detected by photomultipliers tubes. In addition, fluorochromes (fluorescent dyes) which are also detected by photomultipliers tubes can be added to the cells, changing their fluorescence intensity depending on specific cell compartments and/or cell functions changes. Thus, certain cell components and/or cell functions can be selectively monitored and/or assayed throughout any biotransformation process. In this way, cells can be differentiated based on their structure and/or metabolic activity, based on FC analysis. This is not possible when using conventional microbiology methods for microbial cultures monitoring such as optical density or dry cell weight as such techniques do not give any information on cell physiological status [36]. When applying FC to low-cost feedstocks bioconversion processes, it provides crucial information for understanding the inhibitory mechanism and the cellular stress/adaptation response to these feedstocks.

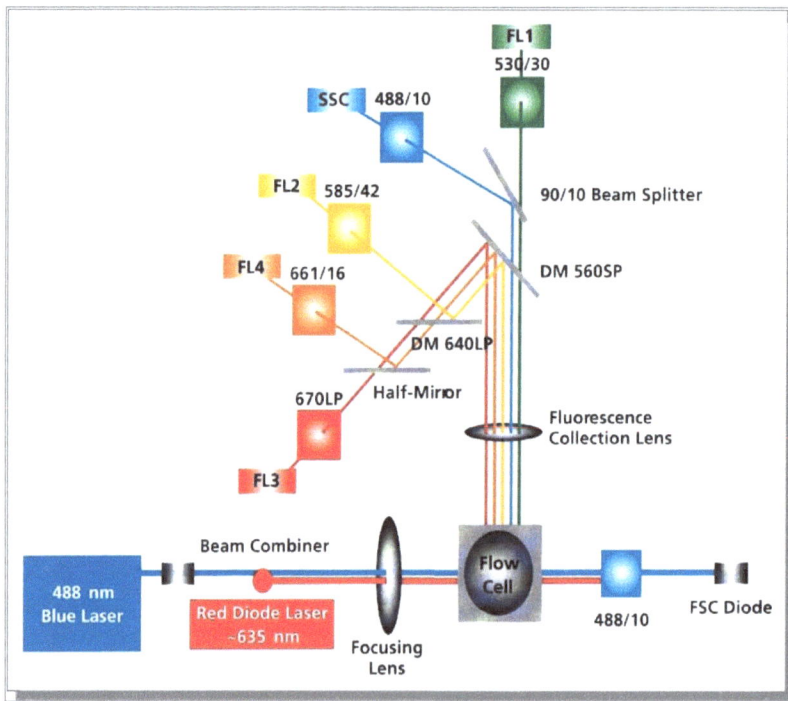

Fig. (2). Scheme of the FACScalibur benchtop flow cytometer (courtesy of BD Biosciences). A set of lenses focus the light source (laser or mercury lamp). The fluidic system transports particles in the sheath fluid and individual cells pass through the light beam at high speed, being possible to detect light scattering and fluorescence intensity. Optical filters combined with light detectors [photodiodes (FSC) or photomultiplier tubes (SSC, FL1, FL2, FL3, FL4)] allow sensitive detection of the signals of interest. Forward scatter (FSC) measured in the plane of the beam gives information on cell size; Side scatter (SSC) measured at 90° to the laser beam reports on cell granularity and internal complexity.

Reactive Oxygen Species (ROS)

ROS are generated in cells as metabolic byproducts as a stress response, and their intracellular accumulation can be increased by environmental stress conditions, genetic mutations and cell ageing, being known to cause DNA, proteins, lipids and cytoskeleton damage, and to induce programmed cell death (Allen *et al.* 2010) [11]. Therefore, ROS accumulation as a cell response to the presence of lignocellulosic hydrolysates inhibitors is an indicative that cells are under stress conditions. Fluorescent probes have been used with microscope image techniques to detect intracellular ROS accumulation [37]. However, such techniques are tedious to perform and, for statistical accuracy, rely on an equal distribution of cells within the counting chamber, a situation almost impossible to occur. Moreover, fluorescence microscopy is usually limited to the measurement of ten to hundreds of cells and it is the user who makes the decision on whether a cell has taken up the stain and hence, on its physiological state. This leads to inaccuracy and inconsistencies between temporally spaced samples when different users have made the measurement [36].

ROS detection can be carried out in a faster way, and more accurately, using FC coupled with fluorescent stains. ROS accumulation in *S. carlsbergensis* cells when grown in the presence of increasing concentrations of several inhibitors such as acetic acid [38], formic acid [39], HMF [40] and furfural [41] was performed, using the fluorochrome Dihydrorhodamine 123 (DHR). This dye freely enters the cell membrane and is oxidized by ROS to produce the fluorescent product Rhodamine 123, which characterizes the degree of ROS accumulation in individual cells, being the results available in less than 15 min. Other ROS detection probes have been used such as Dihydroethidium (DHE) [42, 43]. Kim and Hahn (2013) [44] studied the mechanisms by which furfural and HMF induce oxidative stress in *S. cerevisiae* cells, having demonstrated that thiol-reactivity of furfural and HMF contributes for depleting cellular glutathione and to accumulate ROS in the yeast cells which were detected by FACS using the stain 2' 7'-dichlorodihydrofluorescein diacetate (H_2DCF-DA).

Cell Membrane Damage

The cell membrane (plasma membrane) is a thin semi-permeable membrane that surrounds the cytoplasm of a cell. Its function is to protect the integrity of the cell interior by allowing certain substances into the cell, while keeping other substances out and helping support the cell and maintain its shape. Regular passive and active transport across an intact cytoplasmic membrane generate an electrochemical gradient, the membrane potential, which is one of the most frequently used parameters to determine cellular metabolic activity during processes involving microorganisms. A compromised (injured, permeabilised) membrane cannot generate the electrochemical gradient and hence a membrane potential.

Cell membranes appear to be the main sites of hydrophobic inhibitors due to their affinity by the hydrophobic membrane sites. FC allows detection of cell membrane damage in a few minutes, through potential and membrane integrity evaluation, using specific fluorescent dyes.

Huang *et al.* (2014) [15] studied the effect of ten typical organic acids and five aldehydes present in lignocellulosic hydrolysates on the cell membrane integrity of oleaginous yeast *Trichosporon fermentans* using FC coupled with propidium iodide (PI). This stain only penetrates permeabilized membranes and binds to DNA, allowing the differentiation between cells with intact membrane (not stained with PI) from cells with permeabilised membrane (stained with PI). The authors concluded that the organic acids affected the cell membrane integrity of *T. fermentans* more significantly than aldehydes albeit the latter are more toxic to *T. fermentans*. The impact of organic acids (acetic and formic acids) and aldehydes (HMF and furfural) on *S. carlsbergensis* membrane potential and membrane integrity was also evaluated by FC [38 - 41], using PI and the stains 3,3-dihexylocarbocyanine iodide [$DiOC_6(3)$], a lipophilic cationic probe that is accumulated in the cells if the membrane is polarised, being excluded if the membrane is depolarised. In these works, ROS accumulation as a result of the inhibitor presence was also evaluated by FC, and the authors concluded that aldehydes were more toxic to *S. carlsbergensis* cells than the organic acids, having demonstrated that aldehydes induced cell membrane permeabilization beyond cell membrane depolarization, while the effect organic acids was mainly cell membrane depolarization. The different stress response between *T. fermentans* and *S. carlsbergensis* to the organic acids and aldehydes may be due to the fact that the former is an oil producer, while the latter is an ethanologenic microorganism, being subjected to the additional stress effect of the produced ethanol which intensifies the synergies between the inhibitors [41].

FC has also been used to compare *Rhodosporidium toruloides* cell viability when grown on carob pulp syrup and sugarcane molasses (both low-cost agro-food industrial byproducts) for lipids and carotenoids microbial production [45]. Both feedstocks are rich in sugars (sucrose) that can be used as carbon source in microbial fermentations, but contain inhibitor phenolic compounds. FC analysis revealed that the membrane damage was higher when the yeast cells were grown in sugarcane molasses, comparing with the growth on carob pulp syrup, probably due to the higher inhibitors concentration in the sugarcane molasses. In a detailed study on the effect of the acetic acid on the yeast *Saccharomyces cerevisiae*, Dong *et al.* (2017) [46] conducted transcriptomic and metabolomic analyses to investigate the global yeast response under acetic acid stress at different growth stages, providing a novel and comprehensive understanding of the yeast stress responses and programmed cell death induced by acetic acid. The authors used FC to evaluate acetic acid-induced cell death using the Annexin V-FITC/PI apoptosis kit, and mitochondrial degradation by the mitochondrial matrix-targeted green fluorescent protein (GFP), having concluded that for high acetic acid concentrations (> 150 mM), the severe acidification led to intracellular damage and degradation, thus inducing programmed cell death.

FC can also be applied to low-cost feedstocks bioconversion processes in which microbial consortia are used, to assess cell membrane integrity. Baptista *et al.* (2016) [47] and Marques *et al.* (2017) [48] have used this approach to detect membrane integrity of cells from a microbial consortium used to digest anaerobically the residues, wastes and effluents generated in the yeast *Rhodosporidium toruloides* biomass biorefinery process (to obtain carotenoids, biodiesel and biogas) using individual wastes/effluents [47] and mixed wastes/effluents [48]. FC analysis of cells from the microbial consortium stained with the dyes SYBR Green (to discriminate cells from background) and PI (used to differentiate cells with intact membrane from cells with injured membrane) revealed that the proportion of cells with injured membrane was lower in the wastes/effluents mixtures co-digestion than in the individual waste/effluent digestion, indicating that co-digestion mitigates the negative effect of potential toxic compounds present in the effluents and wastes or produced during the process digestion with the microbial consortium.

Intracellular pH

The intracellular pH plays a central role in the cell homeostasis, being an important cell parameter when studying the impact of toxic compounds on the cell physiology, such as weak organic acids, in order to understand the cell inhibition/adaptation mechanism to these compounds. When microorganisms grow in the presence of weak acids, the intracellular pH (pHi) decreases as a result of the weak acid dissociation inside the cytoplasm due to the fact that pHi is naturally neutral, which is much higher than the pKa of weak acids. The decrease in the pHi affects the membrane potential as already mentioned, which contributes for the growth inhibition.

Dyes such as carboxyfluorescein diacetate (cFDA), biscarboxyethyl carboxy-fluorescein (BCECF), and SNARF dyes exhibit pH dependent spectral shifts in excitation, emission or both, thus ratiometric analysis of fluorescence between the spectral shift renders an estimate of pHi [49]. Many of these dyes and, in particular their ester forms, such as carboxyfluorescein diacetate succinimidyl ester (cFDA-SE), are cell permeant, which, once inside the cell, are converted into an impermeant form by nonspecific esterase hydrolysis, thus capturing the dye in the cell. The probe fluorescence depends on the pH value. When measuring pHi, the cells are loaded with the probe and then exposed to different extracellular conditions. The fluorescent signal from the probe inside is proportional to pHi.

The genetically encoded ratiometric fluorescent probe pHluorin also allows pHi detection by FC, permitting the pHi assessment of large populations of cells that express pHluorin. Narayanan *et al.* (2017) [43] used this method to characterize populations with high tolerance to lignocellulosic inhibitors, together with ROS detection, as above referred.

Enzymatic Activity

Cell viability detection can also be carried out by evaluating cellular enzymatic activity (specifically esterase activity) which may be detected by FC, being a suitable method to detect cell stress response. Nevertheless, it has been scarcely used to detect cell stress response to lignocellulosic/wastes/effluents inhibitors. Fluorescein and fluorescein derivatives dyes have been used to assess the cell esterase activity during bioprocesses. In general, a nonfluorescent permeant substrate [cFDA and fluorescein diactetate (FDA)] is taken up by the cell and converted inside the cell to a fluorescent substance (ideally impermeant). Enzyme activity detection is also used to assay membrane integrity if cells have a compromised membrane, as the impermeant end products will be quickly lost to the extracellular space, thus resulting to a decreased fluorescent signal. Jamers *et al.* (2009) [50] studied the effect of cadmium (a heavy metal often present in wastes) on *Chlamydomonas reinhardtii* cells using fluorescein diacetate (FDA) and dihydrorhodamine 123 (DHR123) were used to detect esterase activity and ROS accumulation, respectively. The authors observed an increase in esterase activity with the cadmium concentration increase, as reflected in fluorescein fluorescence which might be related to the activation and upregulation of detoxification processes, in which esterases may take part. They also reported intracellular ROS accumulation as judged by the increase in the DHR 123 fluorescence confirming that this metal exerts its toxicity through the generation of reactive oxygen species in microalgae.

Cell Size and Internal Complexity

FSC and SSC scatter light signals detected by FC have been used to monitor cell size and internal complexity changes in the presence of inhibitor compounds. For example, a gradual increase in the mean values of the FSC and SSC signals was detected during *S. carlsbergensis* alcoholic fermentations in the presence of furfural, which were associated with mitochondria aggregation and swelling, as well as vacuole fragmentation and aggregation [41].

Intracellular Products

Beyond the ability to treat effluents and wastes, a few oleaginous microorganisms are recognised as an important renewable source of pigments and lipids that can be converted into biodiesel [51]. During any bioprocess development, it is crucial to have information, near real time, on the products synthesis, so that the process control strategy can be changed to achieve the highest product yields.

Most of the recent works reporting the carotenoid and lipid production by microorganisms uses traditional gravimetric methods for assessment of carotenoids and lipids. These methods are time consuming, use high amounts of toxic organic solvent and require significant amounts of biomass for carotenoid and lipid extraction and quantification [52]. Importantly, carotenoid and lipid analyses usually takes a few days and, sometimes, data is only available after the cultivation is over, too late to introduce changes to the process control strategy.

FC can be used to *at-line* detect intracellular products during bioprocess development, measuring the cell intrinsic autofluorescence for pigments detection, or using specific fluorescent dyes that bind to lipidic compounds, emitting a fluorescence signal that is proportional to the compound concentration.

Freitas *et al.* (2014a) [45] assayed the total carotenoid content in the yeast *Rhodosporidium toruloides* NCYC 921 grown in the low cost agro-food by-products sugarcane molasses and carob pulp syrup, both used as carbon sources in the yeast medium culture, by measuring the yeast cells autofluorescence along the cultivation, and found that the carob pulp syrup induces higher carotenoid production that the sugarcane molasses.

Microbial lipids can also be quickly detected by FC, as demonstrated by Lynch *et al.* (2015) [53] who have quickly detected lipid production by several microalgal species using FC with the fluorescent stain BODYPI. In addition, since lipids are essential carbon and energy reserves for yeast cells, allowing survival under starvation conditions, these intracellular compounds can also be an indicator of the microbial metabolism health. Lopes da Silva *et al.* (2017b) [41] assayed *S. carlsbergensis* lipid production in the presence of HMF and furfural by FC, using the dye Nile Red, which is excited at 488 nm, and emits yellow fluorescence when is dissolved in neutral lipids and red fluorescence when dissolved in polar lipids. The authors observed that intracellular polar lipids were more affected by the HMF presence than the neutral lipids, probably due to the extensive membrane damage, while both polar and neutral lipids content decreased in the presence of furfural, being this reduction was more notorious during the stationary phase.

STRATEGIES TO SOLVE THE INHIBITORY EFFECT OF TOXIC COMPOUNDS ON MICROBIAL CELLS GROWN IN LOW-COST FEEDSTOCKS

Despite the economic and environmental benefits of using industrial by-products, wastes and effluents for biofuels and high value added products production *via* biochemical conversion, the inhibition action of toxic compounds on micro-organisms is still a significant hurdle. Therefore, strategies to overcome the inhibition should be implemented.

In fact, if a high fraction of dead or injured cells is present during any part of a bioprocess, the yield will decrease as such cells do not participate in the bioconversion [36]. Therefore, it is desired obtaining accurate information on the physiological states of individual cells within a population along the process development.

The inhibitor concentrations depend on the feedstock, as well as on the conditions used during pre-treatments and hydrolysis. Therefore, one possibility is to select less recalcitrant feedstocks and to utilize mild pre-treatment conditions [3]. However, the production of bulk bio-products is yield dependent, which means that low carbon source (sugar) concentrations will result in poor overall product yields, due to the use of insufficient pre-treatment conditions [54].

Another possibility is the detoxification of raw materials (*i.e.*, removal of inhibitors by several methods such as chemical additives, enzymatic treatment, *etc.*) which, according to Almeida *et al.* (2007) [3] should be avoided due to additional process cost and possible loss of fermentable sugars.

An advantageous strategy consists of using microorganisms with improved inhibitor tolerance and/or using a careful fermentation strategy. The fed-batch mode of operation in fermenting hydrolysates/effluents/wastes to produce biofuels and high value added products has been shown to be advantageous in many studies including both conventional fermentations and simultaneous saccharification and fermentation (SSF) due to the fact that the inhibitors concentration can be kept low, which increases the fermentability and products synthesis. To maximize the products synthesis, the hydrolysates/wastes should be fed at the highest possible rate without severely inhibiting the cells. This rate is ideally equal to the inherent detoxification capacity of the cells [3]. *At-line* flow cytometric analysis coupled to a lignocellulosic hydrolysate/effluent/waste feeding control strategy based on the cell physiological status assessed by FC will ensure that the feed rate of inhibitory compounds does not exceed the capacity of the cells to effectively ferment the hydrolysates/wastes. By controlling the feed-rate using a dynamic step response method such as FC, an increased conversion of the hydrolysates/wastes and bioproducts productivity will be obtained without the need for the hydrolysate detoxification step, prior to fermentation. Moreover, FC provides this information, promptly, with a high degree of resolution and allows changing the process control strategy (by changing the operational conditions, such as feed rate, speed rate, aeration rate, medium pH, *etc.*) during its development, which will increase the overall process yield. Such strategy will allow the fast process optimization and process scale-up stage, since the *at-line* information allows products to be harvest at optimum concentrations and the activation of inducible systems to be initiated at the correct time, resulting in high products yields.

CONCLUSIONS

FC is a powerful technique that is especially suitable to detect stress cell response to adverse environments. Although still scarcely applied, it has an enormous potential when associated to the implementation and/or optimization of bioprocesses that use low-cost feedstocks containing inhibitors for biofuels and other bioproducts production. When detecting cellular stress response by FC, it is possible to simultaneously monitor several cellular functions and/or cell compartments, which allow obtaining complete and detailed information on the cell behaviour, thus the bioconversion process development.

Although new developments have been achieved in the FC area over the last decades, it has continued to adopt innovations in order to reduce cost, size and complexity and to increase sensitivity. In the short term, the future of this technology applied to low-cost feedstock conversion to biofuels and high value added products depend on the development of automatic *in situ* FC equipment coupled to the fermentation systems, and the development of user-friendly software. Such tools will allow extracting even more information from the bioprocess development, which in turn will enhance our understanding on how microbes respond to stress conditions such as low-cost feedstocks containing inhibitor compounds.

CONSENT FOR PUBLICATION

Not applicable.

CONFLICT OF INTEREST

The author confirms that he has no conflict of interest to declare for this publication.

ACKNOWLEDGEMENTS

Declared none.

REFERENCES

[1]   Qin, L.; Liu, L.; Zeng, A-P.; Wei, D. From low-cost substrates to Single Cell Oils synthesized by oleaginous yeasts. *Bioresour. Technol.,* **2017**, *245*(Pt B), 1507-1519.
         [http://dx.doi.org/10.1016/j.biortech.2017.05.163] [PMID: 28642053]

[2]   Anwar, Z.; Gulfraz, M.; Irshad, M. Agro-industrial lignocellulosic biomass a key to unlock the future bio-energy: A brief review. *J Radiat. Res. Appl. Sci.,* **2015**, *7*, 153-173.

[3]   Almeida, J.; Modig, T.; Petersson, A.; Gerdal, B.; Lidén, G.; Gorwa-Grauslund, M.F. Increased tolerance and conversion of inhibitors in lignocellulosic hydrolysates by Saccharomyces cerevisiae. *J. Chem. Technol. Biotechnol.,* **2007**, *82*, 340-349.
         [http://dx.doi.org/10.1002/jctb.1676]

[4]   Maiorella, B.; Blanch, H.W.; Wilke, C.R. By-product inhibition effects on ethanolic fermentation by Saccharomyces cerevisiae. *Biotechnol. Bioeng.,* **1983**, *25*(1), 103-121.
         [http://dx.doi.org/10.1002/bit.260250109] [PMID: 18548541]

[5]   Larsson, L.; Palmqvist, E.; Hahn-Hägerdal, B.; Tengborg, C.; Stenberg, K.; Zacchi, G.; Nilvebrant, N-O. The generation of fermentation inhibitors during dilute acid hydrolysis of softwood. *Enzyme Microb. Technol.,* **1999**, *1999*(24), 151-159.
         [http://dx.doi.org/10.1016/S0141-0229(98)00101-X]

[6]   Palmqvist, E.; Grage, H.; Meinander, N.Q.; Hahn-Hägerdal, B. Main and interaction effects of acetic acid, furfural, and p-hydroxybenzoic acid on growth and ethanol productivity of yeasts. *Biotechnol. Bioeng.,* **1999**, *63*(1), 46-55.
         [http://dx.doi.org/10.1002/(SICI)1097-0290(19990405)63:1<46::AID-BIT5>3.0.CO;2-J]   [PMID: 10099580]

[7]   Oliva, J.M.; Sáez, F.; Ballesteros, I.; González, A.; Negro, M.J.; Manzanares, P.; Ballesteros, M. Effect of lignocellulosic degradation compounds from steam explosion pretreatment on ethanol fermentation by thermotolerant yeast Kluyveromyces marxianus. *Appl. Biochem. Biotechnol.,* **2003**, *105*(108), 141-153.
         [http://dx.doi.org/10.1385/ABAB:105:1-3:141] [PMID: 12721481]

[8]   Wikandri, R.; Millat, R.; Syamsiyah, S.; Muriana, R.; Ayuningsih, Y. Effect of furfural, hydroxymethylfurfural and acetic acid on indigenous microbial isolate for bioethanol production. *Agric. J.,* **2012**, *5*(2), 195-109.

[9]   Taherzadeh, M.J.; Gustafsson, L.; Niklasson, C.; Lidén, G. Physiological effects of 5-hydroxymethylfurfural on Saccharomyces cerevisiae. *Appl. Microbiol. Biotechnol.,* **2000**, *53*(6), 701-708.
         [http://dx.doi.org/10.1007/s002530000328] [PMID: 10919330]

[10]   Liu, Z.L.; Slininger, P.J.; Dien, B.S.; Berhow, M.A.; Kurtzman, C.P.; Gorsich, S.W. Adaptive response of yeasts to furfural and 5-hydroxymethylfurfural and new chemical evidence for HMF conversion to 2,5-bis-hydroxymethylfuran. *J. Ind. Microbiol. Biotechnol.,* **2004**, *31*(8), 345-352.
         [http://dx.doi.org/10.1007/s10295-004-0148-3] [PMID: 15338422]

[11]   Allen, S.A.; Clark, W.; McCaffery, J.M.; Cai, Z.; Lanctot, A.; Slininger, P.J.; Liu, Z.L.; Gorsich, S.W. Furfural induces reactive oxygen species accumulation and cellular damage in Saccharomyces cerevisiae. *Biotechnol. Biofuels,* **2010**, *3*, 2-10.
         [http://dx.doi.org/10.1186/1754-6834-3-2] [PMID: 20150993]

[12]   Palmqvist, E.; Hahn-Hägerdal, B. Fermentation of lignocellulosic hydrolysates. I: inhibition and detoxification. *Bioresour. Technol.,* **2000**, *74*, 17-24.
[http://dx.doi.org/10.1016/S0960-8524(99)00160-1]

[13]   Hu, C.; Zhao, X.; Zhao, J.; Wu, S.; Zhao, Z.K. Effects of biomass hydrolysis by-products on oleaginous yeast *Rhodosporidium toruloides. Bioresour. Technol.,* **2009**, *100*(20), 4843-4847.
[http://dx.doi.org/10.1016/j.biortech.2009.04.041] [PMID: 19497736]

[14]   Huang, C.; Wu, H.; Liu, Z.J.; Cai, J.; Lou, W.Y.; Zong, M.H. Effect of organic acids on the growth and lipid accumulation of oleaginous yeast *Trichosporon fermentans. Biotechnol. Biofuels,* **2012**, *5*, 4.
[http://dx.doi.org/10.1186/1754-6834-5-4] [PMID: 22260291]

[15]   Huang, C.; Zhu, D.; Wu, H.; Lou, W.; Zong, M. Evaluating the influence of inhibitors present in lignocellulosic hydrolysates on the cell membrane integrity of oleaginous yeast *Trichosporon fermentans* by flow cytometry. *Process Biochem.,* **2014**, *49*, 395-401.
[http://dx.doi.org/10.1016/j.procbio.2013.12.007]

[16]   Jin, M.; Slininger, P.J.; Dien, B.S.; Waghmode, S.; Moser, B.R.; Orjuela, A.; Sousa, Lda.C.; Balan, V. Microbial lipid-based lignocellulosic biorefinery: feasibility and challenges. *Trends Biotechnol.,* **2015**, *33*(1), 43-54.
[http://dx.doi.org/10.1016/j.tibtech.2014.11.005] [PMID: 25483049]

[17]   Fontanille, P.; Kumar, V.; Christophe, G.; Nouaille, R.; Larroche, C. Bioconversion of volatile fatty acids into lipids by the oleaginous yeast Yarrowia lipolytica. *Bioresour. Technol.,* **2012**, *114*, 443-449.
[http://dx.doi.org/10.1016/j.biortech.2012.02.091] [PMID: 22464419]

[18]   Vajpeyi, S.; Chandran, K. Microbial conversion of synthetic and food waste-derived volatile fatty acids to lipids. *Bioresour. Technol.,* **2015**, *188*, 49-55.
[http://dx.doi.org/10.1016/j.biortech.2015.01.099] [PMID: 25697838]

[19]   Huang, X-F.; Liu, J.N.; Lu, L.J.; Peng, K.M.; Yang, G.X.; Liu, J. Culture strategies for lipid production using acetic acid as sole carbon source by *Rhodosporidium toruloides. Bioresour. Technol.,* **2016**, *206*, 141-149.
[http://dx.doi.org/10.1016/j.biortech.2016.01.073] [PMID: 26851898]

[20]   Lim, S.L.; Chu, W.L.; Phang, S.M. Use of *Chlorella vulgaris* for bioremediation of textile wastewater. *Bioresour. Technol.,* **2010**, *101*(19), 7314-7322.
[http://dx.doi.org/10.1016/j.biortech.2010.04.092] [PMID: 20547057]

[21]   Monteiro, C.M.; Castro, P.M.; Malcata, F.X. Metal uptake by microalgae: underlying mechanisms and practical applications. *Biotechnol. Prog.,* **2012**, *28*(2), 299-311.
[http://dx.doi.org/10.1002/btpr.1504] [PMID: 22228490]

[22]   Miazek, K.; Iwanek, W.; Remacle, C.; Richel, A.; Goffin, D. Effect of metals, metalloids and metallic nanoparticles on microalgae growth and industrial product biosynthesis: A review. *Int. J. Mol. Sci.,* **2015**, *16*(10), 23929-23969.
[http://dx.doi.org/10.3390/ijms161023929] [PMID: 26473834]

[23]   Carey, D.E.; McNamara, P.J. The impact of triclosan on the spread of antibiotic resistance in the environment. *Front. Microbiol.,* **2015**, *5*, 780.
[http://dx.doi.org/10.3389/fmicb.2014.00780] [PMID: 25642217]

[24]   Yazdani, S.S.; Gonzalez, R. Anaerobic fermentation of glycerol: a path to economic viability for the biofuels industry. *Curr. Opin. Biotechnol.,* **2007**, *18*(3), 213-219.
[http://dx.doi.org/10.1016/j.copbio.2007.05.002] [PMID: 17532205]

[25]   Samul, D.; Leja, K.; Grajek, W. Impurities of crude glycerol and their effect on metabolite production. *Ann. Microbiol.,* **2014**, *64*, 891-898.
[http://dx.doi.org/10.1007/s13213-013-0767-x] [PMID: 25100926]

[26]   André, A.; Diamantopoulou, P.; Philippoussis, A.; Sarris, D.; Komaitis, M.; Papanikolaou, S. Biotechnological conversions of biodiesel derived water glycerol into added-value compounds by

higher fungi: production of biomass, single cell oil and oxalic acid. *Ind. Crops Prod.,* **2010**, *31*, 407-416.
[http://dx.doi.org/10.1016/j.indcrop.2009.12.011]

[27]   Chatzifragkou, A.; Papanikolaou, S. Effect of impurities in biodiesel-derived waste glycerol on the performance and feasibility of biotechnological processes. *Appl. Microbiol. Biotechnol.,* **2012**, *95*(1), 13-27.
[http://dx.doi.org/10.1007/s00253-012-4111-3] [PMID: 22581036]

[28]   Konstantinovic, S.; Danilovic, B.; Ciric, J.; Ilic, S.; Savic, D.; Veljkovic, V. Valorization of crude glycerol from biodiesel production. *Chem. Ind. Chem. Eng. Q.,* **2016**, *22*(4), 461-489.
[http://dx.doi.org/10.2298/CICEQ160303019K]

[29]   Sarma, S.; Dhillon, G.; Brar, S.; Bihan, Y.; Buelna, G.; Verma, M. Investigation of the effect of different crude glycerol components on hydrogen production by Enterobacter aerogenes NRRL B-407. *Renew. Energy,* **2013**, *60*, 566-571.
[http://dx.doi.org/10.1016/j.renene.2013.06.007]

[30]   Venkataramanan, K.P.; Boatman, J.J.; Kurniawan, Y.; Taconi, K.A.; Bothun, G.D.; Scholz, C. Impact of impurities in biodiesel-derived crude glycerol on the fermentation by *Clostridium pasteurianum* ATCC 6013. *Appl. Microbiol. Biotechnol.,* **2012**, *93*(3), 1325-1335.
[http://dx.doi.org/10.1007/s00253-011-3766-5] [PMID: 22202963]

[31]   Pott, R.W.; Howe, C.J.; Dennis, J.S. Photofermentation of crude glycerol from biodiesel using *Rhodopseudomonas palustris*: comparison with organic acids and the identification of inhibitory compounds. *Bioresour. Technol.,* **2013**, *130*, 725-730.
[http://dx.doi.org/10.1016/j.biortech.2012.11.126] [PMID: 23334033]

[32]   Picot, J.; Guerin, C.L.; Le Van Kim, C.; Boulanger, C.M. Flow cytometry: retrospective, fundamentals and recent instrumentation. *Cytotechnology,* **2012**, *64*(2), 109-130.
[http://dx.doi.org/10.1007/s10616-011-9415-0] [PMID: 22271369]

[33]   Wilkerson, M.J. Principles and applications of flow cytometry and cell sorting in companion animal medicine. *Vet. Clin. North Am. Small Anim. Pract.,* **2012**, *42*(1), 53-71.
[http://dx.doi.org/10.1016/j.cvsm.2011.09.012] [PMID: 22285157]

[34]   Stauber, J.L.; Franklin, N.M.; Adams, M.S. Applications of flow cytometry to ecotoxicity testing using microalgae. *Trends Biotechnol.,* **2002**, *20*(4), 141-143.
[http://dx.doi.org/10.1016/S0167-7799(01)01924-2] [PMID: 11906740]

[35]   Gomes, A.; Fernandes, E.; Lima, J.L. Fluorescence probes used for detection of reactive oxygen species. *J. Biochem. Biophys. Methods,* **2005**, *65*(2-3), 45-80.
[http://dx.doi.org/10.1016/j.jbbm.2005.10.003] [PMID: 16297980]

[36]   Lopes da Silva, T.; Roseiro, J.C.; Reis, A. Applications and perspectives of flow cytometry applied to microbial biofuels. *TIBTEC,* **2012**, *30*, 225-232.
[http://dx.doi.org/10.1016/j.tibtech.2011.11.005]

[37]   Nebe-von-Caron, G.; Stephens, P.J.; Hewitt, C.J.; Powell, J.R.; Badley, R.A. Analysis of bacterial function by multi-colour fluorescence flow cytometry and single cell sorting. *J. Microbiol. Methods,* **2000**, *42*(1), 97-114.
[http://dx.doi.org/10.1016/S0167-7012(00)00181-0] [PMID: 11000436]

[38]   Freitas, C.; Neves, E.; Reis, A.; Passarinho, P.C.; da Silva, T.L. Effect of acetic acid on *Saccharomyces carlsbergensis* ATCC 6269 batch ethanol production monitored by flow cytometry. *Appl. Biochem. Biotechnol.,* **2012**, *168*(6), 1501-1515.
[http://dx.doi.org/10.1007/s12010-012-9873-7] [PMID: 22971830]

[39]   Freitas, C.; Neves, E.; Reis, A.; Passarinho, P.C.; da Silva, T.L. Use of multi-parameter flow cytometry as tool to monitor the impact of formic acid on *Saccharomyces carlsbergensis* batch ethanol fermentations. *Appl. Biochem. Biotechnol.,* **2013**, *169*(7), 2038-2048.
[http://dx.doi.org/10.1007/s12010-012-0055-4] [PMID: 23359009]

[40]   Lopes da Silva, T.; Baptista, C.; Reis, A.; Passarinho, P.C. R.; Reis, A.; Passarinho, P.C. Using flow cytometry to evaluate the stress physiological response of the yeast *Saccharomyces carlsbergensis* ATCC 6269 to the presence of 5-Hydroxymethylfurfural during ethanol fermentations. *Appl. Biochem. Biotechnol.,* **2017**, *181*(3), 1096-1107. a
[http://dx.doi.org/10.1007/s12010-016-2271-9] [PMID: 27757805]

[41]   Lopes da Silva, T.; Santo, R.; Reis, A.; Passarinho, P.C. Effect of furfural on *Saccharomyces carlsbergensis* growth, physiology and ethanol production. *Appl. Biochem. Biotechnol.,* **2017**, *182*(2), 708-720. b
[http://dx.doi.org/10.1007/s12010-016-2356-5] [PMID: 27987192]

[42]   Landolfo, S.; Politi, H.; Angelozzi, D.; Mannazzu, I. ROS accumulation and oxidative damage to cell structures in *Saccharomyces cerevisiae* wine strains during fermentation of high-sugar-containing medium. *Biochim. Biophys. Acta,* **2008**, *1780*(6), 892-898.
[http://dx.doi.org/10.1016/j.bbagen.2008.03.008] [PMID: 18395524]

[43]   Narayanan, V.; Schelin, J.; Gorwa-Grauslund, M.; van Niel, E.W.; Carlquist, M. Increased lignocellulosic inhibitor tolerance of *Saccharomyces cerevisiae* cell populations in early stationary phase. *Biotechnol. Biofuels,* **2017**, *10*, 114.
[http://dx.doi.org/10.1186/s13068-017-0794-0] [PMID: 28484514]

[44]   Kim, D.; Hahn, J. Roles of Yap1 transcription factor and antioxidants in yeast tolerance to furfural and 5-hydroxymethylfurfural that function as thiol-reactive electrophiles generating oxidative stress. *Appl. Environ. Microbiol.,* **2013**, *79*(16), 5069-5077.
[http://dx.doi.org/10.1128/AEM.00643-13] [PMID: 23793623]

[45]   Freitas, C.; Parreira, T. B., Roseiro, J., Reis, A., Lopes da Silva, T. Selecting low-cost carbon sources for carotenoid and lipid production by the pink yeast *Rhodosporidium toruloides* NCYC 921 using flow cytometry. *Bioresour. Technol.,* **2014**, *158*, 335-359. a
[http://dx.doi.org/10.1016/j.biortech.2014.02.071]

[46]   Dong, Y.; Hu, J.; Fan, L.; Chen, Q. RNA-Seq-based transcriptomic and metabolomic analysis reveal stress responses and programmed cell death induced by acetic acid in Saccharomyces cerevisiae. *Sci. Rep.,* **2017**, *7*, 42659.
[http://dx.doi.org/10.1038/srep42659] [PMID: 28209995]

[47]   Batista, A.P.; López, E.P.; Dias, C.; Lopes da Silva, T.; Marques, I.P. Wastes valorization from Rhodosporidium toruloides NCYC 921 production and biorefinery by anaerobic digestion. *Bioresour. Technol.,* **2017**, *226*, 108-117.
[http://dx.doi.org/10.1016/j.biortech.2016.11.113] [PMID: 27992793]

[48]   Marques, I.P.; Batista, A.P.; Coelho, A.; Lopes da Silva, T. Co-digestion of *Rhodosporidium toruloides* biorefinery wastes for biogas production. *Process Biochem.,* **2017**, •••
[http://dx.doi.org/10.1016/j.procbio.2017.09.023]

[49]   Tracy, B.P.; Gaida, S.M.; Papoutsakis, E.T. Flow cytometry for bacteria: enabling metabolic engineering, synthetic biology and the elucidation of complex phenotypes. *Curr. Opin. Biotechnol.,* **2010**, *21*(1), 85-99.
[http://dx.doi.org/10.1016/j.copbio.2010.02.006] [PMID: 20206495]

[50]   Jamers, A.; Lenjou, M.; Deraedt, P.; Bockstaele, D.; Blust, R.; Coen, W. Flow cytometric analysis of cadmium-exposed green alga *Chlamydomonas reinhardtii* (Chlorophyceae). *Eur. J. Phycol.,* **2009**, *44*, 541-550.
[http://dx.doi.org/10.1080/09670260903118214]

[51]   Dias, C.; Sousa, S.; Caldeira, J.; Reis, A.; Lopes da Silva, T. New dual-stage pH control fed-batch cultivation strategy for the improvement of lipids and carotenoids production by the red yeast *Rhodosporidium toruloides* NCYC 921. *Bioresour. Technol.,* **2015**, *189*, 309-318.
[http://dx.doi.org/10.1016/j.biortech.2015.04.009] [PMID: 25898094]

[52]   Freitas, C.; Nobre, B.; Gouveia, L.; Roseiro, J.; Reis, A.; Lopes da Silva, T. New at-line flow

cytometric protocols for determining carotenoid content and cell viability during *Rhodosporidium toruloides* NCYC 921 batch growth. *Process Biochem.,* **2014**, *49*, 554-562. b
[http://dx.doi.org/10.1016/j.procbio.2014.01.022]

[53]   Lynch, F.; Santana-Sánchez, A.; Jamsa, M.; Sivonen, K.; Aro, E.; Allahverdiyeva, Y. Screening native isolates of cyanobacteria and green alga for integrated wastewater treatment, biomass accumulation and neutral lipid production. *Algal Res.,* **2015**, *11*, 411-420.
[http://dx.doi.org/10.1016/j.algal.2015.05.015]

[54]   Jönsson, L.J.; Alriksson, B.; Nilvebrant, N-O. Bioconversion of lignocellulose: inhibitors and detoxification. *Biotechnol. Biofuels,* **2013**, *6*(1), 16.
[http://dx.doi.org/10.1186/1754-6834-6-16] [PMID: 23356676]

CHAPTER 3

Recent Trends in the Application of Ionic Liquids for Micro Extraction Techniques

I. Rykowska[*], I. Nowak and W. Wasiak

Faculty of Chemistry, Adam Mickiewicz University, Umultowska 89b, 61-614 Poznań, Poland

Abstract: In recent years, the requirements for separation and preconcentration procedures have undergone numerous changes. A general trend is not only to improve the analytical performance of microextraction techniques but also to endeavor to satisfy the requirements of green chemistry. As a result, new modified and derived methods have been developed. One of the most popular techniques that meet the expectations of analysts is a dispersive liquid-liquid microextraction (DLLME). Owing to its rapidity, low costs, simplicity of operation, high recovery, as well as low consumption of both organic or inorganic compounds the method has been widely accepted as a miniaturized sample preparation technique. Despite the advances mentioned, DLLME still requires expensive and hazardous organic solvents, the use of multistep procedures leading to high risk of analyte losses, and has low selectivity and sample clean-up efficiency. That is why much attention has been paid to the development of green activities such as replacing toxic organic solvents and automating extraction techniques. In this context, a new group of solvents namely ionic liquids (ILs) has been utilized in combination with micro extraction procedures.

ILs as molten salts are made of cations and anions. The novelty in relation to ILs is their application as the new non-molecular class of solvents characterized by a low melting point temperature arbitrary fixed at or below 100°C, as opposed to inorganic salts which are solid with melting point well above 500°C (NaI, NaBr, NaCl, 661, 747, 801°C, respectively). The term IL covers inorganic as well as organic salts.

The potential of ionic liquids is based on their unique properties such as a low melting point, a negligible vapor pressure, high thermal stability, a significant viscosity, miscibility with water and other solvents. The ionic liquids environment is very different from that of polar or non-polar organic solvents. Besides the non-molecular nature of ILs, the significant advantages are their non-measurable vapor pressure at room temperature and appreciable liquid ranges. The desired physical and chemical properties of ILs may be controlled by selecting cation/anion combination or by incorporating specific functional groups in the IL molecule. Consequently, combinations of a variety of cation and anions lead to a tremendous number of ionic liquids and that is why ILs are often referred to as designer solvents.

[*] **Corresponding author I. Rykowska:** Faculty of Chemistry, Adam Mickiewicz University, Umultowska 89b, 61-614 Poznań, Poland; Tel: +48 61 829 1763; E-mail: obstiwo@amu.edu.pl

Atta-ur-Rahman & Sibel A. Ozkan (Eds.)

The unique properties of ionic liquids raised growing interest of scientists and engineers regarding an application for these compounds for the extraction purposes. Ionic Liquids and Polymeric Ionic Liquids (PILs) have been used in a Single-Drop Microextraction (SDM), a Liquid-Phase Microextraction (LPM), a Solid-Phase Microextraction (SPME), a Dispersive Liquid-Liquid Microextraction (DLLME), a Hollow Fiber-Supported Liquid Membrane Extraction (HFSLME) and in a Solid Phase Extraction (SPE), till the present day.

The chapter presents extensive theoretical and practical information on the possible application of ionic liquids in various separation micro extraction techniques. It was shown that ILs might be successfully used in chemistry, medicine, environmental research and in other areas where is a need for selective isolation and enrichment of analytes. A systematic review has also been performed involving utilization of ILs in DLLME for the determination of organic compounds and metals in a variety of samples. It has been presented that the DLLME method has been successfully applied for the extraction and determination of a broad spectrum of organic and inorganic analytes from a variety of samples such as environmental, water, food, cosmetics, and biological.

Keywords: Dispersive Liquid-Liquid Microextraction, Hollow Fiber-Supported Liquid Membrane Extraction, Ionic liquids, Microextraction, Single-Drop Microextraction, Solid-Phase Microextraction.

INTRODUCTION

Public concern in protecting the environment causes that much emphasis is placed on the development of accurate, sensitive, rapid, secure and automated analytical techniques. Even if we continuously apply more and more advanced technologies, still most analytical techniques are not able to deal with a complex environment matrix. Due to this fact, the sample preparation remains a critical step in the overall analytical procedure. The primary goal of sample preparation is to purify the sample, increase the concentration of the analytes, and possibly modify the sample to adapt it to the requirements of the test equipment. It is widely observed that the less preparatory steps, the more reliable and less error prone the result of the analysis is. The additional challenge of modern analytical chemistry is to design methods that minimize or even exclude the use of hazardous substances. All together, these requirements lead to the searching of new techniques of the extraction that conforms to the "green" chemistry.

Two factors are considered as strongly affecting the extraction efficiency: suitable extractant, and rightly carried out the transfer of the analyte to the extraction phase. The first factor is associated with the chemical and physical properties of the extraction phase. The extractant must be a liquid and has to interact actively with analytes. It may be either a pure solvent or a solution of a chemical compound with high affinity to the analyte. The similarity of the extractant phase to the different types of chemical compounds determines the selectivity of the extraction process. To this goal, it is preferred to use solvents of high polarity, so the interaction with mainly polar analytes increases.

The second factor applies to ensure the best interaction of the two phases - an aqueous sample matrix and the extractant. High degree of the dispersion of the extraction phase in the sample matrix is a critical feature of this procedure. The extraction efficiency increases with the distribution of reagents and with the prolongation of contact time of both phases. Several methods lead to good reproducibility and credibility of the results and make the procedure successful.

For instance, suitable extraction efficiency may be accomplished by some physical operations, including the following:

- Intense mixing, vortexing or manual shaking of the system
- Performing a heating-cooling step of the system. One, however, has to bear in mind that in the CPE method careful attention is desirable for adverse effects which may occur. A micellar system formed by the nonionic surfactant may be destroyed while exposed to high temperature. When this happens, the separation of the surfactant containing the analyte is hardly possible;
- An application of ultrasound.

Another factor increasing the achievement of extraction procedure is an introduction of hydrophilic dispersive solvent to the sample matrix for it provides for receiving an adequate emulsion. Facilitating this process by physical methods, such as ultrasound is recommended.

Formation of water-insoluble extractant phase in an ion-exchange reaction is another way of ensuring high efficiency of the extraction process. In this method, an ionic liquid soluble in water [1] and an ion-exchange reagent are added to the aqueous solution in sequence. Thus, a hydrophobic IL is formed as an extraction solvent. A novel created IL is able to adsorb a variety of analytes present in the sample matrix.

A well performed cleaning up, and concentration of trace target organic analytes from aqueous media require some defined conditions. One of the most important is a proper selection of the extractant. Several parameters characterize an adequate extractant. It should be a liquid under standard conditions, has a low vapor pressure, negligible miscibility with water, and a chemical structure providing a high partition coefficient of the analyte between water and the extractant. Furthermore, due to the fact that the most of the analytes of concern to modern science are polar, the extractant should be highly polar as well. All above-enumerated features fit well with the description of some ionic liquids (ILs).

In recent years, a large number of citations present in the field of analytical chemistry and concerning ionic liquids is observed [2 - 5]. ILs are effectively isolated and reused to significantly reduce the cost of their application. ILs are often called modeling solvents, due to the fact that a variety of ions is available to obtain an IL with the desired physical and chemical properties such as melting point, viscosity, density, and miscibility with water and other solvents.

ILs are characterized by tunable physical and chemical properties, to mention volatility, hydrophilicity, hydrophobicity, thermal stability, acidity and basicity, electrical conductivity, a solubility of both organic and inorganic compounds, sustaining activity of transition metals and enzymes, lubricants for polymers, minerals and metal surfaces, catalytic and extraction properties, *etc*. Compared to conventional solvents, ionic liquids may be superior in usability and more attractive in the process of sample preparation [5].

Recently, a growing interest is observed in new extraction methods with lower requirements for sample volume, simpler equipment and handling, and less reagent consumption. This trend resulted in the development of a series of micro extraction methods based on extraction phases in the microliter order and implied a wide range of applications for different analytes and samples [6 - 11].

IONIC LIQUIDS: NEW GENERATION SOLVENTS

Ionic liquids discovered in 1914 by P. Walden are organic salts with melting points at or below 100°C. ILs are characterized by negligible vapor pressure, high thermal stability and more than twice viscosity compared to conventional organic solvents. Moreover, they are suitable solvents for a wide range of substances. ILs shall be deemed to meet the rules for "green chemistry" liquids [12].

These salts are typically composed of an organic cation and an organic or inorganic anion. Due to the nature of the cation, a simplified classification of the ILs has been introduced, based on ammonium, imine, and phosphonium salts. Ammonium ionic liquids are salts in which the cation is based on quaternary ammonium (nitrogen atom of sp^3 hybridization). In turn, the imine liquids are characterized by the presence of the imino moiety (the nitrogen atom of sp^2 hybridization) substituted by hydrogen atoms or alkyl, alkoxyalkyl or tioalkyl groups. Phosphonium ionic liquids (PILs) have some advantages over imidazolium and pyridinium ILs. They are characterized by high thermal stability, and the kinetics of the salt formation is distinguishingly fast. PILs also have no acidic proton, which makes them durable towards nucleophilic and necessary conditions, and they have a lower density than water, which provides potential benefits for some applications [13]. Phosphonium cation-based ionic liquids are a readily available family of ILs that in some applications offer superior properties as compared to nitrogen cation-based ILs.

The anion of the ionic liquid can be both organic and inorganic. Inorganic counter ions include simple ions such as chlorides, bromides, iodides, nitrite and nitrate ions, sulfate ions, as well as some complex ions such as BF_4^-, PF_6^-, $AlCl_4^-$, $SnCl_3^-$, $ZnCl_3^-$, $CuCl_2^-$, $Al_2Cl_7^-$, $Al_3Cl_{10}^-$, $Fe_2Cl_7^-$ (Table **1**).

Table 1. Structures of the cations and anions of the ionic liquids employed to improve microextraction techniques

Cations	Anions	
 1-Alkyl-3-methylimidazolium	Cl⁻ Chloride	 Tetrafluoroborate
 Pyridinium based ILs	 Heksafluorophosphate	 Trifluoromethanesulfonate
 Phosphonium based ILs	Tris(pentafluoroethyl)trifluorophosphate	
 Polymeric ILs 1-Vinyl-3-alkylimidazolium	Bis(trifluoromethyl)sulfonylimide	

Among the organic counterions, one should mention acetates, fluoroacetates, lactates, salicylates, benzoates, sulfimines and thiazoles.

Most often happens that both an anion and a cation are large ions. It means that their ion centers are spaced, and the interaction between them is mainly of electrostatic nature. Moreover, the symmetry of the cation is low, which in most cases results in decreasing the melting point of the ionic liquid based on such cation [14]. Synthesis of ionic liquids is relatively simple and mostly confined to one or two-step reaction [14 - 16]. The availability of a large and diverse number of functional groups allows designing ionic liquids with various compositions, which in turn gives a broad spectrum of synthetic pathways leading to ILs rich in physical and chemical properties. One estimated that it is possible to form approximately 10^{18} different ionic liquids, most of them undiscovered till present day.

Undertaking the separation techniques using ionic liquids, one should pay particular attention to their miscibility with the sample matrix [17] wherein; the type of an anion mostly determines this capacity. It can be assumed that, for example, halides, nitrate, ethoxide, alkylimidazolium trifluoroacetate salts are completely miscible with water, while all alkylimidazolium tetrafluoroborates mix thoroughly with acetone and dichloromethane, in turn, 1-ethyl-3-methyl-imidazolium ethylsulfate is miscible with water and most polar organic solvents.

The properties of ionic liquids depend mainly on their chemical structure. The effectiveness of an IL depends on its structure and physicochemical properties. The physicochemical properties could be tuned by the selection of the appropriate anion, cation, and the length of substituent alkyl chain.

Melting points (MPs) of ILs depend on the cationic moieties compare to the anionic moieties. MPs rely on the degree of non-symmetricity of the cation moieties. For higher symmetric anions higher melting points were observed. MPs are increased with the decrease of the alkyl-chain lengths. However, at a specific range, *e.g.*, ~8–10 carbon atoms, Van der Waals forces an increase of the melting points.

Stability of an IL concerns many categories such as thermal, chemical, or structural stability. The thermal stability, usually the most important factor, for a typical IL varies from 200°C to 400°C [18]. It was reported that ionic liquids with low nucleophilic or coordinating anions usually exhibited the highest thermal stability [18]. ILs also show good chemical and physical stability.

Ionic liquids are green solvents – they show low toxicity and they are biocompatible with traditional solvents. The *toxicity* of ILs depends on the material structure and its moieties. Potential toxicity of ILs is still in the debate [19], and the result of some research is controversial.

ILs has been applied to many micro extraction techniques. They are characterized by a good affinity to organic or inorganic compounds, low *flammability*, and excellent *electrical conductivity*. The rate of diffusion and mass transport is reduced by their high *viscosity*. ILs are tunable for miscible/immiscible and liquid analytes due to the high *density*. The distinguished *surface tension* allows massive, stable drops to be obtained at the tip of a needle or a capillary. Low *vapor pressure* ensures that the evaporation does not reduce drop volumes during the extraction process [20]. Last but not least, ionic liquids are able to dissolve a wide range of different compounds. They served as "green" solvents and offered new "green" analytical tools. The more detailed ILs' physicochemical properties, the solvent classification, and the green contributions can be found in Poole and Lenca (2015) review paper [21].

Single Drop Micro Extraction (SDME)

SDME (Single Drop Micro Extraction) is an example of one of the most commonly used LPME-related techniques [22]. It is based on the process of extracting the analytes in a drop of a solvent, suspended in a needle tip of a micro syringe. The drop may be directly immersed either in the aqueous sample (DI-SDME) or its headspace (HS-SDME) [23 - 25]. Schematic diagrams of DI-SDME and HS-SDME systems are presented in Fig. (**1**).

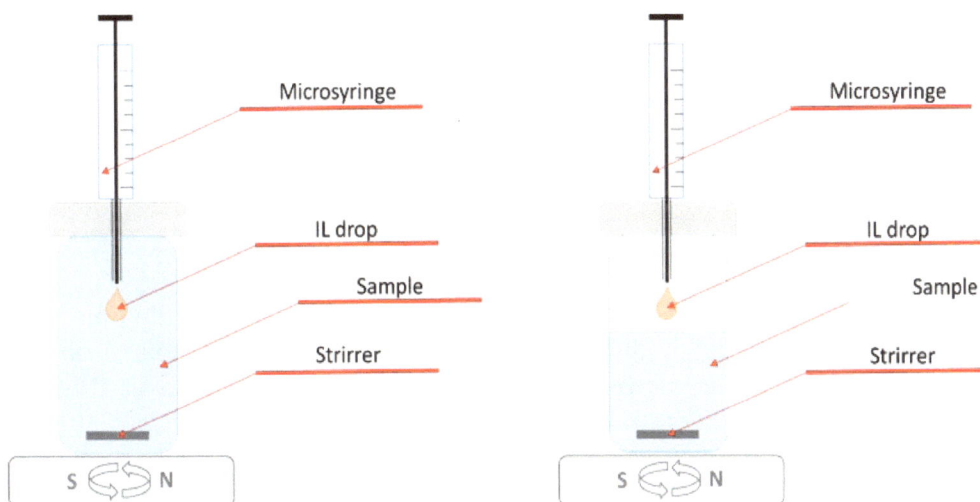

Fig. (1). Schematic diagrams of DI-SDME and HS-SDME systems.

Even if several organic solvents were used as extractants in SDME, severe instability of the drop and reduced precision of the method have been reported. These disadvantages were caused by low viscosity and evaporation of the organic extractant. In contrast, ILs offer high viscosity (which facilitates the formation of a larger-volume drop), low vapor pressure and excellent thermal stability, avoiding evaporation and irreproducible losses of the solvent, as well as its immiscibility with water. Also, ILs may be readily prepared from relatively inexpensive materials and tuned by a combination of different anions and cations for task-specific extraction of analytes from various samples. These properties make ILs perfectly suitable for SDME in the so-called "ionic liquid-based SDME" (SDME) and have led to the development of several applications in which its IL-related equivalent replaces conventional SDME technique.

Indirect immersion SDME (DI-SDME), the extraction solvent is usually water immiscible, since water is the most common sample matrix. On the contrary, in headspace SDME (HS-SDME), to ensure minimal loss of the solvent, hydrophobic organic solvents are commonly chosen as extraction solvents in the early stages of SDME method development [26 - 28].

In 2003 Liu and coworkers demonstrated that ILs may act as excellent extraction solvents for SDME due to their tunable structures [29]. Compared to traditional organic solvent droplets such as 1-octanol, hexane, toluene, and chloroform that volatilize at elevated temperatures and/or low pressures, ILs are capable of making more abundant and more stable droplets in SDME due to their low vapor pressures [30]. Furthermore, the chemical structure of ILs may be easily tuned to provide better selectivity for specific analytes and generate ILs with higher viscosities to improve droplet stability.

A summary of applications in which individual ILs have been employed in SDME is shown in Table **2** [29 - 73].

Hollow-fiber-based Liquid-phase Micro Extraction (HF-LPME)

In conventional SDME, a small organic solvent drop held at the needle tip can be unstable and may dislodge during extraction. To eliminating this problem, hollow-fiber liquid-phase micro extraction technique was proposed in 1999 by Pedersen-Bjergaard and Rasmussen [74]. In this technique, a piece of a polypropylene hollow fiber, containing an acceptor solution in its lumen, is immersed in a vigorously stirred aqueous sample solution for the extraction of target analytes. After extraction, the acceptor solution is removed by a micro syringe and further analyzed by appropriate analytical technique.

Table 2. Representative references on the use of ionic liquids (ILs) in single-drop microextraction (SDME).

Extraction mode	IL	Analyte	Sample	Instrumentation	LOD [µg/L]	Recovery [%]	Ref.
IL–DI-SDME	[C8Mim][PF6]	PAHs	Water	HPLC-VWD-FLD	-	-	[29]
IL–DI-SDME	[HMIm][PF6]	Phenols	Water	HPLC-FLD	0.3÷7.8	90÷113	[31]
IL–DI-SDME	[OMIm][PF6]	Formaldehyde	Shitake mushroom	HPLC-PDA	5	80÷102	[32]
IL–HS-SDME	[C4Mim][PF6]	Chloroanilines	Water	HPLC-PDA	0.5÷1.0	81.9÷99.6	[33]
IL–HS-SDME	[BMIm][PF6]	DDT and metabolites	Water	HPLC-UV	0.05÷0.08	86.8÷102.0	[34]
IL–DI-SDME	[HMIm][PF6]	Benzophenone-3	Urine	HPLC-PDA	1.3	-	[35]
IL–HS-SDME	[HMIm][PF6]	Chlorobenzenes	Water	HPLC-PDA	0.016÷0.039	81.7÷105.5	[36]
IL–HS-SDME	[C4Mim][PF6]	Chloroanilines	Water	HPLC-PDA	0.102÷0.203	60.8÷120.6	[37]
IL–HS-SDME	[C4Mim][PF6]	Phenols	Water	HPLC-UV	0.3÷0.5	89.4÷114.2	[38]
IL–HS-SDME	[BMIm][PF6]	Aromatic amines	Water	HPLC-UV	0.09÷0.38	81.9÷99.1	[39]
IL–HS-SDME	[OMIm][PF6]	BTEXs	Water	GC-MS	0.022÷0.091	88.9÷103.1	[40]
IL–HS-SDME	[OMIm][PF6]	Phenols	Water	G-C-FID	0.1÷0.4	81÷111	[41]
IL–HS-SDME	[BMIm][PF6]	Cobalt, mercury, lead	Waters, biological samples	ETV-ICP-MS	0.015÷0.098	92.8÷108.7	[42]
IL–HS-SDME	[C4Mim][PF6]	Dichloromethane, p-xylene, n-undecane	Water	GC-MS	5.6÷15.6	-	[43]
IL–HS-SDME	[C8Mim][PF6]	Trihalomethanes	Water	GC-MS	0.5-0.9	91.6-101.7	[44]
IL–HS-SDME	[C8MIm][PF6]	Chlorobenzenes	Dye wastewater	GC-FID	0.1÷0.5	88.9÷110.9	[45]
IL–HS-SDME	[C4Mim][PF6]	Organotin	Water	HPLC-FLD	0.62÷0.95	86.9÷92.1	[46]
IL–DI-SDME	[C6Mim][FAP]	PAHs	Water	HPLC-UV	0.1÷0.6	79÷114	[47]

(Table 2) cont......

Extraction mode	IL	Analyte	Sample	Instrumentation	LOD [µg/L]	Recovery [%]	Ref.
IL–HS-SDME	[OMIm][PF6]	Halocompounds	Tap, swimming-pool, river and drinking water	IMS	0.1÷0.9	95.9÷101.7	[48]
IL–DI-SDME	[C$_4$MIm][PF$_6$]	Pb	Water	ETAAS	0.015	-	[49]
IL–DI-SDME	[HMIm][PF$_6$]	Mercury species	Water	HPLC-PDA	1.0÷22.8	83÷123	[50]
IL–HS-SDME	[HMIm][PF$_6$]	Chlorobenzenes	Water	GC-MS	0.001-0.004	90÷115	[51]
IL–DI-SDME	[BMIm][PF$_6$]	Manganese	Water	ETAAS	0.024	95÷105	[52]
IL–DI-SDME	[C$_6$MIm][PF$_6$]	Aromatic amines	Water	HPLC-UV	1.0-2.5	81.7÷97.9	[53]
IL–DI-SDME	[C$_8$MIm][PF$_6$]	Carbonyl compounds	Water	HPLC-UV	0.04÷2.03	84.2÷106.9	[54]
IL–DI-SDME	[C$_4$MIm][PF$_6$]	Phenols	Water	CE	0.005÷0.5	107÷257	[55]
IL–HS-SDME	[C$_4$MIm][PF$_6$]	Organochlorine pesticides	Soil	GC-ECD	0.1÷0.5 µg kg^{-1}	-	[56]
IL–HS-SDME	[C$_4$MIm][PF$_6$]	Dichlorobenzenes	Soil	GC-FID	0.001-0.002 µg kg^{-1}	-	[57]
IL– HS-SDME	[C$_4$MIm] [Cl]-SDS [C$_4$MIm][Cl]-[C$_{10}$MIm] [Br]	Aromatic compounds	Water	HPLC-UV	0.3÷260.3 0.1÷260.3	-	[58]
SDME	[HMIm][PF$_6$]	UV filters	Water	LC-UV	0.06÷3.0	-	[59]
SDME	CYPHOS® IL 101	Lead	Water	ETAAS	0.0032	-	[60]
In situ-SDME	Dithizone-CCl	Mercury	Water	CCD	0.2	-	[61]
IL-HS-SDME	CYPHOS® IL 101	Hg species	Seawater, fish tissues, human hair, and win	ETAAS	0.01	95÷105	[62]

(Table 2) cont....

Extraction mode	IL	Analyte	Sample	Instrumentation	LOD [μg/L]	Recovery [%]	Ref.
IL-DI-SDME	[C₆Mim][PF₆]	Co	Tap, subterranean canal, river, underground water	ETAAS	0.04	-	[63]
SDME	Imidazolium-based	2,4,6-tr:choloroanisole	Water and wine	IMS	0.0001	-	[64]
SDME	-	Copper	Water and food	UV-VIS	0.15	-	[65]
IL-DI-SDME	[C₄Mim][PF₆]	Colchicine	Water	CE	0.25 mg L^{-1}	-	[66]
IL-HS-SDME	[C4Mim][PF6]	MMT	Tap water, gasoline sample	ETAAS	0.010	81.2÷101.0	[67]
IL-DI-SDME	[C₄Mim][PF₆]-Pt	Pathogenic bacteria	Water	MALDI-TOF MS	-	-	[68]
IL-HS-SDME		Musk fragrances	Water	GC-IT-MS/MS	0.010÷0.030	-	[69]
SDME		Sulfonamides	Water	HPLC	1÷1500	-	[70]
IL-HS-SDME	[C₈Mim][PF₆]	NMs, PCMs, and PCM degradation Compound	Sewage sludge	GC-IT-MS/MS	0.5÷1.5	-	[71]
SDME		Cadmium	Water	W-coil ET-AAS	0.015	-	[72]
vacuum MIL-H-SDME	[aliquat⁺]₂[MnCl₄²⁻] [P₆,₆,₆,₁₄⁺]₂[MnCl₄²⁻] [Dy(hfacac)₄⁻][P₆,₆,₆,₁₄⁺] [Dy(hfacac)₃]	Short chain free fatty acid	Milk	GC-MS	-	79.5÷111.0	[73]

Two main extraction modes may be pointed out depending on the solvents involved:

- *Two-phase mode:* the extractant, usually an organic solvent, is contained in the lumen of the fiber and its pores for the extraction of the analytes from an aqueous media;
- *Three-phase mode:* the analytes are transferred to an intermediate solvent that is immobilized in the fiber pores and subsequently back-extracted into a different acceptor phase contained in the fiber lumen.

The schematic illustration of HF-LPME device is shown in Fig. (**2**).

To obtain efficient extractions, the critical aspect in HF-LPME is the selection of the solvent. While organic solvents are typically used in the scope of this technique, recently ILs have been investigated, showing their great potential while impregnated within the pores of the fiber. Similar to previously mentioned extraction techniques, ILs have some unique properties that make them attractive alternatives to traditional organic acceptor phases.

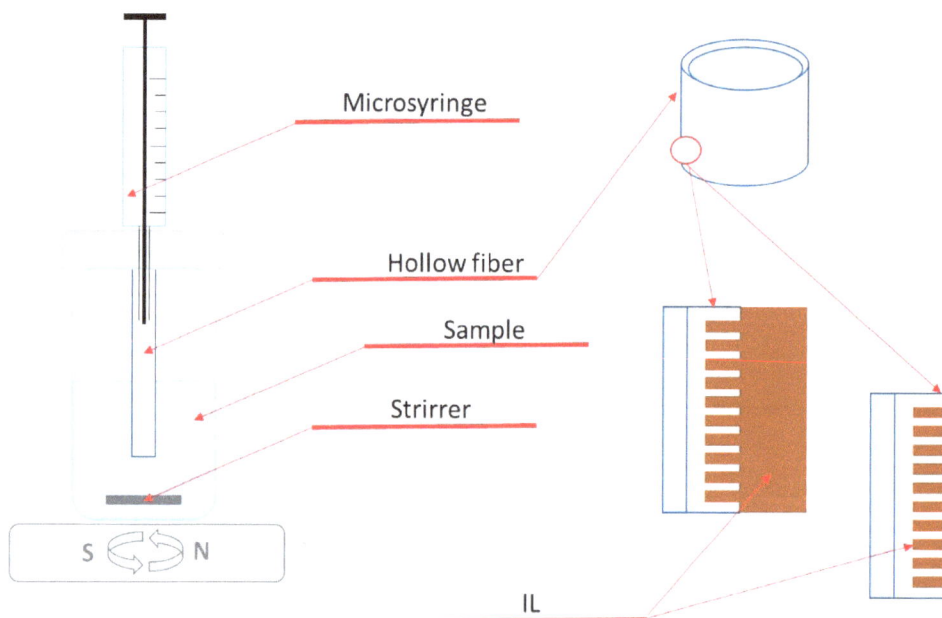

Fig. (2). The schematic diagram of HF-LPME.

The first application of ILs in HF-LPME was described by Peng and co-workers in 2007 [75]. They developed IL supported three-phase HF-LPME mode coupled with HPLC-UV to determine four chlorophenols in environmental water samples.

[C_8MIm][PF_6] ionic liquid was immobilized in the pores of a polypropylene hollow fiber, and an alkaline aqueous solution was used as acceptor phase.

Several HF-LPME techniques have been recently proposed for the extraction of metal ions by IL solvents immobilized within the fiber pores. Chen and coworkers studied the extraction of cadmium(II) using the [C_4MIm][PF_6] IL as the extraction solvent [76]. In this method, the IL was impregnated into the pores of the fiber to facilitate the transfer of the metal ions from the donor phase to an acceptor phase that contained EDTA for chelation and trapping of the metal. The same [C_4MIm][PF_6] IL was employed in another study for the extraction of monomethyl mercury and inorganic mercury using an HF-LPME method [77]. The extracted mercury species were subjected to HPLC–ICP-MS. Pimparu *et al.* used the Aliquat 336 IL within the pores of a hollow fiber for the extraction of chromium(VI) [78].

Wang and coworkers developed a method for the selective extraction of lead(II) found in medicinal plants using a crowned IL as the extractant [79]. The authors found that adding EDTA to the acceptor phase in the lumen of the fiber improved extraction of lead(II) ions.

In 2014, Wang *et al.* [80] suggested a three-phase HF-LPME method for the extraction of kanamycin sulfate combined with electrochemiluminescence detection. In this procedure, 1-octyl-3-methylimidazolium hexafluorophosphate ([OMIm][PF_6]) as an extraction solvent and a hollow fiber supported liquid membrane between the sample solution containing analyte and an aqueous solution (pH 10) as acceptor phase were used. In 2014, an HF-LPME method for the simultaneous extraction and preconcentration of Ag, Al, As, Mn and Ti as ammonium pyrrolidine dithiocarbamate (APDC) complexes in [C_6MIm][PF_6] ionic liquid were proposed by Nomngongo *et al.* [81] They used multivariate techniques for the optimization of analytical parameters. The gasoline samples were digested by using microwave-assisted digestion system before applying the HF-LPME. IL as the extraction solvent phase was impregnated into hollow fiber membrane pores of the hollow fiber wall. The target analytes extracted in the IL phase were then transferred to an aqueous phase by adding different concentrations of nitric acid. They sonicated the mixture for 10 min. After centrifugation, the upper nitric acid phase was collected for the determination of the analyte concentrations with the ICP-OES.

A summary of applications in which ILs have been applied to HF-LPME is shown in Table **3** [82 - 86].

Table 3. Representative references on the use of ionic liquids (ILs) in hollow fiber microextraction (HF-LPME).

Extraction mode	IL	Analyte	Sample	Instrumentation	LOD [µg/L]	Recovery [%]	Ref.
IL-three phase HFLPME	[BMIm]Cl)	PAHs	Water	HPLC	0.00025	90.9÷109.7	[82]
IL-HF-LPME	[HMIm][FAP]	Ultraviolet filters	Water	HPLC	0.3÷0.5	92.0÷115.0	[83]
IL-HF-LPME	[OMIm]PF$_6$	Kanamycin sulfate	Water and milk	ECL	0.67	-	[84]
IL-HF-LPME	[C$_8$MIm][PF$_6$]	Neutral red dye	Free soft drink samples	UV-VIS or ECL	0.36	-	[85]
IL-HF-LPME	[C$_6$MIm][PF$_6$]	Ag, Al, As, Mn, and Ti	Diesel and gasoline	CP-OES	0.04÷0.09	96.0÷101.0	[86]

Dispersive Liquid-liquid Micro Extraction (IL-DLLME)

Dispersive liquid-liquid micro extraction (IL-DLLME) was introduced in 2006 by Rezaee *et al.* [87] for the preconcentration of organic analytes from aqueous matrices [88]. The technique becomes more and more popular, expanding to the new application and research areas, including not only chemistry and biochemistry, but also genetics and molecular biology, chemical engineering, environmental science, medicine, engineering, agricultural and biological sciences, pharmacology, toxicology and pharmaceutics, social sciences, and many others.

Schematic diagram of DLLME procedure is shown in Fig. (**3**). DLLME is based on a ternary component solvent system, in which extraction and disperser solvents are rapidly introduced into the aqueous sample to form a cloudy solution. Extraction equilibrium is quickly achieved, due to the extensive surface contact among the droplets of the extraction solvent and the sample. After centrifugation, the extraction solvent is typically sedimented at the bottom of the tube (if the density is above that of water) and taken with a micro syringe for its later chromatographic analysis.

DLLME, according to the current trends of modern analytical chemistry, is simple, inexpensive, environmentally friendly, and could offer high enrichment factors.

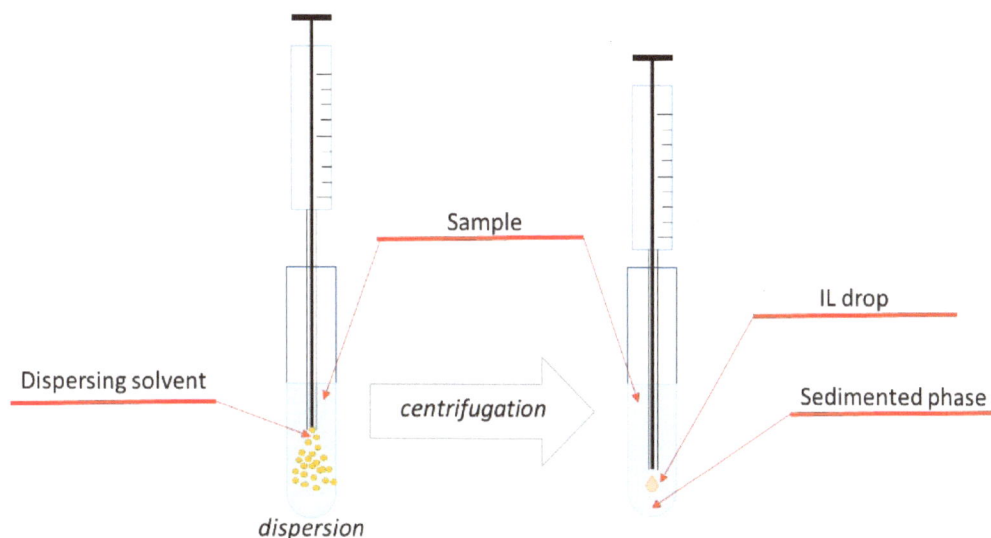

Fig. (3). Schematic diagram of DLLME procedure.

The essence of this method is the right choice of the extraction solvent. The proper dispersive solvent has to be miscible with both extraction and aqueous phases. It creates a turbid solution in which the interactions of the two phases increase, thus leading to high extraction efficiency [88]. However, in the literature, several limitations in the application of DLLME are pointed out [89]. It is essential that the extraction solvent has a higher density than water and form cloudy solution in the presence of a dispersive solvent. Similarly, another difficulty of DLLME lies in the requirements related to the dispersive solvent.

Initially, the ionic liquids have been applied to DLLME technique as IL-DLLME by Zhou and coworkers, and by Baghdadi and Shemirani [90, 91]. In turn, Liu *et al*. were the pioneers of the conventional IL-DLLME application [92]. Their method was used for the preconcentration and separation of heterocyclic insecticides in water prior to HPLC-DAD determination. IL ($[C_6MIm][PF_6]$) was also used as the extractor together with methanol as the dispersive solvent [93]. In recent years many modifications to the method have been introduced, *e.g.*, sample cooling, ultrasound, microwaves and centrifugation support or further-reaching changes such as an *in situ* IL-DLLME method.

Fig. (**4**) presents different variants of DLLME based on the use of ionic liquids proposed in the last ten years [94].

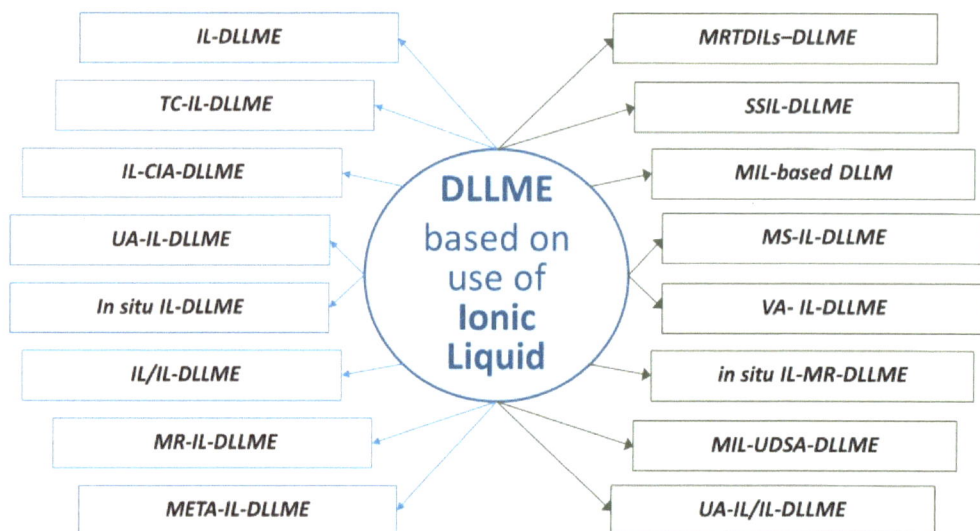

Fig. (4). DLLME modifications based on the use of ionic liquids [94].

As Trujillo-Rodríguez, Rocío-Bautista and Pino & Afonso (2013) in review [95] have shown an enormous number of works had been proposed for the IL-DLLME method in the period of 2008-2013 as an effective alternative to conventional DLLME. In this paper four main operation modes have been distinguished: conventional IL-DLLME, temperature-controlled IL-DLLME, ultrasound-assisted, microwave-assisted or Vortex-assisted IL-DLLME, and *in situ* IL-DLLME. The most-valuable publications related to IL-DLLME have been specified below, starting from the conventional applications of IL-DLLME.

A summary of applications in IL-DLLME is shown in Table **4** [96 - 127].

In situ Ionic Liquids Dispersive Liquid-liquid Micro Extraction (*In situ* IL-DLLME)

The *in situ* ionic liquids dispersive liquid-liquid micro extraction method (*in situ* IL-DLLME) based on ILs is also known as *in situ* solvent formation micro extraction. The technique was proposed by Bahdadi and Shemirani in 2009, as an *in situ* solvent formation micro extraction (ISFME). They applied ISFME to the quantitative determination of some metals [128]. A similar approach has been proposed the same year by Yao and Anderson for the determination of aromatic hydrocarbons [129, 130]. In the *in situ* IL-DLLME method the hydrophilic IL is introduced into an aqueous solution containing the analytes. Then an ion-exchange reagent which promotes a metathesis reaction is added forming an insoluble, hydrophobic IL [129].

Table 4. Representative references on the use of ionic liquids (ILs) in IL-DLLME.

Extraction mode	IL	Analyte	Sample	Instrumentation	LOD [ng L⁻¹]	Recovery [%]	Ref.
IL-DLLME	$[C_6MIm][NTf_2]$, $[C_8MIm][NTf_2]$, $[C_{10}MIm][NTf_2]$	Parabens: (MeP), (EtP), (PrP), (i-BuP) (BuP)	swimming pool water	TD-GC-MS	QOD 4.3÷8.1	-	[96]
IL DLLME	$[C_6MIm][PF_6]$	Heterocyclic insecticides	water	HPLC-DAD	0.53÷1.28	79.0÷110.0	[97]
IL DLLME	$[C_8MIm][PF_6]$	Pesticides	water	HPLC-UV	0.1÷5.0	99.9÷115.4	[98]
IL DLLME	$[C_6MIm][PF_6]$	Tetrabromobisphenol A	water	HPLC-ESI-MS/MS	60	84.6÷89.4 102.5÷110.8	[99]
IL DLLME	$[C_6MIm][PF_6]$	Persistent organic pollutants	water	HPLC-UV	0.33÷0.63	94.4÷115.3	[100]
IL DLLME	$[NH_2C_6][MPyr-FAPa]$	Formaldehyde	wastewater	spectrophotometry	0.1÷55.1	96.0÷104.0	[101]
IL DLLME	$[C_8MIm][PF_6]$	Emerging contaminants	water	HPLC-UV	0.03÷2.0	91.0÷110.0	[102]
IL DLLME	$[C_8MIm][PF_6]$	Polycyclic aromatic hydrocarbons	water	HPLC-FD	0.23÷0.35	90.3-103.8	[103]
IL DLLME	$[C_8MIm][PF_6]$	Antimicrobials	water	HPLC-ESI-MS/MS	0.28÷0.83	88.0÷111.0	[104]
IL DLLME	$[C_8MIm][PF_6]$	Pyrethroid pesticides	water	HPLC-UV	0.12÷0.22	95.0÷110.8	[105]
IL DLLME	$[C_6MIm][PF_6]$	Hexabromocyclododecane diastereomers	water	LC-ESI-MS/MS	0.17÷0.29	88.0÷114.0	[106]
IL DLLME	$[P_{66614}][NTf_2]$	Fenoxaprop-p-ethyl, quizalofop-p-tefuryl, propaquizafop haloxyfop-p-methyl	soy milk and soy sauce	HPLC-DAD	120÷340·10³	-	[107]
IL DLLME	$[C_6MIm][PF_6]$	Pesticides	aqueous extract of bananas	HPLC-DAD	320÷4 660	69.0÷97.0	[108]

(Table 4) cont......

Extraction mode	IL	Analyte	Sample	Instrumentation	LOD [ng L^{-1}]	Recovery [%]	Ref.
IL DLLME	[C$_6$Mim][PF$_6$]	Multi-class pesticides	aqueous extract of grapes or plums	HPLC-DAD	0.651÷5.44	72.0÷100.0 66.0÷105.0	[109]
IL DLLME	[C$_8$Mim][PF$_6$]	Parabens	aqueous extract of pancakes	HPLC-UV	1.0÷1.5	60.1÷79.5	[110]
IL DLLME	[C$_4$Mim][PF$_6$]	Drugs	urine	HPLC-UV	8.3÷32	-	[111]
IL DLLME	[C$_8$Mim][PF$_6$]	Drugs	aqueous extract of human urine	HPLC-DAD	1 500÷3 300	-	[112]
IL DLLME	[C$_6$Mim][PF$_6$]	Emodin and metabolites	rats urine	HPLC-UV	500÷1 000	87.1÷105.0	[113]
IL DLLME	[C$_8$Mim][PF$_6$]	3-OHBaP	human urinary	HPLC-HRMS/MS	0.2	-	[114]
IL DLLME	[C$_8$Mim][PF$_6$]	Rhodamine B	water or aqueous solution of soap, matches tips, pencil colored or textile dyes mixture	FO-LADS	1050	-	[115]
IL DLLME	[C$_8$Mim][PF$_6$]	Phenols	aqueous cosmetics	CE-UV	5 000÷100·10^3	81.6÷119.4	[116]
IL DLLME	[C$_6$Mim][PF$_6$]	Coumarins	aqueous extract of roots	HPLC-UV	13÷ 660	86.4÷101.6	[117]
IL DLLME	[C$_8$Mim][PF$_6$]	Ursolic acid	organic extract of force loquat capsule	HPLC-UV	-	95.9÷105.5	[118]
IL DLLME	[C$_6$Mim][PF$_6$]	Ta (I), Ta (III) species	water	ETAAS	3.3	-	[119]
IL DLLME	[C$_6$Mim][PF$_6$]	Hg (II), MeHg$^+$, EtHg$^+$	aqueous solution of liquid cosmetic	HPLC-ICPMS	1.3÷7.2	-	[120]

(Table 4) cont.....

Extraction mode	IL	Analyte	Sample	Instrumentation	LOD [ng L⁻¹]	Recovery [%]	Ref.
IL DLLME	[C₈MIm][PF₆]	As (III)	wines	ETAAS	5	-	[121]
IL DLLME	[C₆MIm][FAP]	Cr (VI), Co (II), Cu (II), Ni (II)	water	UPLC-UV	300÷2 000	80.0	[122]
IL DLLME	[C₆Py][PF₆]	Zn (II)	aqueous solution of deproteinized milk	FAAS	220	-	[123]
IL DLLME	[C₆MIm][PF₆]	Dy (III), Sm (III), Eu (III), Gd (III)	aqueous extract of uranium dioxide powder	ICP-OES	340÷1 290	-	[124]
IL DLLME	[C₆MIm][NTf₂]	Cu (II)	water	FAAS	450	-	[125]
IL DLLME	[C₆MIm][PF₆]	Co (II)	water, urine and saliva	ETAAS	3.8	-	[126]
IL DLLME	CYPHOS® IL101	Se (IV)	water and aqueous garlic extract	ETAAS	15	-	[127]

A novel dispersion liquid-liquid micro extraction method based on the solidification of sedimentary ionic liquids (SSIL-DLLME), in which an *in situ* metathesis reaction forms a hydrophobic IL has been proposed by Hu *et al.* [131]. The method was developed to determine four pyrethroid insecticides (i.e., permethrin, cyhalothrin, fenpropathrin, and transfluthrin) in water followed by a separation based on high-performance liquid chromatography. In this developed *in situ* DLLME method the extraction efficiency and yield have been noticeably increased. After centrifugation, tributyldodecylphosphonium hexafluorophosphate $[P_{44412}][PF_6]$ used as an extraction solvent, was easily collected by solidification at the bottom of the tube.

One of the first proposals based on the combination of *in situ* IL-DLLME and WCAES was reported by Vidal *et al.* [132]. This technique was presented as a novel and eco-friendly analytical method for total chromium determination and chromium speciation in water samples. In this paper tungsten coil, atomic emission spectrometry (WCAES) has been combined with the *in situ* ionic liquid dispersive liquid-liquid micro extraction (*in situ* IL-DLLME). Tap, bottled mineral, and natural mineral water samples were analyzed at 60 mg L^{-1} spiking level of total Cr content at two Cr(VI)/Cr(III) ratios, and relative recovery values ranged between 88% and 112% showing that the matrix has a negligible effect.

In turn, H. Yua *et al.* [133] applied three structurally different ionic liquids (ILs), namely 1-butyl-3-methylimidazolium chloride ([BMIm][Cl]), 1-(6-hydroxyethyl)-3-methylimidazolium chloride ([HeOHMIm][Cl]) and 1-benzyl-3-(2-hydroxy-ethyl)imidazolium bromide ([BeEOHIm][Br]), as extraction solvents using *in situ* dispersive liquid-liquid micro extraction (*in situ* DLLME) for the preconcentration of two microcystin variants, microcystin-RR (MC-RR) and microcystin-LR (MC-LR) from aqueous samples.

Li *et al.* [134] applied six ionic liquids () as solvents in the extraction and preconcentration of deoxyribonucleic acid (DNA) using an *in situ* dispersive liquid-liquid micro extractions (DLLME) approach. The Authors investigated an effect of different IL substituents and a functional group on the extraction efficiency of DNA. Most analyses have been carried out using HPLC because direct GC analysis can cause accumulation of the nonvolatile IL in the GC inlet.

Authors of [135] have developed *in situ* IL-DLLME method coupled to headspace gas chromatography (HS-GC) employing electron capture (ECD) and mass spectrometry (MS) detection for the analysis of polychlorinated biphenyls (PCBs) and acrylamide at trace levels from milk and coffee samples. It was unveiled that the optimized *in situ* DLLME method is characterized by good analytical precision, excellent linearity, and provided detection limits down to the low ppt level for PCBs and the low ppb level for acrylamide in aqueous samples. The utility of the *in situ* DLLME was verified in quantitative studies of acrylamide in brewed coffee samples.

In turn, Cachol *et al*. [136] proposed a new procedure based on direct microvial thermal desorption injection. In this paper, direct analysis of ionic liquid extracts by gas chromatography and mass spectrometry (GC-MS) was proposed. The *in situ* ionic liquid dispersive liquid-liquid micro extractions (*in situ* IL DLLME) has been recommended as a useful tool for the quantification of bisphenol A (BPA), bisphenol Z (BPZ) and bisphenol F (BPF). The authors subjected 15 different plastic containers to the migration of the test solutions containing the analytes at concentrations ranging between 0.07 and 37 μg L^{-1}. BPA proved to be the most common and at higher concentrations than the other analytes.

Ionic Liquid Dispersive Liquid-liquid Micro Extraction with Magnetic Retrieval (MR-IL-DLLME)

Another new modification of an *in situ* IL-DLLME is the usage of magnetic nanoparticles of iron oxide (Fe_3O_4) for recovery of the *in situ* created hydrophobic ionic liquid [137]. This technique, known as ionic liquid dispersive liquid-liquid micro extraction (IL-DLLME with magnetic retrieval (MR-IL-DLLME), has been developed and used to analyze five benzoylurea insecticides (BUs) in environmental water samples [137]. MNPs (magnetic nanoparticles), in this case, iron oxides were used, have a large surface area and can be easily isolated from a sample solution by applying an external magnetic field [137 - 139]. MNPs acting as sorbents retrieve the IL that contained the analytes. The large interfacial area between the IL and the sample solution contributes to a faster mass transfer [140, 141]. The application of MNPs in *in situ* IL-DLLME stands for another rapid, simple, effective and eco-friendly micro extraction technique.

The *in situ* IL-DLLME combined with ultra-small Fe_3O_4 magnetic nanoparticles was presented in work [142]. The developed technique was acclaimed as a pretreatment method to detect pyrethroid pesticides in water samples. New anion-exchange reagents including Na[DDTC] and Na[N(CN)₂] were optimized for *in situ* extraction pyrethroids. The proposed method is nanometer-level micro extraction (average size 80 nm) with the advantages of simplicity, rapidity, and sensitivity.

Shi *et al*. [143] developed a novel dispersive liquid-liquid micro extraction method based on amine-functionalized Fe_3O_4 magnetic nanoparticles for the determination of six phenolic acids in vegetable oils by high-performance liquid chromatography. Amine-functionalized Fe_3O_4 was synthesized by a one-pot solvothermal reaction between Fe_3O_4 and 1,6-hexanediamine and characterized by transmission electron microscopy and Fourier transform infrared spectrophoto-metry. A trace amount of phosphate buffer solution (extractant) was adsorbed on bare Fe_3O_4-NH_2 nanoparticles by hydrophilic interaction to form the "magnetic extractant". Rapid extraction could be achieved while the "magnetic extractant" on amine-functionalized Fe_3O_4 nanoparticles were dispersed in the sample solution by vortexing. After extraction, the "magnetic extractant" was collected by application of an external magnet. Finally, the founded method was successfully applied for the determination of six phenolic acids in eight kinds of vegetable oils.

Magnetic Stirrer Induced Dispersive Ionic-liquid Micro Extraction (MS-IL-DLLME)

A subsequent version of the dispersive liquid-liquid micro extraction was a magnetic stirrer induced dispersive ionic-liquid micro extraction (MS-IL-DLLME) described by Naeemullah *et al.* [144]. The authors quantified the trace level of vanadium in real water and food samples by graphite furnace atomic absorption spectrometry (GF-AAS) developing MS-IL-DLLME. In this extraction method, a magnetic stirrer was applied to obtain a dispersive medium of 1-butyl-3-methylimidazolium hexafluorophosphate $[C_4MIm][PF_6]$ in an aqueous solution of real water samples and digested food samples. This method was successfully applied to real water and acid digested food samples. Such approach significantly improved the capability of recovery of vanadium-4-(2-pyridylazo) resorcinol (PAR) chelate. The variables playing a vital role in desired micro extraction methods were optimized to obtain the maximum recovery of the analytes under study.

Zhang *et al.* in 2012 [137] presented a different approach for the magnetic retrieval of an ionic liquid in IL-DLLME. After standard steps of the conventional IL-DLLME, a magnetic sorbent was dispersed into the sample by vortexing. Subsequently, the IL adsorbed on the magnetic sorbent was recovered using an external magnet. Defectively, the solvent dispersion and recovery steps do not run synchronously.

Magnetic Effervescent Tablet-assisted Ionic Liquid Dispersive Liquid-liquid Microextraction (META-IL-DLLME)

A solution to this problem was proposed by Yang *et al.* [145]. They worked out a novel analytical technique denoted magnetic effervescent tablet-assisted ionic liquid dispersive liquid-liquid micro extraction (META-IL-DILMA). This method combined ionic liquid-based dispersive liquid-liquid microextraction with the magnetic retrieval of the extractant. A magnetic effervescent tablet composed of Fe_3O_4 magnetic nanoparticles, sodium carbonate, sodium dihydrogen phosphate and 1-hexyl-3-methylimidazolium bis(trifluoromethanesulfonimide) was used for extractant dispersion and retrieval. The combination of effervescence dispersion and magnetic recovery leads to the reduction of constraints in the classic IL-DLLME. The extraction solvent was retrieved through iron oxide nanoparticles, and this step was accompanied by a dispersion that was chemically assisted by an effervescent reaction. As a result, the distribution and collection of the green extractant could be completed almost simultaneously. This optimized method was applied in the analysis of four fungicides (azoxystrobin, triazolone, cyprodinil, trifloxystrobin) in environmental water samples. The recoveries ranged between 70.7% and 105%. META-IL-DLLME was found to be a timesaving, environmentally safe, and efficient field sampling technique.

The same solution was applied by Wang *et al.* [146] to the speciation of Se(IV) and Se(VI). They proposed a novel method based on magnetic effervescent tablet-assisted ionic liquid dispersive liquid-liquid micro extraction (META-IL-DLLME) followed by graphite furnace atomic absorption spectrometry (GF-AAS) determination. In this study, the speciation of selenium in various food and beverage samples has been performed. The authors, an exceptional magnetic effervescent tablet containing CO_2 sources (sodium carbonate and sodium dihydrogenphosphate), ionic liquids and Fe_3O_4 magnetic nanoparticles (MNPs) used to combine extractant dispersion and magnetic recovery procedures into a single step. Using described technique food and beverage samples including black tea, milk powder, mushroom, soybean, bamboo shoots, energy drink, bottled water, carbonated drink and mineral water for the speciation of Se(IV) and Se(VI) with satisfactory relative recoveries (92.0–108.1%) have been examined.

Effervescence-assisted Dispersive Liquid-liquid Micro Extraction (EA-DLLME)

Wu *et al.*, as reported in [147], developed a practical and straightforward novel ionic liquid-based TiO_2 nanofluid, effervescence-assisted dispersive liquid-liquid microextraction (EA-DLLME) method to detect acaricides in honey and tea by high-performance liquid chromatography (HPLC-DAD). In two-steps, effervescence-assisted DLLME procedure the extractant was firstly homogeneously mixed with carbonate and acid. Farther on, the mixture was slowly added to the sample, and the extractant was distributed with the assistance of generated carbon dioxide. Effervescence assistance led to dispersion without the usage of a dispersive solvent. Thus the extraction solvent formed fine droplets in aqueous samples equivalently. Under the optimized conditions, the linear ranges of this proposed method were 0.5–500 g L^{-1} with correlation coefficients in the range of 0.9985–1.0000. The extraction efficiencies for the target analytes varied from 70.70 to 84.58%. The detection and quantitation limits were in the ranges of 0.04–0.18 g L−1 and 0.13–0.60 g L^{-1}, respectively. The intra- and inter-day relative standard deviations (n = 3) were found to range from 2.32 to 5.71%, which showed perfect repeatability. Overall, the EA-DLLME method was reported as time-saving and eco-friendly, with future potential for micro extraction.

Magnetic Ionic Liquids Dispersive Liquid-liquid Micro Extraction (MILs-DLLME)

Nowadays, the combined use of magnetic nanoparticles and ILs has become an innovative area and a hot topic of research in LPME methods [148 - 151]. An original type of magnetic ionic liquids (MILs) with a single component has been developed. MILs interact well with an external magnetic field [151] which makes them useful and efficient substitutes for conventional nonmagnetic extraction solvents in DLLME. In the last decade, the development of MILs encouraged a new wave of research, a wide range of opportunities to create innovative devices and develop processes and products [152 - 155].

Unlike conventional IL-DLLME, the publications describing magnetic retrieval of MILs in IL-based procedures are hardly present [154]. MILs-DLLME allows replacement of a dispersing solvent and centrifugation by usage of Fe_3O_4 magnetic nanoparticles which renders the phase magnetic and leads to a gainful distribution of substances [137].

Recently, a novel class of magnetic ionic liquids (MILs) with a single-component was discovered. In this case, the magnetic property is not introduced as external magnetic supports, but complex ions of metals are used [154, 156, 157].

The first example of MILs is 1-butyl-3-methylimidazolium tetrachloroferrate [C_4MIm][$FeCl_4$]. Although the compound has been known for some time, its magnetic behavior was not described until 2004 [158]. These MILs are based on the anions containing high-spin diiron(III), which were in the forms of tetrachloro- or tetrabromoferrate(III), with varieties of counter cations. Because of their high single-ion magnetic moments, MILs show a good response to an external magnetic field [151] and are very promising to be employed as novel extraction solvents to take the place of routine nonmagnetic ILs of DLLME.

In 2014, Wang *et al.* [159] used MILs in DLLME for the pre-concentration and determination of triazine herbicides in vegetable oils by LC. In this method a MIL, precisely 1-hexyl-3-methylimidazolium tetrachloroferrate [C_6MIm][$FeCl_4$], was used as the extractor phase. The authors reported that the phase separation was shortened by using magnetic separation in this method.

Magnetic Ionic Liquids Dispersive Liquid-liquid Micro Extraction and Stirring-assisted Drop-breakup Micro Extraction (MIL-SAME)

Chatzimitakos *et al.* [160] proposed a novel stirring-assisted drop-breakup microextraction based on the extraction properties of the MIL methyltrioctylammonium tetrachloroferrate $[N_{8881}][FeCl_4]$. The developed procedure makes use of low volumes of readily synthesized ILs based on the magnetic extracting phases avoiding the use of toxic solvents. This approach complies with the principles of green chemistry. Not like in the classical DLLME, in MIL-SADBME, usage of a dispersing solvent and centrifugation are omitted. Here, the extracting phase with the analytes can be harvested by applying an external magnetic field. To demonstrate MIL-SADBME applicability, the proposed micro extraction procedure was studied in conjunction with HPLC for the determination of selected phenols and acidic pharmaceuticals in aqueous matrices, taking into account the primary experimental variables involved. The results obtained were accurate and highly reproducible, thus making it a good alternative approach for the routine analysis of phenols and acidic pharmaceuticals.

Magnetic Ionic Liquid-based Up-and-down-shaker-assisted Dispersive Liquid-liquid Micro Extraction (MIL-UDSA-DLLME)

Among the most recent publications devoted to MILs, last-year report of Wang *et al.* [161] seems to be very promising. The Authors proposed a novel and sensitive magnetic ionic liquid-based up-and-down-shaker-assisted dispersive liquid-liquid micro extraction (MIL-UDSA-DLLME) method for the graphite furnace atomic absorption spectrometric measurement of inorganic selenium speciation from various rice matrices. In the first micro extraction step, the magnetic ionic liquid (1-butyl-3-methylimidazolium tetrachloroferrate, $[C_4MIm][FeCl_4]$ was selected to extract the complex of Se(IV) and 2,3-diaminonaphthalene from aqueous solution by the assistance of up-and-down-shaker vortex agitator. After the micro extraction step, the magnetic ionic liquid containing target analytes was collected at the bottom of the tube by applying an external magnetic field around the test tube. The developed methodology was also successfully used for the speciation of Se(IV) and Se(VI) in different rice samples with the relative recoveries within the acceptable range of 94.9–104.8% for the addition recovery tests.

Magnetic Multiwalled Carbon Nanotubes Extraction Based on Polymeric Ionic Liquids (m-MWCNTs@PIL)

In [162] a first magnetic adsorbent was proposed, namely benzyl groups functionalized imidazolium-based polymeric ionic liquids (PIL)-coated magnetic multiwalled carbon nanotubes (MWCNTs) (m-MWCNTs@PIL), for the extraction of Cu and Zn-superoxide dismutase (Cu, Zn-SOD). The research concerned the extraction performance of the m-MWCNTs@PIL in the magnetic solid-phase extraction (MSPE) procedure coupled with the determination by UV-Vis spectrophotometer. Compared with m-MWCNTs@IL and m-MWCNTs, the mMWCNTs@PIL exhibited the highest extraction capacity of 29.1 mg/g for Cu, ZnSOD. The adsorbed Cu, Zn-SOD remained high specific activity after being eluted from m-MWCNTs@PIL by 1 mol/L NaCl solution. Besides, the mMWCNTs@PIL could be easily recycled and successfully employed in the extraction of Cu, Zn-SOD from real samples. The proposed method has a great potential in the extraction of Cu, Zn-SOD or other analytes from biological samples.

A summary of applications of IL-DLLME related techniques is shown in Table **5** [163 - 222].

Ionic Liquid Solid Phase Extraction (IL-SPE)

Solid phase extraction (SPE), among many techniques developed for the isolation and the pre-concentration of different organic compounds and metal ions, belongs to the most frequently used [223, 224]. Formerly, SPE application area was usually restricted to the pre-concentration of analytes showing strongly hydrophobic properties, whereas the recovery rates for polar compounds were not satisfactory. Nowadays research is concentrated on obtaining better sorbents, which applied to SPE would let to better extraction of given compounds from contaminated samples, as well as to higher recovery rates.

At the first view, one may note that the liquid state of ILs is lost when immobilized onto a solid support, such as for SPE. Nevertheless, under special conditions, multi-modal type interactions can be still exploited. In recent years, ILs have been immobilized onto silica or polymeric supports (known as supported IL phases – SILPs) to take advantage of the chemical functionality that ILs can impart. As a result, new groups of stationary phases, to be applied for different extraction (*e.g.*, SPE and SPME) and separation techniques (*e.g.*, GC, LC and CE), were obtained. SILPs can, therefore, be considered as another class of a sorptive material [10]. Concerning sorptive extraction techniques, SILPs were first applied as an SPME coating in 2005. SILPs as SPE materials emerged later, in 2009. Since then, the number of applications has increased steadily. Mainly due to their customizable structures, IL-modified solid supports emerged as suitable alternatives to conventional materials, to improve the selectivity of SPE methods for a broad range of analytes.

(Table 5) cont.....

Extraction mode	IL	Analyte	Sample	Instrumentation	LOD [ng L⁻¹]	Recovery [%]	Ref.
TC-IL DLLME	[C8MIm][PF6]	phthalate esters and pyrethroid insecticides	water	HPLC-UV	0.34÷0.48	-	[169]
TC-IL DLLME	[C8MIm][PF6]	pyrethroid insecticides	water	HPLC	0.27÷0.68	89.2÷102.7	[170]
TC-IL DLLME	[C6MIm][PF6]	phenols	water	HPLC-DAD	0.24÷0.45	85.8÷117.0	[171]
TC-IL DLLME	[C6MIm][PF6]	dichloro-diphenyltrichloroethane (DDT) and its main metabolite	water	HPLC-UV	0.04÷0.43	-	[172]
TC-IL DLLME	[C8MIm][PF6]	pesticides	water	HPLC-UV	0.58÷0.86	99.6÷106.5	[173]
TC-IL DLLME	[C8MIm][PF6]	phenolic compounds	water	HPLC-VWD	0.68÷1.36	97.1÷108.1	[174]
TC-IL DLLME	[C8MIm][PF6]	phthalate esters	water	HPLC-VWD	0.45÷1.40	90.1÷99.2	[175]
TC-IL DLLME	[C6MIm][PF6]	carbamate pesticides	water	HPLC-VWD	0.23÷0.48	95.5÷104.4	[176]
TC-IL DLLME	[C8MIm][PF6]	endocrine disruptors	water	HPLC-FD	0.1	-	[177]
TC-IL DLLME	[C6MIm][PF6]	hexabromocyclodo- decane diastereomers	water	HPLC-ESIMS/MS	0.17÷0.49	77.2÷99.3	[178]
TC-IL DLLME	[C6MIm][PF6]	ofloxacin antibiotic	aqueous extract of powered tablet, aqueous solution of deproteinized human plasma and human urine	spectrofluorometric	29	91.5÷106.9 89.5÷97.1	[179]
TC-IL DLLME	[C6MIm][PF6] [C6MIm][NTf2]	Hg (II)	water	spectrophotometry	300	-	[180]
TC-IL DLLME	[C6MIm][PF6]	Pb (II)	water	FAAS	9 500	94.8÷104.1	[181]

(Table 5) cont.....

Extraction mode	IL	Analyte	Sample	Instrumentation	LOD [ng L⁻¹]	Recovery [%]	Ref.
TC-IL DLLME	$[C_4MIm][PF_6]$	V (IV) and V (V)	water	ETAAS	4.9	-	[182]
TC-IL DLLME	$[C_6MIm][PF_6]$	Zn (II)	water and aqueous solid food extract	FAAS	180	91.2÷102.2	[183]
TC-IL DLLME	$[C_6MIm][PF_6]$	Ag nanoparticles	water and aqueous extract of hair	GF-AAS	5.2	-	[184]
US-IL DLLME	$[C_6MIm][PF_6]$	aromatic amines	water	HPLC-UV	0.17÷0.29	92.2÷119.3	[185]
US-IL DLLME	$[C_4MIm][PF_6]$	drugs	water	HPLC-UV	15÷26	80.5÷112.0	[186]
US-IL DLLME	$[C_6MIm][PF_6]$	aromatic amines	water	HPLC-UV	0.37	82.0÷94.0	[187]
US-IL DLLME	$[C_6MIm][FAP]$	total content of alkylbenzene sulfonates	water	spectrophotometry	0.2÷5.0	-	[188]
US-IL DLLME	$[C_6MIm][PF_6]$	UV filters	water	HPLC-UV	0.10÷0.24	71.0÷118.0	[189]
US-IL DLLME	$[C_6MIm][PF_6]$	phenylurea herbicides	water	HPLC-UV	0.10÷0.24	-	[190]
US-IL DLLME	$[C_6MIm][PF_6]$	phenol	water	FO-LADS	0.86	99.7÷103.1	[191]
US-IL DLLME	$[C_6MIm][FAP]$	tricyclic antidepressants	water	HPLC-UV	0.3÷1.0	94.3÷114.7	[192]
US-IL DLLME	$[C_8MIm][PF_6]$	pyrethroid pesticides	aqueous solution of honey	HPLC-DAD	0.21÷0.38	101.2÷103.0	[193]
US-IL DLLME	$[C_6MIm][PF_6]$	fungicides	red wine	HPLC-DAD	2.8÷16.8	70.0÷107.5	[194]

(Table 5) cont......

Extraction mode	IL	Analyte	Sample	Instrumentation	LOD [ng L⁻¹]	Recovery [%]	Ref.
US-IL DLLME	[C$_6$Mim][PF$_6$]	celastrol	aqueous solution of human urine	HPLC-DAD	1.6	93.2÷109.3	[195]
US-IL DLLME	[C$_6$Mim][PF$_6$]	Cd (II)	water	ETAAS	7.4	87.2÷106	[196]
US-IL DLLME	[C$_8$Mim][NTf$_2$]	Rh (III)	water and aqueous extract of leaves of roses	FAAS	370	-	[197]
USA-IL-DLLME	[C$_6$Mim][NTf$_2$]	Se (IV)	ice tea, soda and mineral water beer, cow's milk, red wine, mixed fruit juice, date, apple, orange, grapefruit, egg and honey	Atomic absorption spectrometer equipped with a deuterium background correction system and electrothermal atomizer, HGA-800	12	-	[198]
VA-IL DLLME	[C$_4$Mim][PF$_6$]	UV filters	water	HPLC-UV	1.9÷6.4	-	[199]
VA-IL DLLME	[C$_8$Mim][PF$_6$]	fluoroquinolones	ground water	HPLC-FD	0.8÷13	85.0÷107.0	[200]
VA-IL DLLME	[C$_4$Mim][PF$_6$]	glucocorticoids	water	HPLC-UV	4.11÷9.19	≥ 97.2	[201]
VA-IL DLLME	[C$_6$Mim][PF$_6$]	pesticides and metabolites	aqueous extract of soils	HPLC-FD	0.02÷90.2*	-	[202]
VA-IL DLLME	[C$_8$Mim][PF$_6$]	organophosphorus pesticides	aqueous solution of fruit juices	HPLC-UV	0.061÷0.73*	-	[203]

(Table 5) cont.....

Extraction mode	IL	Analyte	Sample	Instrumentation	LOD [ng L⁻¹]	Recovery [%]	Ref.
US-IL DLLME	[C₆MIm][PF₆]	celastrol	aqueous solution of human urine	HPLC-DAD	1.6	93.2÷109.3	[195]
US-IL DLLME	[C₆MIm][PF₆]	Cd (II)	water	ETAAS	7.4	87.2÷106	[196]
US-IL DLLME	[C₈MIm][NTf₂]	Rh (III)	water and aqueous extract of roses	FAAS	370	-	[197]
USA-IL-DLLME	[C₆MIm][NTf₂]	Se (IV)	ice tea, soda and mineral water, beer, cow's milk, red wine, mixed fruit juice, date, apple, orange, grapefruit, egg and honey	Atomic absorption spectrometer equipped with a deuterium background correction system and electrothermal atomizer, HGA-800	12	-	[198]
VA-IL DLLME	[C₄MIm][PF₆]	UV filters	water	HPLC-UV	1.9÷6.4	-	[199]
VA-IL DLLME	[C₈MIm][PF₆]	fluoroquinolones	ground water	HPLC-FD	0.8÷13	85.0÷107.0	[200]
VA-IL DLLME	[C₄MIm][PF₆]	glucocorticoids	water	HPLC-UV	4.11÷9.19	≥ 97.2	[201]
VA-IL DLLME	[C₆MIm][PF₆]	pesticides and metabolites	aqueous extract of soils	HPLC-FD	0.02÷90.2*	-	[202]
VA-IL DLLME	[C₈MIm][PF₆]	organophosphorus pesticides	aqueous solution of fruit juices	HPLC-UV	0.061÷0.73*	-	[203]

(Table 5) cont......

Extraction mode	IL	Analyte	Sample	Instrumentation	LOD [ng L⁻¹]	Recovery [%]	Ref.
VA-IL DLLME	[C$_6$Mim][PF$_6$]	As (III) and As (IV)	urine and blood	FI-HG-AAS	0.02÷10 urine 5 blood	-	[204]
VA-IL DLLME	CYPHOS® IL 101	Co (II)	water and pharmaceutical formulations	ETAAS	8	97.2÷103.0	[205]
MW-IL DLLME	[C$_6$Mim][PF$_6$]	formaldehyde	aqueous solution of beverages	HPLC-UV	0.12	84.9÷95.1	[206]
MW-IL DLLME	[C$_6$Mim][PF$_6$]	aminoglycosides	aqueous extract of milk	HPLC-FD	0.11÷0.50	96.4÷105.4	[207]
in situ IL DLLME	[C$_4$Mim][BF$_4$]/[Li][NTf$_2$]	triazine herbicides	water	HPLC-UV	0.52÷1.3	88.4÷114.0	[208]
in situ IL DLLME	[C$_6$Mim][Cl]/[Li][NTf$_2$]	insecticides aryloxyphenoxypropionate herbicides	water	HPLC-VWD	0.98÷2.54	82.0÷102.0	[209]
in situ IL DLLME	[C$_8$Mim][Cl]/[H][PF$_6$]	medicinal products	aqueous extract of herb	HPLC-UV	80.0÷97.0	-	[210]
in situ IL DLLME	[C$_{16}$C$_4$Im][Br]/[Li][NTf$_2$]	PAHs	aqueous extract of toasted cereals	HPLC-UV-FD	0.03÷83	80.0÷95.0	[211]
in situ IL DLLME		(PCBs) acrylamide	milk, coffee	HS-GC –ECD or MS	5 to 25	-	[135]
in situ IL DLLME	[C$_4$Mim][NTf$_2$]/[C$_4$Mim][BF$_4$]	(CPs)	honey	HPLC	800÷3 200	91.6÷114.3	[212]

(Table 5) cont.....

Extraction mode	IL	Analyte	Sample	Instrumentation	LOD [ng L^{-1}]	Recovery [%]	Ref.
in situ IL DLLME	[C$_{16}$C$_3$(OH)$_2$Im][Br]/[Li][NTf$_2$]	DNA deoxyribonucleic acid	aqueous DNA solution	HPLC-UV	-	-	[134]
ATPS- in situ IL-DLLME	[HMIm][PF$_6$]	curcuminoids	Curcuma Longa	HPLC	3 000	93.0	[213]
in situ IL DLLME	[C$_6$MIm][NTf$_2$], [C$_8$MIm][NTf$_2$], [C$_{10}$MIm][NTf$_2$], [C$_{12}$MIm][NTf$_2$]	bisphenol A (BPA), bisphenol Z (BPZ), bisphenol F (BPF)	different plastic containers,	GC-MS	5.7 7.8 5.4	95÷113	[136]
IL-DLLME/in situ IL-DLLME	[C$_4$MIm][PF$_6$] or [C$_{16}$C$_4$lm][Br]	Cu(II)	water	FAAS	3 300 and 5 100 when using [C$_{16}$C$_4$lm][Br]	105.0÷110.0	[214]
in situ IL DLLME	[C$_6$MIm][BF$_4$]/[Na][PF$_6$]	Cd (II)	saline water	FAAS	70	-	[215]
in situ IL DLLME	[C$_6$MIm][BF$_4$]/[Na][PF$_6$]	Pd (II)	seawater, food additive, and aqueous solutions of tea and blood	spectrophotometry	200	-	[216]
in situ IL DLLME	[C$_6$MIm][BF$_4$]/[Na][PF$_6$]	Ag (I)	photographic and X-ray waste	spectrophotometry	400	-	[217]
MIL-based DLLM	[C$_6$MIm][FeCl$_4$]	triazine herbicides: cyanazine, desmetryn, secbumeton, terbutryn, dimethametryn, dipropetryn	vegetable oils	LC	1 310÷1 490	81.8÷114.2	[218]

(Table 5) cont.....

Extraction mode	IL	Analyte	Sample	Instrumentation	LOD [ng L⁻¹]	Recovery [%]	Ref.
MIL-UDSA-DLLME	$[C_4MIm][FeCl_4]$	Se (IV), Se (VI)	different rice samples	FAAS	18	92.0÷108.1	[146]
MRTILs-DLLME		Au, Ag	water	ETAAS	3.2 and 7.3	96.0÷104.5	[219]
(UA-IL/IL-DLLME)	$[C_6MIm][PF_6]$	sulfamerazine, sulfamethizole, sulfachlorpyridazine sulfamonomethoxinesulfmethoxazole sulfisoxazol	milk powder	LC	2.94* 9.26* 16.7* 5.28* 3.35* 6.66*	90.4÷114.8	[220]
UAE-IL-DLLME	$[C_6MIm][BF_4]$	six alkaloids	M. cordata	UPLC-MS/MS	80	86.4÷112.5	[221]
UA-IL-DLLME	$[C_6MIm][BF_4]$	antidepressant drugs citalopram hydrobromide and nortriptyline hydrochloride	human plasma	HPLC	10 000 for citalopram and 6 000 for nortriptyline	-	[222]

The first study utilizing IL-modified SPE materials was reported by Tian and coworkers [225]. This research concerned silica support coupled with 1-trimethoxy(propyl)silane-3-methylimidazolium chloride to generate IL-modified solid support for the extraction of tanshinones from Salvia miltiorrhiza. Compared to a commercial silica cartridge, the new support was characterized by higher recovery values, due to the stronger interactions of the IL-modified silica material with tanshinones. Recently, IL-SPE has expanded to new types of solid supports functionalized with ILs or PILs to improve the selectivity and extraction efficiency, as well as new configurations (*e.g.*, on-line IL-SPE) to increase sample throughput.

Imidazole and pyridine have been the most common IL groups used in the polymerization. The first work concerning these compounds was published in 2009 by Fontanals *et al.* [226] where N-methylimidazolium trifluoroacetate salt was a part of the polymer. The IL polymer was used as SPE sorbent for the extraction of 10 pharmaceutical products from water samples.

Dai and coworkers applied a PIL based on the 1-vinyl-3-methylimidazolium chloride ([VMIm][Cl]) monomer for the on-line extraction and separation of hydroxybenzoic acids from Typha angustifolia [227]. The PIL was cross-linked with ethylene glycol dimethacrylate and Mobil Composition of Matter No. 48 (MCM-48) at 60°C for 24 h.

Liu *et al.* employed a monolithic PIL column based on the 1-vinyl-3 butylimidazolium chloride ([VC$_4$Im][Cl]) IL for the separation of three calcium antagonists [228]. The column was prepared using *in situ* polymerization between [VC$_4$Im][Cl], glycidyl methacrylate, and ethylene dimethacrylate initiated by AIBN.

Only a few works about modified silica or modified polymers by ILs have been published. The results published from 2009 until today are described by L. Vidal *et al.* in the review [15]. The Authors highlighted some of their recent advances and provided an overview of the current status of ionic liquid-modified materials applied in solid-phase extraction, liquid and gas chromatography and capillary electrochromatography concerning recent applications.

A summary of applications in IL-SPE is shown in Table **6** [229 - 260].

Table 6. Applications of supported ionic-liquid phases (SILPs) to solid-phase extraction (SPE).

Extraction mode	SILP	Analyte	Sample	Instrumentation	LOD [µg/L]	Recovery [%]	Ref.
SPE-RTLs	Sil impregnated with [C$_6$Mim][Br] or [C$_{12}$Mim][Br]	Phthalates	Tap, canal. River and effluent water	HPLC-UV	0.12 and 0.17	85÷107	[229]
SPE-SILP	VBC-DVB[Mim][CF$_3$COO]	Acidic pharmaceuticals	River water and wastewater	HPLC-UV	-	87÷69	[230]
SPE-IL	Silpr[Mim][Cl]	Tanshinones	Salvia miltiorrhiza Bunge	HPLC-UV	76÷93	87.2÷91.5	[231]
SPE-SILP	Silpr[EMIm][Cl] 2-Ethyl-4-methylimidazolium chloride	Liquiritin Glycyrrhizic acid	Licorice	HPLC-UV	365÷464	81,0÷90,3	[232]
SPE-IL	Sil-AgBF$_4$/SiO$_2$[IM][PF$_6$] Sil-AgNO$_3$/SiO$_2$[IM][PF$_6$] Sil-AgBF$_4$/SiO$_2$[IM][BF$_4$] Sil-AgNO$_3$/SiO$_2$[IM][BF$_4$]	Polyunsaturated fatty acid methyl esters	Fish oil	GC-FID	-	93,5	[233]
SPE-IL	[MIm][PF$_6$]/[BF$_4$]	Essential fatty acid methyl esters	Soy-derived biodiesel	GC-FID	-	85,2÷95,7	[234]
SPE-SILP	Sil[C$_4$MIm][PF$_6$]	Cd^{+2}	Tap and lake water	FAAS	0.60	95÷103	[235]
SPE-IL	PVC[Mim][Cl]	Cr+4	Tap and snow water	ETAAS, ICP-MS	0.003	92÷100	[236]
SPE-IL	GMA-EGDMA[MIm][Cl] N-Methylimidazolium	Caffeine Theophylline	Green tea	HPLC	-	Caffeine: 87÷90 Theophylline: 88÷91	[237]
SPE-IL	VBC-DVB[MIN][Br] N-(Propyl-3-amino)imidazolium bromide	Matrine Oxymetrine	Sophora flavescens Ait.	HPLC		Oxymetrine: 94 Matrine: 81	[238]

(Table 6) cont.....

Extraction mode	SILP	Analyte	Sample	Instrumentation	LOD [μg/L]	Recovery [%]	Ref.
SPE-IL	MAA-GMA-EGDMA[BIm][Cl]	β-Sitosterol	Salicornia herbacea L.	LC-UV	0.08	97.20 to 102.93	[239]
SPE-IL	PVP-DVB[Bu][Cl] N-Butyl-pyridinium chloride	Liquiritin Glycyrrhizin	Licorice and medicines	HPLC	-	80÷90	[240]
SPE-IL	[MIm][PF$_6$]	12 sulfonylurea herbicides	Environmental water and soil	HPLC-UV	0.012÷0.142	Water: 53.8÷118.2 Soil: 60.9÷121.3	[241]
SILP	Fe$_3$O$_4$Silpr[MIm][Cl]	Chlorophenols	Water	HPLC	0.20÷0.35	-	[242]
On-line SPE	MIP-ViBuIM[Cl]	Chlorsulfuron	Surface water	HPLC	0.001	81.0÷110.1	[243]
SPE-SIL	Sil[C$_4$MIm][Br]	Pb^{+2}	River water	FAAS	0.7	99÷100	[244]
SPE-SILP	MIP-Silpr[IM][Cl]	Tanshinone I, tanshinone IIA and cryptotanshinone	Salvia Miltiorrhiza Bunge and ginseng drinks	LC	76.85÷93	87.5÷90.0	[245]
SPE-SILP	MIP-VBC-DVB[IM][Cl] MIP-VBC-DVB[MIm][Cl] MIP-VBC-DVB[IM-COOH][Cl] MIP-VBC-DVB[IM-NH$_2$][Cl] MIP-VBC-DVB[IM-CN][Cl]	Tanshinone I, tanshinone IIA and cryptotanshinone	Salvia Miltiorrhiza Bunge and functional drinks	HPLC	-	Not available	[246]
SPE-IL	[IM][Cl] [MIm][Cl] [EMIm]Cl	Lactic acid	Fermentation broth Imidazolium:	HPLC-UV	-	85÷91.9	[247]
SPE-IL	Silpr[IM][Cl] Silpr[MIm][Cl] Silpr[EMIm][Cl]	Organic acids, aromatic compounds	Atmospheric aerosol sample	LC-MS	-	87÷110	[248]

(Table 6) cont.....

Extraction mode	SILP	Analyte	Sample	Instrumentation	LOD [µg/L]	Recovery [%]	Ref.
SPE-SILPs	VBC-DVB[MIm][CF$_3$SO$_3$] VBC DVB[MIm][BF4]	Acidic pharmaceuticals	River water and effluent wastewater	HPLC			[249]
SPE-SILs	Silpr[IM][Cl] Silpr[MIm][Cl] Silpr[EIM][Cl] Silpr[C$_4$IM][Cl] Silpr[IM][BF$_4$] Silpr[IM][PF$_6$] Silpr[IM][Tf$_2$N]	Oxymatrine	*Sophora flavescens* Ait	HPLC-UV	53.2	89.7÷93.4	[250]
SPE-IL	MIP-AllEIM[Br] MIP-AllBuIM[Cl] MIP-AllHeIM[Cl] MIP-AllOcIM[Br]	Phenolic acids	*Salicornia herbacea* L	HPLC-UV	20.0–45.0	84.7÷96.6	[251]
SPE-IL	poly[VMIM$^+$][PF$_6^-$]	DNA	bacterial cell lysate			80.7	[252]
On-line IL-based SPE	poly[VC$_4$IM$^+$][Cl$^-$]	nifedipine, nitrendipine, felodipine	human plasma	HPLC-UV	3; 3; 2	90÷106	[253]
IL-MSPE	Dicationic ammonium-based IL	Bhb	blood	UV-vis spectrophotometer			[254]
IL-MSPE	Betaine-based IL	BSA	blood	UV-vis spectrophotometer	1.87		[255]
IL-MSPE	superparamagnetic attapulgite/Fe$_3$O$_4$/polyaniline (ATP/Fe$_3$O$_4$/PANI)	benzoylurea insecticides (BUs)	environmental water samples	HPLC-DAD	0.02÷0.43	77.37÷103.69	[256]

(Table 6) cont.....

Extraction mode	SILP	Analyte	Sample	Instrumentation	LOD [µg/L]	Recovery [%]	Ref.
PIL-SPE	[C₄IM⁺][AMPS⁻]	OVT, BSA, BHb, Cytc, casein	blood	-	OVT÷ 861.5 BSA÷ 842.2 BHb÷983.4 Cytc÷882.9 casein -605.2 mg g⁻¹	-	[257]
IL-MSPE	[TMG⁺]₂[Glu²⁻]	trypsin	aqueous solution	UV-vis spectrophotometer		Over 90	[258]
PIL-MA-SPE	VMDB/PIL-MA	Parabens and aromatic amines	Lake, river and tap waters	HPLC-DAD	0.5÷200.0	83.1÷108	[259]
IL-MSPE	dicationic ionic liquid-NTf₂ [DICAT-NTf₂]	PAHs	environmental samples	GC-MS	0.0008÷0.2086	80.2÷111.9	[260]

Ionic Liquid Solid-phase Micro Extraction (IL-SPME)

In the early 1990s Pawliszyn and co-workers developed solid-phase micro extraction as miniaturized version of solid-phase extraction [261]. It is a non-exhaustive extraction technique based on the partitioning of analytes among the sample matrix and either liquid or solid stationary phase, in turn coated on a solid support (Fig. **5**). The sorption step is followed by the desorption of the analytes. This technique is fast, simple and conforms to the rules of the "green" chemistry. It combines sampling and sample preparation into one single step.

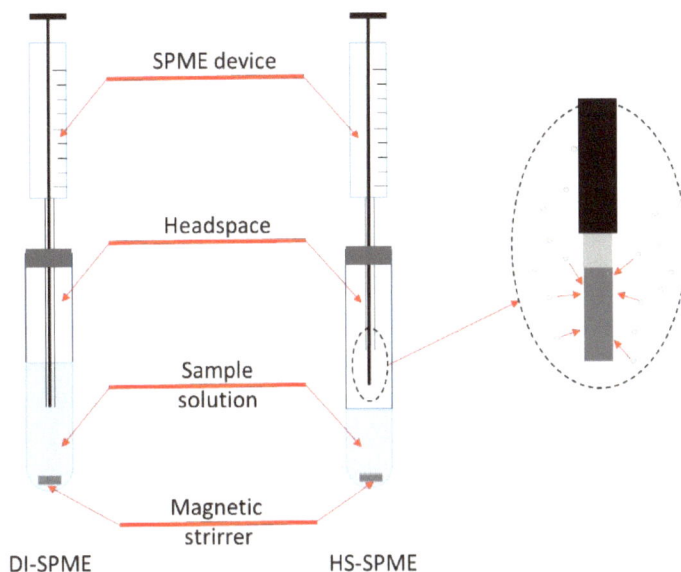

Fig. (5). Schematic diagram of DI- and HS-SPME procedure.

To enhance extraction selectivity and sensitivity and to expand the applicability of SPME for a broader range of analytes, an increasing interest in developing of novel sorbent materials has appeared. Since 2005, due to negligible vapor pressure, good thermal stability and tunable extractability for various compounds, ILs are used as adsorbing coatings for SPME [262]. An extensive review of the application of ILs as sorbent materials for SPME is presented by Anderson and co-workers [263].

Jiang and co-workers [262] used ILs as disposable SPME fiber coatings in the headspace extraction of benzene, toluene, ethylbenzene, and xylenes (BTEXs) from paints. An IL 1-octyl-3-methylimidazolium hexafluorophosphate ($[C_8MIm][PF_6]$) was applied as a SPME coating. The IL was physically adsorbed on fused silica or stainless steel fiber of an SPME device by dipping the fiber into the solution of [C₈MIm][PF₆] in dichloromethane for 1 min. Then the analytes were extracted using HS-SPME mode and desorbed in the injection port of GC at 200°C. Finally, IL-coating was washed out from the fiber with solvents. The coating procedure was repeated before every extraction. This disposable IL-coated fiber showed reproducibility comparable with the commercially available SPME fibers, but its sensitivity was lower because of the relatively thin coating.

Later on, four different ILs including 1-butyl-3- methylimidazolium tetrafluoroborate ([C₄MIm][BF₄]), 1-octyl-3-methylimidazolium tetrafluoroborate ([C₈MIm][BF₄]), [C₈MIm][PF₆] and 1-ethyl-3-methylimidazolium ethylsulphate ([C₂MIm][EtSO₄]) were synthesized and examined as disposable coatings for the HS-SPME of methyl tert-butyl ether in gasoline samples [264].

Hsieh and co-workers [265] introduced SPME fiber pre-coated with Nafion membrane before IL coating to increase the amount of IL adsorbed on this fiber and thus to enlarge extraction efficiency. Nafion is a cation-exchange polymer. It enhances the quantity and stability of the IL surface-adsorbed by means of the electrostatic interactions between IL and the membrane.

Due to enhanced surface interactions, Nafion support produced an even coating of the IL while ILs coated onto the bare silica resulted in a droplet-like coating. After each analysis, IL and Nafion coatings were removed by solvents. Three ILs, namely 1-methyl-3-octylimidazolium trifluoromethanesulfonate ([C₈MIm][TfO]), 1-benzyl-3- methylimidazolium trifluoromethanesulfonate ([BeMIm][TfO]) and 1-methyl-3-phenylpropylimidazolium trifluoromethanesulfonate ([PhproMIm][TfO]), were applied for the fiber coating. Ss target analytes, polycyclic aromatic hydrocarbons (PAHs) were selected.

In 2008, Anderson and co-workers introduced polymeric ILs (PILs) as a base for SPME coatings [266]. The selectivity of PIL-based coatings can be modulated by introducing functional groups to the cationic portion of IL or by incorporating different anions to impart desired solvent characteristics. Three homologous polymeric imidazolium-based IL coatings, namely poly(1-vinyl- 3-hexylimidazolium) bis(trifluoromethyl)sulfonylimide (poly[ViHIm][NTf₂]), poly(1-vinyl-3-dodecylimidazolium) bis(trifluoromethyl)sulfonylimide (poly[ViDDIm][NTf₂]), and poly-1-vinyl-3-hexadecylimidazolium bis(trifluoromethyl)sulfonylimide (poly[ViHDIm][NTf₂]) were synthesized [266].

Meng and Anderson synthesized a new generation PIL, poly(1-4-vinylbenzyl)-3-hexadecylimidazolium bis(trifluoromethyl)sulfonylimide (poly[ViBHDIm][NTf₂]) [267]. This PIL was functionalized with benzyl groups capable of imparting π-π interactions. Due to these interactions and high hydrophobicity, poly[ViBHDIm][NTf₂] exhibited high selectivity towards PAHs from aqueous samples when used as a sorbent coating in DI-SPME. As it was found, the benzyl-functionalized poly[ViBHDIm][NTf$_2$] PIL exhibited higher extraction efficiency compared to PIL lacking such functionalization as well as to commercial PDMS sorbent coating. Also, the poly[ViBHDIm][NTf$_2$] coating exhibited long lifetimes.

Recently, ILs have been employed in the preparation of sol-gel materials in which ILs serve as solvents, pore templates, drying control chemical additives and catalysts. It was reported that ILs reduced cracking and shrinking of sol-gel coatings. For the first time, IL-mediated sol-gel materials for CME of nonpolar and moderately polar analytes were suggested in 2009 [268].

Table **7** summarizes some representative examples of IL-based SPME technique [262 - 316].

Table 7. Representative references on the use of ionic liquids (ILs) in solid-phase microextraction (SPME).

Extraction mode	IL	Analyte	Sample	Instrumentation	LOD [µg/L]	Recovery [%]	Ref.
IL-HS-SPME	[OMIm][PF$_6$]	BTEXs	Paints	GC-FID	100÷800	70÷114	[262]
IL-HS-SPME	[OMIm][TfO]	PAHs	Waters	GC-MS	0.004÷0.005	80÷110	[265]
HS-SPME	[BMIm][CO$_2$CF$_3$]	Acetone	Humana plasma	-	3.3÷7.6	88÷115	[269]
IL-HS-SPME	Poly[ViHim][NTf$_2$] Poly[ViDDim][NTf$_2$] Poly[ViHDim][NTf$_2$]	Esters	Wines	GC-FID	2.5÷50.0	61.9÷115.1	[266]
IL-HS-SPME	[BMIm][PF$_6$]	PAHs	Mosquito coils fumes	GC-FID	-	-	[270]
IL-HS-SPME	Poly[ViHDim][NTf$_2$]	Aliphatic hydrocarbons, esters	-	GC-FID	300÷600	69.9÷106	[271]
IL-HS-SPME	[EeMIm][NTf$_2$]	Methamphetamine amphetamine	Urine	GC-MS	0.1, 0.5	89.0÷113.8	[272]
IL-HS-SPME	[BMIm][PF$_4$]	Heterocyclic aromatic amines	Meat extracts	HPLC-PLD	0.0003÷0.075	68.5÷118	[273]
HS-SPME	[HPh][BF$_4$], [Bpy][PB$_4$]	Dodecanal, decanol, heptanophenone, pyrene, phenanthrene	-	GC-FID	0.0032÷0.0174	-	[268]
HS-SPME	[OMIm][Cl], [HPh][BF$_4$]	Alcohols, aldehydes, aromatic amines, ketones, PAHs, phenols, acids	-	GC-FID	0.0019÷0.0331	-	[274]
DI-SPME	Poly([ViHDIM][NTf$_2$])	PAHs, substituted phenols	Water	GC-MS	0.005÷4.4	75.8÷119	[275]
DI-SPME	Poly([VBHDIM][NTf$_2$])	PAHs	Water	GC-FID	0.003÷0.07	-	[267]
HS-SPME	Poly([VHIM][NTf$_2$]) Poly([VHIM][taurate])	CO$_2$	-	GC-TC	-	-	[276]
DI-SPME	[AMIm][N(SO$_2$CF$_3$)$_2$]	Phenolic environmental estrogens	Water	GC-FID	0.0030÷0.1248	83.1÷104.1	[277]

(Table 7) cont.....

Extraction mode	IL	Analyte	Sample	Instrumentation	LOD [µg/L]	Recovery [%]	Ref.
	$[(C_2OHIM)_2 T]_2[NTf_2]$ $[(C_2OHIM)_3 T]_2[TfO]$	Alcohols, acetone, ethyl acetate, and acetonitrile	Water		0.047÷0.086	97.4÷104.7	[278]
HS-SPME	Poly([ViHIM][Cl])	Phenols, volatile fatty acids, alcohols	Water	GC-FID	0.02÷7.5	-	[279]
HS-SPME	$[C_8MIm][BF_4]$	Methyl tert-butyl ether	Gasoline	GC-FID	0.09	90-96	[264]
HS-SPME	Poly([ViHIM][NTf_2]) 50%/poly([ViHIM][Cl]) 50%	Alcohols, aromatic hydrocarbons	Water	GC-FID	0.4÷129	-	[280]
HS-SPME	$[MTPIM][NTf_2]$	Methyl tert-butyl ether	Gasoline	GC-FID	0.1	91.6÷95.8	[281]
HF-SPME	$[C_4MIm][OH]$-mediated sol-gel sorbent	Diazinon, fenitrothion, malathion, fenvalerate, phosalone, and tridemorph	Human hair and water samples	HPLC–DAD	0.004÷0.095 (in water) 0.003÷0.08 (in hair)	86.2÷98.8 (in water)	[282]
HS-SPME	Nanostructured polyaniline–IL $[C_4MIm][PF6]$ composite	Organochlorine pesticides	Lake water	GC-ECD	0.00012÷0.00031	88.9÷112.9	[283]
HS-SPME	$[C_4MIm][BF_4]$	benzene derivatives	waste water and tap water	GC	0.0093÷0.0481	87.1÷108.1	[284]
DI-SPME	$[A(Benzo_{15}C_5)HIM]$ $[PF_6]$	Alcohols, phthalate esters, PEEs, fatty acids, and aromatic amines	Water	GC-FID	-	-	[285]
HF-SPME	Butyl sulfone pyridinium cation- and Keggin anion-based IL as porogen for sol-gel sorben	Organophosphorus pesticides	Human hair samples	Sol-gel chemistry	0.00034÷0.84	83÷92	[286]
DI-SPME	$[VC_6IM][Cl]$ copolymerized with $[(VIM)_2C_8]_2[Br]$ and $[(VIM)_2C1_2]_2[Br]$	Alcohols, aldehydes, and esters	Well and river water	GC-MS	0.01÷2.5	58÷135.8	[287]
DI-SPME	Copolymerized $[VC_{16}IM][PF_6]$	Pyrethroids	Vegetables	GC–ECD	0.21÷0.49	67.4÷94.01	[288]
DI-SPME	$[VC_4IM][Cl]$ copolymerized with 1-vinyl-3-(3triethoxysilylpropyl)-4,5-dihydroimidazolium chloride	PAHs	Water	GC-MS	0.05÷0.25	49.6÷105	[289]
DI-SPME	$Poly[VC_8IM][SS]$	Anilines, phenols, and phthalate esters	Water	GC	0.01÷10	73.4÷113	[290]
DI-SPME	$Poly[VC_8IM][NapSO_3]$	PAHs	Hair spray and nail polish	GC	0.005÷0.05	-	[291]
SPME	$[AMIM][NTf_2]$-OHTSO	Phthalate esters	Agricultural plastic films	GC	0.003÷0.063	90.2÷111.4	[292]
SPME	$[TESPMIm][PF_6]$, $[TESPMIm][BF_4]$, and $[TESPMIm][NTf_2]$ based sol-gel sorbent	PEEs, fatty acids, aromatic amines, alcohols, phthalate esters, and PAHs	Water	GC	-	-	[293]
HS-SPME	IL coated SPME fiber	Chlorophenols	Landfill leachate	GC-MS	0.008	87	[294]

(Table 7) cont.....

Extraction mode	IL	Analyte	Sample	Instrumentation	LOD [µg/L]	Recovery [%]	Ref.
HS-SPME	PDMS, PDMS-IL-HT, PDMS-IL-LT	Benzene, toluene, ethylbenzene, and xylenes	Water	GC-FID	0.0001÷0.5	91.2–103.3	[295]
SPME	Poly[VC$_8$IM][PF$_6$]	Phenols and PAHs River	River water samples	GC	0.01÷0.06	83.35÷119.24	[296]
HS-SPME	[AMIM][BF$_4$]	Organophosphate	Lake, waste, and tap water, sewage treatment plant effluent	GC-FID	0.0007÷0.012	75.2÷101.8	[297]
DLLME-SPME	IL-zeolite imidazolate framework 4	Antidepressant drugs	Water samples	HPLC	0.3÷1	94.3÷114.7	[298]
SPME	[VBC$_{16}$IM][NTf$_2$] copolymerized with [(VIM)$_2$C$_{12}$]$_2$[NTf$_2$] and [(VBIM)$_2$C$_{12}$]$_2$ [NTf$_2$]	PCBs	Seawater and bovine milk	GC/ECD	0.001÷0.0025	61.2÷135.7	[299]
HS-SPME	IL-mediated MWCNTspoly (dimethylsiloxane) fiber	Polycyclic aromatic hydrocarbons	Urine samples	GC	0.0005÷0.004	89.3÷107.2	[300]
IL-SPME	Crown ether functionalized ionic liquids based fiber	PAEs	Water samples	GC-FID	-	-	[301]
IL-magnetic SPME	ILs Fe$_3$O$_4$ nanoparticles	Cadmium	Water, vegetables, and hair samples	FAAS	0.40	98.1÷101	[302]
IL-MD-SPME	Magnetic ILs	Sulfonamides	Butter samples	HPLC	1.2÷2.17 ng/g	73.25÷103.85	[303]
L-SPME	Magnetic ILs	Triazine herbicides	Vegetable oils	HPLC	1.31–1.49	81.8–114.2	[304]
In-tube SPME	Ionic liquid-modified organic polymer monolith	Acidic food additives	Coca–Cola	HPLC	1.2÷13.5	85.4÷98.3	[305]
HS-SPME	MWCNTs-PILs fiber	2-naphthol	Pomelo and orange	GC	0.20÷0.94 ng/g	1.9÷110	[306]
MD-SPME	Magnetic stirrer induced ILs	Vanadium	Food and water samples	AAS	0.018÷0.125	97÷99	[307]
IL-magnetic SPME	ILs Fe$_3$O$_4$ nanoparticles	Cadmium	Fruit and vegetables	FAAS	0.32	96÷100	[308]
PIL-SPME	Poly[VHIM$^+$][Cl$^-$]	DNA	Bacterial plasmid DNA	-	-	-	[309]
SPME	[(VIM)$_2$C$_{12}$]$_2$[NTf$_2$] In [VBHDIM][NTf$_2$]	Acrylamide	Coffee and coffee powder	GC-MS	100÷1000	-	[310]
HF-PIL-SPME	Poly[APVBzIM$^+$][SS$^-$]	Estrogens	Milk	HPLC	1÷2	85.5÷112	[311]
DI-PIL-SPME	Poly[AMIM$^+$][NTf$_2$$^-$]	Steroid sex hormones	Water, urine	HPLC-DAD	0.027÷–0.12	75.6÷116	[312]
PIL-SPME	Poly[VC9COOHIM+][Br−]	DNA	Crude cell lysate				[313]
SPME	Poly (TA-co-DB/ED) 1-trimethyl-(4-vinylbenzyl) aminium chloride (TA)	Endocrine disrupting chemicals	Human urine and environmental water	HPLC-DAD	0.011÷0.065	70.6÷119.0	[314]
SPME	IL/polyacrylate/NH$_2$/oligo dT$_{20}$	mRNA	RNA	RT-qPCR	5.0	-	[315]
HS-SPME	PTh-IL-Clay polythiophene – ionic liquid clay	Pesticides	Fruit juice	GC-ECD	0.002÷0.667	-	[316]

SUMMARY

Regarding the analysis of trace amounts of chemicals, especially in environmental samples, we may observe that the most critical task is the extraction of the analyte using an efficient and possibly cheap extraction technique. Among several proposals, ionic liquids deserve particular attention. Growing potential of ILs, as well as their applicability at room temperature, seem to form a significant alternative to environmentally-harmful, conventional organic solvents and traditional analytic techniques. Due to unique properties and advantages of ILs, described in that chapter, the ionic liquids have attracted global attention and gained popularity in many areas of analytical chemistry including modern sample preparation techniques.

Growing application of ILs results in continuous and significant improvements in traditional analytical methods and techniques dealing with a vast spectrum of the analytes and samples. Replacing conventional solvents with ILs brings us closer to the green chemistry and results in a substantial decrease in costs, amount of work and time. Also, the performance of IL-based techniques is usually better in comparison with the traditional micro extraction methods. We do expect that the trend of the application of ILs to improve many analytical techniques and procedures will continue, resulting in better and better methods and broader application areas.

ABBREVIATIONS

Ionic liquids	
$[Aliquat]_2[MnCl_4]$	AliquatTetrachloromanganate(II)
[BMIm][Cl]	1-Butyl-3-methylimidazolium chloride
$[BMIm][PF_6]$	1-Butyl-3-methyl-imidazolium hexafluorophosphate
$[BMIm][BF_4]$	1-Butyl-3-methylimidazolium tetrafluoroborate
$[BMIm][CO_2CF_3]$	1-Butyl-3-methylimidazolium trifluoroacetate
$[BMIm][NTf_2]$	1-Butyl-3-methylimidazolium bis(trifluoromethyl)sulfonylimide
[BMIm][Cl]	1-Butyl-3-methylimidazolium chloride
[BeEOHIm][Br]	1-Benzyl-3-(2-hydroxyethyl)imidazoliumbromide
$[C_4MIm][BF_4]$	1-Butyl-3-methylimidazolium tetrafluoroborate
$[C_6MIm][NTf_2]$	1-Hexyl-3-methylimidazolium bis(trifluoromethylsulfonyl)imide
$[C_8MIm][NTf_2]$	1-Methyl-3-octylimidazolium bis[trifluoromethylsulfonyl]imide
$[C_{10}MIm][NTf_2]$	1-Decyl-3-methylimidazolium bis(trifluoromethylsulfonyl)imide)
$[C_4MIm][PF_6]$	1-Butyl-3-methylimidazolium hexafluorophosphate
$[C_6MIm][PF_6]$	1-Hexyl-3-methylimidazolium hexafluorophosphate
$[C_8MIm][PF_6]$	1-Octyl-3-methylimidazolium hexafluorophosphate

Abbrev cont.....

Ionic liquids	
[C$_6$MIm][FAP]	1-Hexyl-3-methylimidazolium tris(pentafluoroethyl)trifluorophosphate
[C$_6$Py][PF$_6$]	1-Hexylpyridinium hexafluorophosphate
[DMIm][BF$_4$]	1-Decyl-3-methylimidazoliumtetrafluoroborate
[EMIm][Cl]	2-Ethyl-4-methylimidazolium chloride
[HMIm][PF$_6$]	1-Hexyl-3-methyl-imidazolium hexafluorophosphate
[HMIm][BF$_4$]	1-Hexyl-3-methyl-imidazolium tetrafluoroborate
[HMIm][NTf$_2$]	1-Hexyl-3-methyl-imidazolium bis(trifluoromethyl)sulfonylimide
[HMIm][FAP]	1-Hexyl-3-methylimidazolium tris(pentafluoroethyl)trifluorophosphate)
[HeOHMIm][Cl]	1-(6-Hydroxyethyl)-3-methylimidazoliumchloride
[HPy][PF$_6$]	1-Hexylpyridinium hexafluorophosphate
[HPh][BF$_4$]	Trihexyltetradecylphosphonium
[HMIm][OTf]	1-Hexyl-3-methylimidazolium triflate
MAA-GMA-EGDMA[BIm][Cl]	Methacrylic acid-Glycidyl methacrylate-Ethylene glycol dimethacrylate---butyl-3-methylimidazolium chloride
MIP-VBC-DVB[Im][Cl]	Imidazolium chloride
MIP-VBC-DVB[MIm][Cl]	N-Methylimidazolium chloride
MIP-VBC-DVB[Im-COOH][Cl]	Carboxyl-imidazolium chloride
MIP-VBC-DVB[Im-NH$_2$][Cl]	N-(Propyl-3-amino)imidazolium
MIP-VBC-DVB[Im-CN][Cl]	Cyanoethyl-imidazolium chloride
[OMIm][Cl]	1-Octyl-3-methyl-imidazolium chloride
[OMIm][PF$_6$]	1-Octyl-3-methyl -imidazolium hexafluorophosphate
[OMIm][TfO]	1-Octyl-3-methyl -imidazolium trifluoromethanesulfonate
[EeMIm][NTf$_2$]	1-Etoxy-alkyl-3-methylimidazolium ionic liquid 1-Ethoxyethyl-3-methylimidazloium bis(trifluoromethyl)sulfonylimide
[NH$_2$C$_6$][MPyrr-FAPa]	1-(6-aminohexyl)-1-methylpyrrolidinium tris(pentafluoroethyl)trifluorophosphate
Poly[ViHIm][NTf$_2$]	Poly (1-vinyl-3-hexylimidazolium bis[(trifluoromethyl)sulfonyl]imide)
Poly[ViDDIm][NTf$_2$]	Poly (1-dodecyl-3-vinylimidazolium bis[(trifluoromethyl)sulfonyl]imide)
Poly[ViHDIm][NTf$_2$]	Poly (1-vinyl-3-hexadecylimidazolium bis[(trifluoromethyl)sulfonyl]imide)
[P$_{44412}$][NTf$_2$]	Tributyldodecylphosphonium bis(trifluoromethylsulfonyl)imide
[P$_{44412}$][PF$_6$]	Tributyldodecylphosphonium hexafluorophosphate
[P$_{44412}$][DDTC]	Tributyldodecylphosphonium diethyldithiocarbamate
[P$_{4448}$][N(CN)$_2$]	Tributyloctylphosphonium dicyanamide
[P$_{44412}$][N(CN)$_2$]	Tributyldodecylphosphonium dicyanamide
[P$_{44416}$][N(CN)$_2$]	Tributylhexadecylphosphonium dicyanamide
PVC[MIm][Cl]	Methylimidazolium chloride
[BPy][PB$_4$]	N-butyl-4-methylpyridinium tetrafluoroborate

Abbrev cont.....

Ionic liquids	
[P$_{6,6,6,14}$]$_2$[MnCl$_4$]	[P$_{6,6,6,14}$]$_2$[MnCl$_4$] trihexyl(tetradecyl)phosphonium tetrachloromanganate(II)
[P$_{6,6,6,14}$][Dy(hfacac)$_4$]	Trihexyl(tetradecyl)phosphoniumtetrakis(hexafluoroacetylaceto)dysprosate(III)
[P$_{6,6,6,14}$][Dy(hfacac)$_3$]	Tihexyl(tetradecyl)phosphonium tris(hexafluoroacetylaceto)manganate(II)
[P$_{66614}$][NTf$_2$]	Trihexyltetradecyl-phosphonium bis(trifluoromethylsulfonyl)amide
CYPHOS® IL 101	Tetradecyl(trihexyl)phosphonium chloride
[A(Benzo15C5)HIm][PF$_6$]	1-Allyl-3-(6′-oxo-benzo-15-crown-5 hexyl) imidazolium hexafluorophosphate
[SEt$_3$][NTf]	Triethylsulfonium bis(trifluoromethylsulfonyl)imide
Silpr[Im][Cl]	Imidazolium
Silpr[MIm][Cl]	N-Methylimidazolium chloride
Silpr[EMIm][Cl]	2-Ethyl-4-methylimidazolium chloride
VBC-DVB[MIm][CF$_3$COO]	N-Methylimidazolium trifluoroacetate

Microextraction techniques	
SDME	Single drop micro extraction
DI-SDME	Direct immersion single drop micro extraction
HS-SDME	Headspace single drop micro extraction
IL-SDME	Ionic liquid based single drop micro extraction
HF-LPME	Hollow fiber based liquid phase micro extraction
HF-SLME	Hollow Fiber-Supported Liquid Membrane Extraction
DLLME	Dispersive liquid-liquid micro extraction
IL-DLLME	Ionic liquid based dispersive liquid-liquid micro extraction
TC-IL-DLLME	Temperature-controlled ionic liquid dispersive liquid phase micro extraction
in situ IL-DLLME	*in situ* ionic liquid dispersive liquid–liquid micro extraction
MR-IL-DLLME	Magnetic retrieval ionic liquid dispersive liquid–liquid micro extraction
META-IL-DLLME	Magnetic effervescent tablet-assisted ionic liquid dispersive liquid-liquid micro extraction
EA-DLLME	Effervescence-assisted dispersive liquid–liquid micro extraction
MIL-SADBME	Magnetic ionic liquids stirring-assisted drop-breakup micro extraction
MIL-UDSA-DLLME	Magnetic ionic liquid-based up-and-down-shaker-assisted dispersive liquid-liquid micro extraction
SSIL-DLLME	Solidification of sedimentary ionic liquids dispersive liquid-liquid micro extraction
MNPs	Magnetic nanoparticles
CV-ILAHS-SDME	Cold vapor ionic liquid-assisted headspace single drop micro extraction
vacuum MIL-HS-SDME	A vacuum headspace single-drop micro extraction method based on the use of magnetic ionic liquids

Abbrev cont.....

Microextraction techniques	
MWCNTs	Magnetic multiwalled carbon nanotubes
SPE	Solid Phase Extraction
MSPE	Magnetic solid-phase extraction
IL-SPE	Ionic liquids Solid Phase Extraction
SPME	Solid-Phase Micro extraction
DI-SPME	Direct immersion- Solid Phase Microextraction
HS-SPME	Headspace–Solid Phase Microextraction
ILs	Ionic Liquids
PILs	Polymeric Ionic Liquids
MILs	Magnetic ionic liquids
SILPs	Polymer-supported ionic liquid phases
CPE	Cloud-point extraction
MPs	Melting points
LC	Liquid Chromatography
HPLC	High-performance liquid chromatography
HPLC-DAD	High-performance liquid chromatography with a diode-array detector
HPLC-VWD	High-performance liquid chromatography with a variable wavelength detector
HPLC-FD	High-performance liquid chromatography with a fluorescence detection
HPLC-HRMS/MS	High performance liquid chromatography-high resolution mass spectrometry/ mass spectrometry
HPLC-ESIMS/MS	High-performance liquid chromatography combined with electrospray mass spectrometry/ mass spectrometry
UPLC	Ultra-Performance Liquid Chromatography
GC	Gas chromatography
HS-GC	Headspace gas chromatography
TD-GC-MS	Thermal desorption-gas chromatography mass spectrometry
CE	Capillary electrophoresis
CE-UV	Capillary Electrophoresis-Ultraviolet
ECD	Electron Capture Detector
MS	Mass spectrometry-based detector
IMS	Ion Mobility Spectrometer
ICP-MS	Inductively coupled plasma – mass spectrometry
CCD	Charge Coupled Device
ICP-OES	Inductively Coupled Plasma - Optical Emission Spectrometers
FAAS	Flame atomic absorption spectrometry

Abbrev cont.....

Microextraction techniques	
AAS	Atomic absorption spectrometry
FI-HG-AAS	Flow Injection Hydride Generation Atomic Absorption Spectrometry
GF-AAS	Graphite furnace atomic absorption spectrometry
WCAES	Tungsten coil atomic emission spectrometry
PDMS	Polydimethylsiloxane
EDTA	Ethylenediaminetetraacetic acid
RT-qPCR	quantitative reverse transcriptase real-time PCR

Analytes	
PAHs	Polycyclic aromatic hydrocarbons
DDT	Dichlorodiphenyltrichoroethane
BTEXs	Benzene, toluene, ethylbenzene and xylene isomers
THMs	Trihalomethanes (normally, chloroform, dichlorobromomethane, chlorodibromomethane and bromoform)
MMT	Methylcyclopentadienyl-manganese tricarbonyl
NMs	Nitro musks
PCMs	Polycyclic musks
3-OHBaP	3-Hydroxybenzo[a]pyrene
CPs	Chlorophenols
BPA	Bisphenol A
BPZ	Bisphenol Z
BPF	Bisphenol F
BHb	Bovine haemoglobin
BSA	Bovine serum albumin
BUs	Benzoylurea insecticides
OVT	Ovotransferrin
Cyt	Cytochrome *c*
MC-RR	Microcystin-RR
MC-LR	Microcystin-LR

CONSENT FOR PUBLICATION

Not applicable.

CONFLICT OF INTEREST

The author confirms that he has no conflict of interest to declare for this publication.

ACKNOWLEDGEMENTS

Declared none.

REFERENCES

[1]   Ho, T.D.; Yu, H.; Cole, W.T.S.; Anderson, J.L.; Majors, R.E. Ionic liquids and their applications in sample preparation. *LC GC Eur.,* **2013**, *15*, 101-109.

[2]   Berthod, A.; Ruiz-Ángel, M.J.; Carda-Broch, S. Ionic liquids in separation techniques. *J. Chromatogr. A,* **2008**, *1184*(1-2), 6-18.
[http://dx.doi.org/10.1016/j.chroma.2007.11.109] [PMID: 18155711]

[3]   Nerín, C.; Salafranca, J.; Aznar, M.; Batlle, R. Critical review on recent developments in solventless techniques for extraction of analytes. *Anal. Bioanal. Chem.,* **2009**, *393*(3), 809-833.
[http://dx.doi.org/10.1007/s00216-008-2437-6] [PMID: 18931996]

[4]   Sun, P.; Armstrong, D.W.; Bashammakh, A.S.; Alwael, H.; El-Shahawi, M.S. Ionic liquids in analytical chemistry. *Anal. Chim. Acta,* **2010**, *661*(1), 1-16.
[http://dx.doi.org/10.1016/j.aca.2009.12.007] [PMID: 20113709]

[5]   Liu, R.; Liu, J.F.; Yin, Y.G.; Hu, X.L.; Jiang, G.B. Ionic liquids in sample preparation. *Anal. Bioanal. Chem.,* **2009**, *393*(3), 871-883.
[http://dx.doi.org/10.1007/s00216-008-2445-6] [PMID: 18958452]

[6]   Ocaña-González, J.A.; Fernández-Torres, R.; Bello-López, M.A.; Ramos-Payán, M. New developments in microextraction techniques in bioanalysis. A review. *Anal. Chim. Acta,* **2016**, *905*, 8-23.
[http://dx.doi.org/10.1016/j.aca.2015.10.041] [PMID: 26755133]

[7]   Clark, K.D.; Emaus, M.N.; Varona, M.; Bowers, A.N.; Anderson, J.L. Ionic liquids: solvents and sorbents in sample preparation. *J. Sep. Sci.,* **2018**, *41*(1), 209-235.
[http://dx.doi.org/10.1002/jssc.201700864] [PMID: 28926208]

[8]   Zhao, Q.; Anderson, J.L. Task-specific microextractions using ionic liquids. *Anal. Bioanal. Chem.,* **2011**, *400*(6), 1613-1618.
[http://dx.doi.org/10.1007/s00216-011-4848-z] [PMID: 21400073]

[9]   Vičkačkaite, V.; Padarauskas, A. Ionic liquids in microextraction techniques. Review article. *Cent. Eur. J. Chem.,* **2012**, *10*(3), 652-674.

[10]   Fontanals, N.; Borrull, F.; Marcé, R.M. Ionic liquids in solid-phase extraction. *Trends Analyt. Chem.,* **2012**, *41*, 15-26.
[http://dx.doi.org/10.1016/j.trac.2012.08.010]

[11]   Pomati, F.; Castiglioni, S.; Zuccato, E.; Fanelli, R.; Vigetti, D.; Rossetti, C.; Calamari, D. Effects of a complex mixture of therapeutic drugs at environmental levels on human embryonic cells. *Environ. Sci. Technol.,* **2006**, *40*(7), 2442-2447.
[http://dx.doi.org/10.1021/es051715a] [PMID: 16646487]

[12]   Spietelun, A.; Marcinkowski, Ł.; de la Guardia, M.; Namieśnik, J. Green aspects, developments and perspectives of liquid phase microextraction techniques. *Talanta,* **2014**, *119*, 34-45.
[http://dx.doi.org/10.1016/j.talanta.2013.10.050] [PMID: 24401382]

[13]   Farzad Atefi, F.; Garcia, M.T.; Singer, R.D.; Scammells, P.J. Phosphonium ionic liquids: design,

synthesis and evaluation of biodegradability. *Green Chem.,* **2009**, *11*, 1595-1604.
[http://dx.doi.org/10.1039/b913057h]

[14]   Tan, Z.; Liu, J.; Pang, L. Advances in analytical chemistry using the unique properties of ionic liquids. *Trends Analyt. Chem.,* **2012**, *39*, 218-227.
[http://dx.doi.org/10.1016/j.trac.2012.06.005]

[15]   Vidal, L.; Riekkola, M.L.; Canals, A. Ionic liquid-modified materials for solid-phase extraction and separation: a review. *Anal. Chim. Acta,* **2012**, *715*, 19-41.
[http://dx.doi.org/10.1016/j.aca.2011.11.050] [PMID: 22244164]

[16]   Ho, T.D.; Canestraro, A.J.; Anderson, J.L. Ionic liquids in solid-phase microextraction: a review. *Anal. Chim. Acta,* **2011**, *695*(1-2), 18-43.
[http://dx.doi.org/10.1016/j.aca.2011.03.034] [PMID: 21601027]

[17]   Liu, J.F.; Li, N.; Jiang, G.B.; Liu, J.M.; Jönsson, J.A.; Wen, M.J. Disposable ionic liquid coating for headspace solid-phase microextraction of benzene, toluene, ethylbenzene, and xylenes in paints followed by gas chromatography-flame ionization detection. *J. Chromatogr. A,* **2005**, *1066*(1-2), 27-32.
[http://dx.doi.org/10.1016/j.chroma.2005.01.024] [PMID: 15794551]

[18]   Maton, C.; De Vos, N.; Stevens, C.V. Ionic liquid thermal stabilities: decomposition mechanisms and analysis tools. *Chem. Soc. Rev.,* **2013**, *42*(13), 5963-5977.
[http://dx.doi.org/10.1039/c3cs60071h] [PMID: 23598738]

[19]   Couling, D.J.; Bernot, R.J.; Docherty, K.M.; Dixon, J.K.; Maginn, E.J. Assessing the factors responsible for ionic liquid toxicity to aquatic organisms *via* quantitative structure-property relationship modelling. *Green Chem.,* **2006**, *8*, 82-90.
[http://dx.doi.org/10.1039/B511333D]

[20]   Abdelhamid, H.N. Ionic liquids for mass spectrometry: Matrices, separation and microextraction. *Trends Analyt. Chem.,* **2016**, *77*, 122-138.
[http://dx.doi.org/10.1016/j.trac.2015.12.007]

[21]   Poole, C.F.; Lenca, N. Green sample-preparation methods using room temperature ionic liquids for the chromatographic analysis of organic compounds. *Trends Analyt. Chem.,* **2015**, *71*, 144-156.
[http://dx.doi.org/10.1016/j.trac.2014.08.018]

[22]   Aguilera-Herrador, E.; Lucena, R.; Cardenas, S.; Valcarcel, M. The roles of ionic liquids in sorptive microextraction techniques. *Trends Analyt. Chem.,* **2010**, *29*(7), 7-, 602-616.
[http://dx.doi.org/10.1016/j.trac.2009.11.009]

[23]   Jeannot, M.A.; Cantwell, F.F. Solvent microextraction into a single drop. *Anal. Chem.,* **1996**, *68*(13), 2236-2240.
[http://dx.doi.org/10.1021/ac960042z] [PMID: 21619310]

[24]   Liu, H.; Dasgupta, P.K. Analytical chemistry in a drop. Solvent extraction in a microdrop. *Anal. Chem.,* **1996**, *68*(11), 1817-1821.
[http://dx.doi.org/10.1021/ac960145h] [PMID: 21619093]

[25]   He, Y.; Lee, H.K. Liquid-Phase Microextraction in a Single Drop of Organic Solvent by Using a Conventional Microsyringe. *Anal. Chem.,* **1997**, *69*, 4634-4640.
[http://dx.doi.org/10.1021/ac970242q]

[26]   Jeannot, M.; Cantwell, F. Mass transfer characteristics of solvent extraction into a single drop at the tip of a syringe needle. *Anal. Chem.,* **1997**, *69*, 235-239.
[http://dx.doi.org/10.1021/ac960814r]

[27]   Theis, A.L.; Waldack, A.J.; Hansen, S.M.; Jeannot, M.A. Headspace solvent microextraction. *Anal. Chem.,* **2001**, *73*(23), 5651-5654.
[http://dx.doi.org/10.1021/ac015569c] [PMID: 11774903]

[28]   Tankeviciute, A.; Kazlauskas, R.; Vickackaite, V. Headspace extraction of alcohols into a single drop.

Analyst (Lond.), **2001**, *126*, 1674-1677.
[http://dx.doi.org/10.1039/b103493f]

[29]   Liu, J.F.; Jiang, G.B.; Chi, Y.G.; Cai, Y.Q.; Zhou, Q.X.; Hu, J-T. Use of ionic liquids for liquid-phase microextraction of polycyclic aromatic hydrocarbons. *Anal. Chem.*, **2003**, *75*(21), 5870-5876.
[http://dx.doi.org/10.1021/ac034506m] [PMID: 14588028]

[30]   Kokosa, J.M. Recent trends in using single-drop microextraction and related techniques in green analytical methods. *Trends Analyt. Chem.*, **2015**, *71*, 194-204.
[http://dx.doi.org/10.1016/j.trac.2015.04.019]

[31]   Liu, J.F.; Chi, Y.G.; Jiang, G.B.; Tai, C.; Peng, J.F.; Hu, J.T. Ionic liquid-based liquid-phase microextraction, a new sample enrichment procedure for liquid chromatography. *J. Chromatogr. A*, **2004**, *1026*(1-2), 143-147.
[http://dx.doi.org/10.1016/j.chroma.2003.11.005] [PMID: 14763740]

[32]   Liu, J.F.; Peng, J.F.; Chi, Y.G.; Jiang, G.B. Determination of formaldehyde in shiitake mushroom by ionic liquid-based liquid-phase microextraction coupled with liquid chromatography. *Talanta*, **2005**, *65*(3), 705-709.
[http://dx.doi.org/10.1016/j.talanta.2004.07.037] [PMID: 18969856]

[33]   Peng, J.F.; Liu, J.F.; Jiang, G.B.; Tai, C.; Huang, M.J. Ionic liquid for high temperature headspace liquid-phase microextraction of chlorinated anilines in environmental water samples. *J. Chromatogr. A*, **2005**, *1072*(1), 3-6.
[http://dx.doi.org/10.1016/j.chroma.2004.11.060] [PMID: 15881452]

[34]   Ye, C.L.; Zhou, Q.X.; Wang, X.M. Headspace liquid-phase microextraction using ionic liquid as extractant for the preconcentration of dichlorodiphenyltrichloroethane and its metabolites at trace levels in water samples. *Anal. Chim. Acta*, **2006**, *572*(2), 165-171.
[http://dx.doi.org/10.1016/j.aca.2006.05.052] [PMID: 17723474]

[35]   Vidal, L.; Chisvert, A.; Canals, A.; Salvador, A. Sensitive determination of free benzophenone-3 in human urine samples based on an ionic liquid as extractant phase in single-drop microextraction prior to liquid chromatography analysis. *J. Chromatogr. A*, **2007**, *1174*(1-2), 95-103.
[http://dx.doi.org/10.1016/j.chroma.2007.07.077] [PMID: 17720175]

[36]   Vidal, L.; Domini, C.E.; Grané, N.; Psillakis, E.; Canals, A. Microwave-assisted headspace single-drop microextration of chlorobenzenes from water samples. *Anal. Chim. Acta*, **2007**, *592*(1), 9-15.
[http://dx.doi.org/10.1016/j.aca.2007.03.066] [PMID: 17499064]

[37]   Vidal, L.; Psillakis, E.; Domini, C.E.; Grané, N.; Marken, F.; Canals, A. An ionic liquid as a solvent for headspace single drop microextraction of chlorobenzenes from water samples. *Anal. Chim. Acta*, **2007**, *584*(1), 189-195.
[http://dx.doi.org/10.1016/j.aca.2006.10.053] [PMID: 17386603]

[38]   Ye, C.; Zhou, Q.; Wang, X.; Xiao, J. Determination of phenols in environmental water samples by ionic liquid-based headspace liquid-phase microextraction coupled with high-performance liquid chromatography. *J. Sep. Sci.*, **2007**, *30*(1), 42-47.
[http://dx.doi.org/10.1002/jssc.200600256] [PMID: 17313140]

[39]   Zhou, Q.X.; Ye, C.L. Ionic liquid for improved single-drop micro extraction of aromatic amines in water samples. *Mikrochim. Acta*, **2008**, *162*, 153-159.
[http://dx.doi.org/10.1007/s00604-007-0857-1]

[40]   Aguilera-Herrador, E.; Lucena, R.; Cárdenas, S.; Valcárcel, M. Ionic liquid-based single-drop microextraction/gas chromatographic/mass spectrometric determination of benzene, toluene, ethylbenzene and xylene isomers in waters. *J. Chromatogr. A*, **2008**, *1201*(1), 106-111.
[http://dx.doi.org/10.1016/j.chroma.2008.06.010] [PMID: 18586254]

[41]   Zhao, F.Q.; Li, J.; Zeng, B.Z. Coupling of ionic liquid-based headspace single-drop microextraction with GC for sensitive detection of phenols. *J. Sep. Sci.*, **2008**, *31*(16-17), 3045-3049.
[http://dx.doi.org/10.1002/jssc.200800308] [PMID: 18704999]

[42]   Xia, L.B.; Li, X.; Wu, Y.L.; Hu, B.; Chen, R. Ionic liquids based single drop micro extraction combined with electrothermal vaporization inductively coupled plasma mass spectrometry for determination of Co, Hg and Pb in biological and environmental samples. *Spectrochim. Acta A Mol. Biomol. Spectrosc.,* **2008**, *63*, 1290-1296.
[http://dx.doi.org/10.1016/j.sab.2008.09.018]

[43]   Aguilera-Herrador, E.; Lucena, R.; Cardenas, S.; Valcárcel, M. Direct coupling of ionic liquid based single-drop microextraction and GC/MS. *Anal. Chem.,* **2008**, *80*(3), 793-800.
[http://dx.doi.org/10.1021/ac071555g] [PMID: 18154274]

[44]   Aguilera-Herrador, E.; Lucena, R.; Cárdenas, S.; Valcárcel, M. Determination of trihalomethanes in waters by ionic liquid-based single drop microextraction/gas chromatographic/mass spectrometry. *J. Chromatogr. A,* **2008**, *1209*(1-2), 76-82.
[http://dx.doi.org/10.1016/j.chroma.2008.09.030] [PMID: 18817919]

[45]   Zhao, F.; Lu, S.; Du, W.; Zeng, B. Ionic liquid-based headspace single-drop micro extraction coupled to gas chromatography for the determination of chlorobenzene derivatives. *Mikrochim. Acta,* **2008**, *165*, 29-33.
[http://dx.doi.org/10.1007/s00604-008-0092-4]

[46]   Sheikhloie, H.; Saber-Trhrani, M.; Abrumand-Azar, P.; Waqif-Husein, S. Analysis of tributyltin and triphenyltin in water by ionic liquid-headspace single-drop micro extraction then HPLC with fluorimetric detection. *Acta Chromatogr.,* **2009**, *21*, 577-589.
[http://dx.doi.org/10.1556/AChrom.21.2009.4.5]

[47]   Yao, C.; Pitner, W.R.; Anderson, J.L. Ionic liquids containing the tris(pentafluoroethyl)trifluorophosphate anion: a new class of highly selective and ultra hydrophobic solvents for the extraction of polycyclic aromatic hydrocarbons using single drop microextraction. *Anal. Chem.,* **2009**, *81*(12), 5054-5063.
[http://dx.doi.org/10.1021/ac900719m] [PMID: 19419149]

[48]   Aguilera-Herrador, E.; Lucena, R.; Cárdenas, S.; Valcárcel, M. Ionic liquid-based single drop microextraction and room-temperature gas chromatography for on-site ion mobility spectrometric analysis. *J. Chromatogr. A,* **2009**, *1216*(29), 5580-5587.
[http://dx.doi.org/10.1016/j.chroma.2009.05.071] [PMID: 19523638]

[49]   Manzoori, J.L.; Amjadi, M.; Abulhassani, J. Ultra-trace determination of lead in water and food samples by using ionic liquid-based single drop microextraction-electrothermal atomic absorption spectrometry. *Anal. Chim. Acta,* **2009**, *644*(1-2), 48-52.
[http://dx.doi.org/10.1016/j.aca.2009.04.029] [PMID: 19463561]

[50]   Pena-Pereira, F.; Lavilla, I.; Bendicho, C.; Vidal, L.; Canals, A. Speciation of mercury by ionic liquid-based single-drop microextraction combined with high-performance liquid chromatography-photodiode array detection. *Talanta,* **2009**, *78*(2), 537-541.
[http://dx.doi.org/10.1016/j.talanta.2008.12.003] [PMID: 19203620]

[51]   Chisvert, A.; Román, I.P.; Vidal, L.; Canals, A. Simple and commercial readily-available approach for the direct use of ionic liquid-based single-drop microextraction prior to gas chromatography determination of chlorobenzenes in real water samples as model analytical application. *J. Chromatogr. A,* **2009**, *1216*(9), 1290-1295.
[http://dx.doi.org/10.1016/j.chroma.2008.12.078] [PMID: 19144344]

[52]   Manzoori, J.L.; Amjadi, M.; Abulhassani, J. Ionic liquid-based single drop microextraction combined with electrothermal atomic absorption spectrometry for the determination of manganese in water samples. *Talanta,* **2009**, *77*(4), 1539-1544.
[http://dx.doi.org/10.1016/j.talanta.2008.09.045] [PMID: 19084676]

[53]   Sarafraz-Yazdi, A.; Mofazzeli, F. Ionic liquid-based submerged single drop micro extraction: a new method for the determination of aromatic amines in environmental water samples. *Chromatographia,* **2010**, *72*, 867-873.

[http://dx.doi.org/10.1365/s10337-010-1731-6]

[54]   Liu, Q.; Liu, Y.; Chen, S.; Liu, Q. Ionic liquid for single-drop microextraction followed by high-performance liquid chromatography-ultraviolet detection to determine carbonyl compounds in environmental waters. *J. Sep. Sci.,* **2010**, *33*(15), 2376-2382.
[http://dx.doi.org/10.1002/jssc.201000051] [PMID: 20574954]

[55]   Wang, Q.; Qiu, H.; Li, J.; Liu, X.; Jiang, S. On-line coupling of ionic liquid-based single-drop microextraction with capillary electrophoresis for sensitive detection of phenols. *J. Chromatogr. A,* **2010**, *1217*(33), 5434-5439.
[http://dx.doi.org/10.1016/j.chroma.2010.06.059] [PMID: 20621300]

[56]   Zhang, J.; Lee, H.K. Headspace ionic liquid-based microdrop liquid-phase microextraction followed by microdrop thermal desorption-gas chromatographic analysis. *Talanta,* **2010**, *81*(1-2), 537-542.
[http://dx.doi.org/10.1016/j.talanta.2009.12.039] [PMID: 20188959]

[57]   Dong, S.Y.; Yang, Z.; Huang, T.L. Ionic Liquid-Based Headspace Liquid-Phase Micro-Extraction for the Determination of Dichlorobenzene Isomers in Soils. *Chromatographia,* **2010**, *72*, 1137-1141.
[http://dx.doi.org/10.1365/s10337-010-1772-x]

[58]   Yao, C.; Twu, P.; Anderson, J.L. Headspace single drop micro extraction using micellar ionic liquid extraction solvents. *Chromatographia,* **2010**, *72*, 393-402.
[http://dx.doi.org/10.1365/s10337-010-1675-x]

[59]   Vidal, L.; Chisvert, A.; Canals, A.; Salvador, A. Ionic liquid-based single-drop microextraction followed by liquid chromatography-ultraviolet spectrophotometry detection to determine typical UV filters in surface water samples. *Talanta,* **2010**, *81*(1-2), 549-555.
[http://dx.doi.org/10.1016/j.talanta.2009.12.042] [PMID: 20188961]

[60]   Martinis, E.M. Bert_on, P.; Altamirano, J. C.; Hakala, U.; Wuilloud, R. G. Tetradecyl(trihexyl)phosphonium chloride ionic liquid single-drop micro extraction for electrothermal atomic absorption spectrometric determination of lead in water samples. *Talanta,* **2010**, *80*, 2034-2040.
[http://dx.doi.org/10.1016/j.talanta.2009.11.012] [PMID: 20152449]

[61]   Yang, F.; Liu, R.; Tan, Z.; Wen, X.; Zheng, C.; Lv, Y. Sensitive determination of mercury by a miniaturized spectrophotometer after *in situ* single-drop microextraction. *J. Hazard. Mater.,* **2010**, *183*(1-3), 549-553.
[http://dx.doi.org/10.1016/j.jhazmat.2010.07.059] [PMID: 20696521]

[62]   Martinis, E.; Wuilloud, R. Cold vapor ionic liquid-assisted headspace single-drop micro extraction: a novel preconcentration technique for mercury species determination in complex matrix samples. *J. Anal. At. Spectrom.,* **2010**, *25*, 1432-1439.
[http://dx.doi.org/10.1039/c004678g]

[63]   Amjadi, M.; Manzoori, J.L.; Abulhassani, J. Ionic liquid-based, single-drop microextraction for preconcentration of cobalt before its determination by electrothermal atomic absorption spectrometry. *J. AOAC Int.,* **2010**, *93*(3), 985-991.
[PMID: 20629404]

[64]   Marquez-Sillero, I.; Aguilera-Herrador, E. C_ardenas, S.; Valc_arcel, M. Determination of 2,4,6-tricholoroanisole in water and wine samples by ionic liquid-based single-drop micro extraction and ion mobility spectrometry. *Anal. Chim. Acta,* **2011**, *702*, 199-204.
[http://dx.doi.org/10.1016/j.aca.2011.06.046] [PMID: 21839198]

[65]   Wen, X.; Deng, Q.; Guo, J. Biomolecular Spectroscopy Ionic liquid-based single drop micro extraction of ultra-trace copper in food and water samples before spectrophotometric determination. *Spectrochim. Acta A Mol. Biomol. Spectrosc.,* **2011**, *79*, 1941-1945.
[http://dx.doi.org/10.1016/j.saa.2011.05.095] [PMID: 21697004]

[66]   Wang, Q.; Qiu, H.; Li, J.; Han, H.; Liu, X.; Jiang, S. Novel approach to improve the detection of colchicine *via* online coupling of ionic liquid-based single-drop microextraction with capillary

electrophoresis. *J. Sep. Sci.,* **2011**, *34*(5), 594-600.
[http://dx.doi.org/10.1002/jssc.201000686] [PMID: 21268261]

[67]   Rahmani, M.; Kaykhaii, M. Determination of methylcyclopentadienylmanganese tricarbonyl in gasoline and water *via* ionic-liquid headspace single drop micro extraction and electrothermal atomic absorption spectrometry. *Mikrochim. Acta,* **2011**, *174*, 413-419.
[http://dx.doi.org/10.1007/s00604-011-0648-6]

[68]   Ahmad, F.; Wu, H.F. Characterization of pathogenic bacteria using ionic liquid *via* single drop microextraction combined with MALDI-TOF MS. *Analyst (Lond.),* **2011**, *136*(19), 4020-4027.
[http://dx.doi.org/10.1039/c1an15350a] [PMID: 21804984]

[69]   Vallecillos, L.; Pocurull, E.; Borrull, F. Fully automated ionic liquid-based headspace single drop microextraction coupled to GC-MS/MS to determine musk fragrances in environmental water samples. *Talanta,* **2012**, *99*, 824-832.
[http://dx.doi.org/10.1016/j.talanta.2012.07.036] [PMID: 22967629]

[70]   Guo, X.; Yin, D.; Peng, J.; Hu, X. Ionic liquid-based single-drop liquid-phase microextraction combined with high-performance liquid chromatography for the determination of sulfonamides in environmental water. *J. Sep. Sci.,* **2012**, *35*(3), 452-458.
[http://dx.doi.org/10.1002/jssc.201100777] [PMID: 22258811]

[71]   Vallecillos, L.; Borrull, F.; Pocurull, E. Determination of musk fragrances in sewage sludge by pressurized liquid extraction coupled to automated ionic liquid-based headspace single-drop microextraction followed by GC-MS/MS. *J. Sep. Sci.,* **2012**, *35*(20), 2735-2742.
[http://dx.doi.org/10.1002/jssc.201200326] [PMID: 23019133]

[72]   Wen, X.; Deng, Q.; Wang, J.; Yang, S.; Zhao, X. A new coupling of ionic liquid based-single drop microextraction with tungsten coil electrothermal atomic absorption spectrometry. *Spectrochim. Acta A Mol. Biomol. Spectrosc.,* **2013**, *105*, 320-325.
[http://dx.doi.org/10.1016/j.saa.2012.12.040] [PMID: 23318776]

[73]   Trujillo-Rodríguez, M.J. b, V. Pinoa, J. L. Andersonb Magnetic ionic liquids as extraction solvents in vacuum headspace single drop micro extraction. *Talanta,* **2017**, *172*, 86-94.
[http://dx.doi.org/10.1016/j.talanta.2017.05.021] [PMID: 28602308]

[74]   Pedersen-Bjergaard, S.; Rasmussen, K.E. Liquid-liquid-liquid microextraction for sample preparation of biological fluids prior to capillary electrophoresis. *Anal. Chem.,* **1999**, *71*(14), 2650-2656.
[http://dx.doi.org/10.1021/ac990055n] [PMID: 10424162]

[75]   Peng, J.F.; Liu, J.F.; Hu, X.L.; Jiang, G.B. Direct determination of chlorophenols in environmental water samples by hollow fiber supported ionic liquid membrane extraction coupled with high-performance liquid chromatography. *J. Chromatogr. A,* **2007**, *1139*(2), 165-170.
[http://dx.doi.org/10.1016/j.chroma.2006.11.006] [PMID: 17113589]

[76]   Chen, H.; Han, J.; Wang, Y.; Hu, Y.; Ni, L.; Liu, Y.; Kang, W.; Liu, Y. Hollow fiber liquid-phase micro extraction of cadmium(II) using an ionic liquid as the extractant. *Mikrochim. Acta,* **2014**, *181*, 1455-1461.
[http://dx.doi.org/10.1007/s00604-014-1274-x]

[77]   Wang, Z.; Xu, Q.; Li, S.; Luan, L.; Li, J.; Zhang, S.; Dong, H.; Ding, Y.J.; Chen, L.X. Hollow fiber supported ionic liquid membrane micro extraction for speciation of mercury by high performance liquid chromatography-inductively coupled plasma mass spectrometry. *Anal. Methods,* **2015**, *7*, 1140-1146.
[http://dx.doi.org/10.1039/C4AY02408G]

[78]   Pimparu, R.; Nitiyanontakit, S.; Miró, M.; Varanusupakul, P. Dynamic single-interface hollow fiber liquid phase microextraction of Cr(VI) using ionic liquid containing supported liquid membrane. *Talanta,* **2016**, *161*, 730-734.
[http://dx.doi.org/10.1016/j.talanta.2016.09.036] [PMID: 27769473]

[79]   Wang, Y.; Liu, Y.; Han, J.; Wang, L.; Chen, T.; Ni, L. Selective extraction and preconcentration of

trace lead(ii) in medicinal plant-based ionic liquid hollow fiber liquid phase micro extraction system using dicyclohexyl-18-crown-6 as membrane carrier. *Anal. Methods,* **2015**, *7*, 2339-2346.
[http://dx.doi.org/10.1039/C4AY02625J]

[80]   Wang, H.; Wu, W.W.; Wei, D.Y.; Guo, Z.Y.; Wang, S. Hollow fiber supported ionic liquid membrane micro extraction for preconcentration of kanamycin sulfate with electrochemiluminescence detection. *J. Electroanal. Chem. (Lausanne Switz.),* **2014**, *735*, 136-141.
[http://dx.doi.org/10.1016/j.jelechem.2014.10.017]

[81]   Nomngongo, P.N.; Ngila, J.C.; Msagati, T.A.M.; Moodley, B. Chemometric optimization of hollow fiber-liquid phase micro extraction for preconcentration of trace elements in diesel and gasoline prior to their ICP-OES determination. *Microchem. J.,* **2014**, *114*, 141-147.
[http://dx.doi.org/10.1016/j.microc.2013.12.013]

[82]   Liu, W.; Wei, Z.; Zhang, Q.; Wu, F.; Lin, Z.; Lu, Q.; Lin, F.; Chen, G.; Zhang, L. Novel multifunctional acceptor phase additive of water-miscible ionic liquid in hollow-fiber protected liquid phase microextraction. *Talanta,* **2012**, *88*, 43-49.
[http://dx.doi.org/10.1016/j.talanta.2011.08.017] [PMID: 22265467]

[83]   Ge, D.; Lee, H.K. Ionic liquid based hollow fiber supported liquid phase microextraction of ultraviolet filters. *J. Chromatogr. A,* **2012**, *1229*, 1-5.
[http://dx.doi.org/10.1016/j.chroma.2011.12.110] [PMID: 22307149]

[84]   Wang, H.; Wu, W.W.; Wei, D.Y.; Guo, Z.Y.; Wang, S. Hollow fiber supported ionic liquid membrane micro extraction for preconcentration of kanamycin sulfate with electrochemiluminescence detection. *J. Electroanal. Chem. (Lausanne Switz.),* **2014**, *735*, 136-141.
[http://dx.doi.org/10.1016/j.jelechem.2014.10.017]

[85]   Wang, S.R.; Wang, S. Ionic liquid-based hollow fiber-supported liquid-phase microextraction enhanced electrically for the determination of neutral red. *Yao Wu Shi Pin Fen Xi,* **2014**, *22*(4), 418-424.
[http://dx.doi.org/10.1016/j.jfda.2014.03.006] [PMID: 28911455]

[86]   Nomngongo, P.N.; Ngila, J.C.; Msagati, T.A.M.; Moodley, B. Chemometric optimization of hollow fiber-liquid phase micro extraction for preconcentration of trace elements in diesel and gasoline prior to their ICP-OES determination. *Microchem. J.,* **2014**, *114*, 141-147.
[http://dx.doi.org/10.1016/j.microc.2013.12.013]

[87]   Rezaee, M.; Assadi, Y.; Milani Hosseini, M.R.; Aghaee, E. Ahmadi; F.; Berijani S. Determination of organic compounds in water using dispersive liquid-liquid micro extraction. *J. Chromatogr. A,* **2006**, *1116*, 1-9.
[http://dx.doi.org/10.1016/j.chroma.2006.03.007] [PMID: 16574135]

[88]   Herrera-Herrera, A.V.; Asensio-Ramos, M.; Hernàndez-Borges, J.; Rodriguez-Delgado, M.A. Dispersive liquid-liquid micro extraction for determination of organic analytes. *Trends Analyt. Chem.,* **2010**, *29*(7), 728-751.
[http://dx.doi.org/10.1016/j.trac.2010.03.016]

[89]   Yan, H.; Wang, H. Recent development and applications of dispersive liquid-liquid microextraction. *J. Chromatogr. A,* **2013**, *1295*, 1-15.
[http://dx.doi.org/10.1016/j.chroma.2013.04.053] [PMID: 23680388]

[90]   Zhou, Q.; Bai, H.; Xie, G.; Xiao, J. Trace determination of organophosphorus pesticides in environmental samples by temperature-controlled ionic liquid dispersive liquid-phase microextraction. *J. Chromatogr. A,* **2008**, *1188*(2), 148-153.
[http://dx.doi.org/10.1016/j.chroma.2008.02.094] [PMID: 18346747]

[91]   Baghdadi, M.; Shemirani, F. Cold-induced aggregation microextraction: a novel sample preparation technique based on ionic liquids. *Anal. Chim. Acta,* **2008**, *613*(1), 56-63.
[http://dx.doi.org/10.1016/j.aca.2008.02.057] [PMID: 18374702]

[92]   Liu, Y.; Zhao, E.; Zhu, W.; Gao, H.; Zhou, Z. Determination of four heterocyclic insecticides by ionic

liquid dispersive liquid-liquid microextraction in water samples. *J. Chromatogr. A,* **2009**, *1216*(6), 885-891.
[http://dx.doi.org/10.1016/j.chroma.2008.11.076] [PMID: 19118833]

[93]   Erkan Yilmaz, E.; Soylak, M. *Latest trends, green aspects, and innovations in liquid-phase based micro extraction techniques: a review.,* **2016**.

[94]   Rykowska, I.; Ziemblińska, J.; Nowak, I. Modern approaches in dispersive liquid-liquid micro extraction (DLLME) based on ionic liquids: A review. *J. Mol. Liq.,* **2018**, *259*, 319-339.
[http://dx.doi.org/10.1016/j.molliq.2018.03.043]

[95]   Trujillo-Rodríguez, M.J.; Rocío-Bautista, P.; Pino, V.; Afonso, A.M. Ionic liquids in dispersive liquid-liquid micro extraction. *Trends Analyt. Chem.,* **2013**, *51*, 87-106.
[http://dx.doi.org/10.1016/j.trac.2013.06.008]

[96]   Cacho, J.I.; Campillo, N.; Viñas, P.; Hernández-Córdoba, M. Improved sensitivity gas chromatography-mass spectrometry determination of parabens in waters using ionic liquids. *Talanta,* **2016**, *146*, 568-574.
[http://dx.doi.org/10.1016/j.talanta.2015.09.022] [PMID: 26695305]

[97]   Liu, Y.; Zhao, E.; Zhu, W.; Gao, H.; Zhou, Z. Determination of four heterocyclic insecticides by ionic liquid dispersive liquid-liquid microextraction in water samples. *J. Chromatogr. A,* **2009**, *1216*(6), 885-891.
[http://dx.doi.org/10.1016/j.chroma.2008.11.076] [PMID: 19118833]

[98]   He, L.; Luo, X.; Xie, H.; Wang, C.; Jiang, X.; Lu, K. Ionic liquid-based dispersive liquid-liquid microextraction followed high-performance liquid chromatography for the determination of organophosphorus pesticides in water sample. *Anal. Chim. Acta,* **2009**, *655*(1-2), 52-59.
[http://dx.doi.org/10.1016/j.aca.2009.09.044] [PMID: 19925915]

[99]   Zhao, R.S.; Wang, S.S.; Cheng, C.G. Rapid enrichment and sensitive determination of tetrabromobisphenol A in environmental water samples with ionic liquid dispersive liquid-phase micro extraction prior to HPLC–ESIMS–MS. *Chromatogr.,* **2011**, *73*, 793-797.
[http://dx.doi.org/10.1007/s10337-010-1845-x]

[100] Zhao, R.S.; Zhang, L.L.; Wang, X. Dispersive liquid-phase microextraction using ionic liquid as extractant for the enrichment and determination of DDT and its metabolites in environmental water samples. *Anal. Bioanal. Chem.,* **2011**, *399*(3), 1287-1293.
[http://dx.doi.org/10.1007/s00216-010-4364-6] [PMID: 21060999]

[101] Arvand, M.; Bozorgzadeh, E.; Shariati, S.; Zanjanchi, M.A. Ionic liquid-based dispersive liquid-liquid microextraction for the determination of formaldehyde in wastewaters and detergents. *Environ. Monit. Assess.,* **2012**, *184*(12), 7597-7605.
[http://dx.doi.org/10.1007/s10661-012-2521-4] [PMID: 22258742]

[102] Yao, C.; Li, T.; Twu, P.; Pitner, W.R.; Anderson, J.L. Selective extraction of emerging contaminants from water samples by dispersive liquid-liquid microextraction using functionalized ionic liquids. *J. Chromatogr. A,* **2011**, *1218*(12), 1556-1566.
[http://dx.doi.org/10.1016/j.chroma.2011.01.035] [PMID: 21324466]

[103] Pena, M.T.; Casais, M.C.; Mejuto, M.C.; Cela, R. Development of an ionic liquid based dispersive liquid-liquid microextraction method for the analysis of polycyclic aromatic hydrocarbons in water samples. *J. Chromatogr. A,* **2009**, *1216*(36), 6356-6364.
[http://dx.doi.org/10.1016/j.chroma.2009.07.032] [PMID: 19646707]

[104] Zhao, R.S.; Wang, X.; Sun, J.; Hu, C.; Wang, X.K. Determination of triclosan and triclocarban in environmental water samples with ionic liquid/ionic liquid dispersive liquid-liquid micro extraction prior to HPLC-ESI-MS/MS. *Mikrochim. Acta,* **2011**, *174*, 145-151.
[http://dx.doi.org/10.1007/s00604-011-0607-2]

[105] Xu, X.; Su, R.; Zhao, X.; Liu, Z.; Zhang, Y.; Li, D.; Li, X.; Zhang, H.; Wang, Z. Ionic liquid-based microwave-assisted dispersive liquid-liquid microextraction and derivatization of sulfonamides in

river water, honey, milk, and animal plasma. *Anal. Chim. Acta,* **2011**, *707*(1-2), 92-99.
[http://dx.doi.org/10.1016/j.aca.2011.09.018] [PMID: 22027124]

[106] Zhao, R.S.; Wang, X.; Zhang, L.L.; Wang, S.S.; Yuan, J.P. Ionic liquid/ionic liquid dispersive liquid-liquid micro extraction, a new sample enrichment procedure for the determination of hexabromocyclododecane diastereomers in environmental water samples. *Anal. Methods,* **2011**, *3*, 831-836.
[http://dx.doi.org/10.1039/c0ay00708k]

[107] Lubomirsky, E.; Padró, J.M.; Reta, M.R. Development of a dispersive liquid-liquid micro extraction technique for the analysis of aryloxyphenoxy-propionate herbicides in soy-based foods. *Microchem. J.,* **2016**, *129*, 63-70.
[http://dx.doi.org/10.1016/j.microc.2016.06.015]

[108] Ravelo-Pérez, L.M.; Hernández-Borges, J.; Asensio-Ramos, M.; Rodríguez-Delgado, M.A. Ionic liquid based dispersive liquid-liquid microextraction for the extraction of pesticides from bananas. *J. Chromatogr. A,* **2009**, *1216*(43), 7336-7345.
[http://dx.doi.org/10.1016/j.chroma.2009.08.012] [PMID: 19700165]

[109] Ravelo-Pérez, L.M.; Hernández-Borges, J.; Herrera-Herrera, A.V.; Rodríguez-Delgado, M.A. Pesticide extraction from table grapes and plums using ionic liquid based dispersive liquid-liquid microextraction. *Anal. Bioanal. Chem.,* **2009**, *395*(7), 2387-2395.
[http://dx.doi.org/10.1007/s00216-009-3133-x] [PMID: 19779926]

[110] Yang, P.; Ren, H.; Qiu, H.; Liu, X.; Jiang, S. Determination of four trace preservatives in street food by ionic liquid-based dispersive liquid-liquid micro-extraction. *Chem. Pap.,* **2011**, *65*, 747-753.
[http://dx.doi.org/10.2478/s11696-011-0071-9]

[111] Cruz-Vera, M.; Lucena, R.; Cárdenas, S.; Valcárcel, M. One-step in-syringe ionic liquid-based dispersive liquid-liquid microextraction. *J. Chromatogr. A,* **2009**, *1216*(37), 6459-6465.
[http://dx.doi.org/10.1016/j.chroma.2009.07.040] [PMID: 19674753]

[112] Li, Z.; Chen, F.; Wang, X.; Wang, C. Ionic liquids dispersive liquid-liquid microextraction and high-performance liquid chromatographic determination of irbesartan and valsartan in human urine. *Biomed. Chromatogr.,* **2013**, *27*(2), 254-258.
[http://dx.doi.org/10.1002/bmc.2784] [PMID: 22733604]

[113] Tian, J.; Chen, X.; Bai, X. Comparison of dispersive liquid-liquid microextraction based on organic solvent and ionic liquid combined with high-performance liquid chromatography for the analysis of emodin and its metabolites in urine samples. *J. Sep. Sci.,* **2012**, *35*(1), 145-152.
[http://dx.doi.org/10.1002/jssc.201100729] [PMID: 22144110]

[114] Hu, H.; Liu, B.; Yang, J.; Lin, Z.; Gan, W. Sensitive determination of trace urinary 3-hydroxybenzo[a]pyrene using ionic liquids-based dispersive liquid-liquid microextraction followed by chemical derivatization and high performance liquid chromatography-high resolution tandem mass spectrometry. *J. Chromatogr. B Analyt. Technol. Biomed. Life Sci.,* **2016**, *1027*, 200-206.
[http://dx.doi.org/10.1016/j.jchromb.2016.05.041] [PMID: 27294533]

[115] Taziki, M.; Shemirani, F.; Majidi, B. **2012**.

[116] Zhou, C.; Tong, S.; Chang, Y.; Jia, Q.; Zhou, W. Ionic liquid-based dispersive liquid-liquid microextraction with back-extraction coupled with capillary electrophoresis to determine phenolic compounds. *Electrophoresis,* **2012**, *33*(8), 1331-1338.
[http://dx.doi.org/10.1002/elps.201100469] [PMID: 22589114]

[117] Li, L.H.; Zhang, H.F.; Hu, S.; Bai, X.H.; Li, S. Dispersive liquid-liquid micro extraction coupled with high-performance liquid chromatography for determination of coumarin compounds in Radix Angelicae Dahuricae. *Chromatogr.,* **2012**, *75*, 131-137.
[http://dx.doi.org/10.1007/s10337-011-2177-1]

[118] Jia, X.Y.; Li, N.B.; Luo, H.Q. Determination of ursolic acid in force loquat capsule by ultrasonic extraction and ionic liquid-based reverse dispersive LLME. *Chromatogr.,* **2010**, *71*, 839-843.

[http://dx.doi.org/10.1365/s10337-010-1547-4]

[119]   Escudero, L.B.; Berton, P.; Martinis, E.M.; Olsina, R.A.; Wuilloud, R.G. Dispersive liquid-liquid microextraction and preconcentration of thallium species in water samples by two ionic liquids applied as ion-pairing reagent and extractant phase. *Talanta,* **2012**, *88*, 277-283.
[http://dx.doi.org/10.1016/j.talanta.2011.09.068] [PMID: 22265499]

[120]   Jia, X.; Han, Y.; Wei, C.; Duan, T.; Chen, H. Speciation of mercury in liquid cosmetic samples by ionic liquid-based dispersive liquid-liquid micro extraction combined with high-performance liquid chromatography inductively coupled plasma mass spectrometry. *J. Anal. At. Spectrom.,* **2011**, *26*, 1380-1386.
[http://dx.doi.org/10.1039/c0ja00121j]

[121]   Escudero, L.B.; Martinis, E.M.; Olsina, R.A.; Wuilloud, R.G. Arsenic speciation analysis in mono-varietal wines by on-line ionic liquid-based dispersive liquid-liquid microextraction. *Food Chem.,* **2013**, *138*(1), 484-490.
[http://dx.doi.org/10.1016/j.foodchem.2012.10.054] [PMID: 23265515]

[122]   Razmislevičiene, I.; Olšauskaite, V.; Padarauskas, A. Ionic liquid-based dispersive liquid-liquid microextraction combined with ultraperformance liquid chromatography for the determination of metal ions. *Chemija,* **2011**, *22*, 197-203.

[123]   Abdolmohammad-Zadeh, H.; Sadeghi, G.H. A novel microextraction technique based on 1-hexylpyridinium hexafluorophosphate ionic liquid for the preconcentration of zinc in water and milk samples. *Anal. Chim. Acta,* **2009**, *649*(2), 211-217.
[http://dx.doi.org/10.1016/j.aca.2009.07.040] [PMID: 19699396]

[124]   Mallah, M.H.; Shemirani, F.; Maragheh, M.G. Ionic liquids for simultaneous preconcentration of some lanthanoids using dispersive liquid-liquid microextraction technique in uranium dioxide powder. *Environ. Sci. Technol.,* **2009**, *43*(6), 1947-1951.
[http://dx.doi.org/10.1021/es8030566] [PMID: 19368197]

[125]   Khani, R.; Shemirani, F.; Majidi, B. Combination of dispersive liquid-liquid micro extraction and flame atomic absorption spectrometry for preconcentration and determination of copper in water samples. *Desalination,* **2011**, *266*, 238-243.
[http://dx.doi.org/10.1016/j.desal.2010.08.032]

[126]   Berton, P.; Wuilloud, R.G. Highly selective ionic liquid-based microextraction method for sensitive trace cobalt determination in environmental and biological samples. *Anal. Chim. Acta,* **2010**, *662*(2), 155-162.
[http://dx.doi.org/10.1016/j.aca.2010.01.012] [PMID: 20171314]

[127]   Martinis, E.M.; Escudero, L.B.; Berton, P.; Monasterio, R.P.; Filippini, M.F.; Wuilloud, R.G. Determination of inorganic selenium species in water and garlic samples with on-line ionic liquid dispersive microextraction and electrothermal atomic absorption spectrometry. *Talanta,* **2011**, *85*(4), 2182-2188.
[http://dx.doi.org/10.1016/j.talanta.2011.07.065] [PMID: 21872076]

[128]   Baghdadi, M.; Shemirani, F. *In situ* solvent formation microextraction based on ionic liquids: a novel sample preparation technique for determination of inorganic species in saline solutions. *Anal. Chim. Acta,* **2009**, *634*(2), 186-191.
[http://dx.doi.org/10.1016/j.aca.2008.12.017] [PMID: 19185118]

[129]   Atwood, J.L.; Atwood, J.D. *Advances in Chemistry Series No. 150*; American Chemical Society: Washington, DC, **1976**, pp. 112-127.

[130]   Yao, C.; Anderson, J.L. Dispersive liquid-liquid microextraction using an *in situ* metathesis reaction to form an ionic liquid extraction phase for the preconcentration of aromatic compounds from water. *Anal. Bioanal. Chem.,* **2009**, *395*(5), 1491-1502.
[http://dx.doi.org/10.1007/s00216-009-3078-0] [PMID: 19756541]

[131]   Hu, L.; Zhang, P.; Shan, W.; Wang, X.; Li, S.; Zhou, W.; Gao, H. *In situ* metathesis reaction combined

with liquid-phase micro extraction based on the solidification of sedimentary ionic liquids for the determination of pyrethroid insecticides in water samples. *Talanta*, **2015**, *144*, 98-104.

[132] Vidal, L.; Silva, S.G.; Canals, A.; Nóbrega, J.A. Tungsten coil atomic emission spectrometry combined with dispersive liquid-liquid microextraction: A synergistic association for chromium determination in water samples. *Talanta*, **2016**, *148*, 602-608.
[http://dx.doi.org/10.1016/j.talanta.2015.04.023] [PMID: 26653490]

[133] Yu, H.; Clark, K.D.; Anderson, J.L. Rapid and sensitive analysis of microcystins using ionic liquid-based *in situ* dispersive liquid-liquid microextraction. *J. Chromatogr. A*, **2015**, *1406*, 10-18.
[http://dx.doi.org/10.1016/j.chroma.2015.05.075] [PMID: 26087964]

[134] Li, T.; Joshi, M.D.; Ronning, D.R.; Anderson, J.L. Ionic liquids as solvents for *in situ* dispersive liquid-liquid microextraction of DNA. *J. Chromatogr. A*, **2013**, *1272*, 8-14.
[http://dx.doi.org/10.1016/j.chroma.2012.11.055] [PMID: 23261290]

[135] Zhang, C.; Cagliero, C.; Pierson, S.A.; Anderson, J.L. Rapid and sensitive analysis of polychlorinated biphenyls and acrylamide in food samples using ionic liquid-based *in situ* dispersive liquid-liquid microextraction coupled to headspace gas chromatography. *J. Chromatogr. A*, **2017**, *1481*, 1-11.
[http://dx.doi.org/10.1016/j.chroma.2016.12.013] [PMID: 28017564]

[136] Cachol, J.I.; Campillol, N.; Viñasl, P.; Hernández-Córdoba, M. *In situ* ionic liquid dispersive liquid–liquid micro extraction and direct microvial insert thermal desorption for gas chromatographic determination of bisphenol compounds. *Anal. Bioanal. Chem.*, **2016**, *408*, 243-249.
[http://dx.doi.org/10.1007/s00216-015-9098-z] [PMID: 26476920]

[137] Zhang, J.; Li, M.; Yang, M.; Peng, B.; Li, Y.; Zhou, W.; Gao, H.; Lu, R. Magnetic retrieval of ionic liquids: fast dispersive liquid-liquid microextraction for the determination of benzoylurea insecticides in environmental water samples. *J. Chromatogr. A*, **2012**, *1254*, 23-29.
[http://dx.doi.org/10.1016/j.chroma.2012.07.051] [PMID: 22871379]

[138] Bagheri, H.; Zandi, O.; Aghakhani, A. Magnetic nanoparticle-based micro-solid phase extraction and GC–MS determination of oxadiargyl in aqueous samples. *Chromatogr.*, **2011**, *74*, 483-488.
[http://dx.doi.org/10.1007/s10337-011-2083-6]

[139] Song, Y.; Zhao, S.; Tchounwou, P.; Liu, Y.M. A nanoparticle-based solid-phase extraction method for liquid chromatography-electrospray ionization-tandem mass spectrometric analysis. *J. Chromatogr. A*, **2007**, *1166*(1-2), 79-84.
[http://dx.doi.org/10.1016/j.chroma.2007.07.074] [PMID: 17723235]

[140] Zhao, G.; Song, S.; Wang, C.; Wu, Q.; Wang, Z. Determination of triazine herbicides in environmental water samples by high-performance liquid chromatography using graphene-coated magnetic nanoparticles as adsorbent. *Anal. Chim. Acta*, **2011**, *708*(1-2), 155-159.
[http://dx.doi.org/10.1016/j.aca.2011.10.006] [PMID: 22093359]

[141] Román, I.P.; Chisvert, A.; Canals, A. Dispersive solid-phase extraction based on oleic acid-coated magnetic nanoparticles followed by gas chromatography-mass spectrometry for UV-filter determination in water samples. *J. Chromatogr. A*, **2011**, *1218*(18), 2467-2475.
[http://dx.doi.org/10.1016/j.chroma.2011.02.047] [PMID: 21411104]

[142] Fan, C.; Liang, Y.; Dong, H.; Ding, G.; Zhang, W.; Tang, G.; Yang, J.; Kong, D.; Wang, D.; Cao, Y. *In situ* ionic liquid dispersive liquid-liquid microextraction using a new anion-exchange reagent combined Fe_3O_4 magnetic nanoparticles for determination of pyrethroid pesticides in water samples. *Anal. Chim. Acta*, **2017**, *975*, 20-29.
[http://dx.doi.org/10.1016/j.aca.2017.04.036] [PMID: 28552303]

[143] Shi, Z.; Qiu, L.; Zhang, D.; Sun, M.; Zhang, H. Dispersive liquid-liquid microextraction based on amine-functionalized Fe_3O_4 nanoparticles for the determination of phenolic acids in vegetable oils by high-performance liquid chromatography with UV detection. *J. Sep. Sci.*, **2015**, *38*(16), 2865-2872.
[http://dx.doi.org/10.1002/jssc.201500330] [PMID: 26109040]

[144] Naeemullah, G.K.; Kazi, T.G.; Tuzen, M. Magnetic stirrer induced dispersive ionic-liquid

microextraction for the determination of vanadium in water and food samples prior to graphite furnace atomic absorption spectrometry. *Food Chem.,* **2015**, *172*, 161-165.
[http://dx.doi.org/10.1016/j.foodchem.2014.09.053] [PMID: 25442538]

[145] Yang, M.; Wu, X.; Jia, Y.; Xi, X.; Yang, X.; Lu, R.; Zhang, S.; Gao, H.; Zhou, W. Use of magnetic effervescent tablet-assisted ionic liquid dispersive liquid-liquid microextraction to extract fungicides from environmental waters with the aid of experimental design methodology. *Anal. Chim. Acta,* **2016**, *906*, 118-127.
[http://dx.doi.org/10.1016/j.aca.2015.12.019] [PMID: 26772131]

[146] Wang, L. Wu, L.; Cao, J.; Hong, X.; Ye, R.; Chen, W.; Yuan, T. Magnetic effervescent tablet-assisted ionic liquid dispersive liquid–liquid micro extraction of selenium for speciation in foods and beverages. *Food Addit. Contam. Part A Chem. Anal. Control Expo. Risk Assess.,* **2016**, *33*, 1190-1199.
[http://dx.doi.org/10.1080/19440049.2016.1189807] [PMID: 27181611]

[147] Wu, X.; Li, X.; Yang, M.; Zeng, H.; Zhang, S.; Lu, R.; Gao, H.; Xu, D. An ionic liquid-based nanofluid of titanium dioxide nanoparticles for effervescence-assisted dispersive liquid-liquid extraction for acaricide detection. *J. Chromatogr. A,* **2017**, *1497*, 1-8.
[http://dx.doi.org/10.1016/j.chroma.2017.03.005] [PMID: 28366570]

[148] Qiu, H.; Takafuji, M.; Liu, X.; Jiang, S.; Ihara, H. Investigation of pi-pi and ion-dipole interactions on 1-allyl-3-butylimidazolium ionic liquid-modified silica stationary phase in reversed-phase liquid chromatography. *J. Chromatogr. A,* **2010**, *1217*(32), 5190-5196.
[http://dx.doi.org/10.1016/j.chroma.2010.06.013] [PMID: 20580007]

[149] Li, M.; Zhang, J.; Li, Y.; Peng, B.; Zhou, W.; Gao, H. Ionic liquid-linked dual magnetic microextraction: a novel and facile procedure for the determination of pyrethroids in honey samples. *Talanta,* **2013**, *107*, 81-87.
[http://dx.doi.org/10.1016/j.talanta.2012.12.056] [PMID: 23598196]

[150] Yilmaz, E.; Soylak, M. Ionic liquid-linked dual magnetic microextraction of lead(II) from environmental samples prior to its micro-sampling flame atomic absorption spectrometric determination. *Talanta,* **2013**, *116*, 882-886.
[http://dx.doi.org/10.1016/j.talanta.2013.08.002] [PMID: 24148489]

[151] Mallick, B.; Balke, B.; Felser, C.; Mudring, A.V. Dysprosium room-temperature ionic liquids with strong luminescence and response to magnetic fields. *Angew. Chem. Int. Ed. Engl.,* **2008**, *47*(40), 7635-7638.
[http://dx.doi.org/10.1002/anie.200802390] [PMID: 18759243]

[152] Hayashi, S.; Hamaguchi, H.O. Discovery of a magnetic ionic liquid [bmim]FeCl4. *Chem. Lett.,* **2004**, *33*, 1590-1591.
[http://dx.doi.org/10.1246/cl.2004.1590]

[153] Li, M.; De Rooy, S.L.; Bwambok, D.K.; El-Zahab, B.; Ditusa, J.F.; Warner, I.M. Magnetic chiral ionic liquids derived from amino acids. *Chem. Commun. (Camb.),* **2009**, (45), 6922-6924.
[http://dx.doi.org/10.1039/b917683g] [PMID: 19904348]

[154] Deng, N.; Li, M.; Zhao, L.; Lu, C.; de Rooy, S.L.; Warner, I.M. Highly efficient extraction of phenolic compounds by use of magnetic room temperature ionic liquids for environmental remediation. *J. Hazard. Mater.,* **2011**, *192*(3), 1350-1357.
[http://dx.doi.org/10.1016/j.jhazmat.2011.06.053] [PMID: 21783320]

[155] Santos, E.; Albo, J.; Irabien, A. Magnetic ionic liquids: Synthesis, properties and applications. *RSC Advances,* **2014**, *4*, 40008-40018.
[http://dx.doi.org/10.1039/C4RA05156D]

[156] Del Sesto, R.E.; McCleskey, T.M.; Burrell, A.K.; Baker, G.A.; Thompson, J.D.; Scott, B.L.; Wilkes, J.S.; Williams, P. Structure and magnetic behavior of transition metal based ionic liquids. *Chem. Commun. (Camb.),* **2008**, *4*(4), 447-449.
[http://dx.doi.org/10.1039/B711189D] [PMID: 18188463]

[157] Peppel, T.; Köckerling, M.; Geppert-Rybczyńska, M.; Ralys, R.V.; Lehmann, J.K.; Verevkin, S.P.; Heintz, A. Low-viscosity paramagnetic ionic liquids with doubly charged [Co(NCS)4]2- ions. *Angew. Chem. Int. Ed. Engl.,* **2010**, *49*(39), 7116-7119.
[http://dx.doi.org/10.1002/anie.201000709] [PMID: 20715221]

[158] Hayashi, S.; Hamaguchi, H. Discovery of a magnetic ionic liquid [bmim]FeCl4. *Chem. Lett.,* **2004**, *33*, 1590-1591.
[http://dx.doi.org/10.1246/cl.2004.1590]

[159] Wang, Y.; Sun, Y.; Xu, B.; Li, X.; Jin, R.; Zhang, H.; Song, D. Rapid and sensitive analysis of polychlorinated biphenyls and acrylamide in food samples using ionic liquid-based *in situ* dispersive liquid-liquid micro extraction coupled to headspace gas chromatography. *J. Chromatogr. A,* **2014**, *1373*, 9-16.
[http://dx.doi.org/10.1016/j.chroma.2014.11.009] [PMID: 25464995]

[160] Chatzimitakos, T.; Binellas, C.; Maidatsi, K.; Stalikas, C. Magnetic ionic liquid in stirring-assisted drop-breakup microextraction: Proof-of-concept extraction of phenolic endocrine disrupters and acidic pharmaceuticals. *Anal. Chim. Acta,* **2016**, *910*, 53-59.
[http://dx.doi.org/10.1016/j.aca.2016.01.015] [PMID: 26873468]

[161] Wang, X. Selenium Speciation in Rice Samples by Magnetic Ionic Liquid-Based Up-and-Don-Shaker-Assisted Dispersive Liquid-Liquid Microextraction Coupled to Graphite Furnace Atomic Absorption Spectrometry. *Food Chem.,* **2017**, *10*(6), 1653-1660.

[162] Wen, Q.; Wang, Y.; Xu, K.; Li, N.; Zhang, H.; Yang, Q. A novel polymeric ionic liquid-coated magnetic multiwalled carbon nanotubes for the solid-phase extraction of Cu, Zn-superoxide dismutase. *Anal. Chim. Acta,* **2016**, *939*, 54-63.
[http://dx.doi.org/10.1016/j.aca.2016.08.028] [PMID: 27639143]

[163] Padilla-Alonso, D.J.; Garza-Tapia, M.; Chávez-Montes, A.; González-Horta, A.; Waksman de Torresa, N.H.; Castro-Ríos, R. New temperature-assisted ionic liquid-based dispersive liquid-liquid microextraction method for the determination of glyphosate and aminomethylphosphonic acid in water samples. *J. Liq. Chromatogr. Relat. Technol.,* **2017**, *40*(3), 147-155.
[http://dx.doi.org/10.1080/10826076.2017.1295057]

[164] Fan, C.; Liang, Y.; Dong, H.; Ding, G.; Zhang, W.; Tang, G.; Yang, J.; Kong, D.; Wang, D.; Cao, Y. *In situ* ionic liquid dispersive liquid-liquid microextraction using a new anion-exchange reagent combined Fe₃O₄ magnetic nanoparticles for determination of pyrethroid pesticides in water samples. *Anal. Chim. Acta,* **2017**, *975*, 20-29.
[http://dx.doi.org/10.1016/j.aca.2017.04.036] [PMID: 28552303]

[165] Yu, H.; Clark, K.D.; Anderson, J.L. Rapid and sensitive analysis of microcystins using ionic liquid-based *in situ* dispersive liquid-liquid microextraction. *J. Chromatogr. A,* **2015**, *1406*, 10-18.
[http://dx.doi.org/10.1016/j.chroma.2015.05.075] [PMID: 26087964]

[166] Yu, H.; Merib, J.; Anderson, J.L. Faster dispersive liquid-liquid microextraction methods using magnetic ionic liquids as solvents. *J. Chromatogr. A,* **2016**, *1463*, 11-19.
[http://dx.doi.org/10.1016/j.chroma.2016.08.007] [PMID: 27515554]

[167] Zhou, Q.; Bai, H.; Xie, G.; Xiao, J. Trace determination of organophosphorus pesticides in environmental samples by temperature-controlled ionic liquid dispersive liquid-phase microextraction. *J. Chromatogr. A,* **2008**, *1188*(2), 148-153.
[http://dx.doi.org/10.1016/j.chroma.2008.02.094] [PMID: 18346747]

[168] Gao, Y.; Zhou, Q.; Xie, G.; Yao, Z. Temperature-controlled ionic liquid dispersive liquid-phase microextraction combined with HPLC with ultraviolet detector for the determination of fungicides. *J. Sep. Sci.,* **2012**, *35*(24), 3569-3574.
[http://dx.doi.org/10.1002/jssc.201200553] [PMID: 23166097]

[169] Zhou, Q.; Zhang, X.; Xie, G. Simultaneous analysis of phthalate esters and pyrethroid insecticides in water samples by temperature-controlled ionic liquid dispersive liquid-phase micro extraction

combined with high performance liquid chromatography. *Anal. Methods,* **2011**, *3*, 1815-1820.
[http://dx.doi.org/10.1039/c1ay05137g]

[170] Zhou, Q.; Zhang, X.; Xie, G. Preconcentration and determination of pyrethroid micro extraction in combination with high performance liquid chromatography. *Anal. Methods,* **2011**, *3*, 356-361.
[http://dx.doi.org/10.1039/C0AY00570C]

[171] Zhou, Q.; Gao, Y.; Xiao, J.; Xie, G. Sensitive determination of phenols from water samples by temperature-controlled ionic liquid dispersive liquid-phase micro extraction. *Anal. Methods,* **2011**, *3*, 653-658.
[http://dx.doi.org/10.1039/c0ay00619j]

[172] Bai, H.; Zhou, Q.; Xie, G.; Xiao, J. Enrichment and sensitive determination of dichlorodiphenyltrichloroethane and its metabolites with temperature controlled ionic liquid dispersive liquid phase microextraction prior to high performance liquid phase chromatography. *Anal. Chim. Acta,* **2009**, *651*(1), 64-68.
[http://dx.doi.org/10.1016/j.aca.2009.08.011] [PMID: 19733736]

[173] Zhou, Q.; Zhang, X.; Xie, G.; Xiao, J. Temperature-controlled ionic liquid-dispersive liquid-phase microextraction for preconcentration of chlorotoluron, diethofencarb and chlorbenzuron in water samples. *J. Sep. Sci.,* **2009**, *32*(22), 3945-3950.
[http://dx.doi.org/10.1002/jssc.200900444] [PMID: 19877152]

[174] Jiang, X.; Zhang, H.; Chen, X. Determination of phenolic compounds in water samples by HPLC following ionic liquid dispersive liquid-liquid micro extraction and cold-induced aggregation. *Mikrochim. Acta,* **2011**, *175*, 341-346.
[http://dx.doi.org/10.1007/s00604-011-0672-6]

[175] Zhang, H.; Chen, X.; Jiang, X. Determination of phthalate esters in water samples by ionic liquid cold-induced aggregation dispersive liquid-liquid microextraction coupled with high-performance liquid chromatography. *Anal. Chim. Acta,* **2011**, *689*(1), 137-142.
[http://dx.doi.org/10.1016/j.aca.2011.01.024] [PMID: 21338769]

[176] Zhou, Q.; Pang, L.; Xiao, J. Ultra-trace determination of carbamate pesticides in water samples by temperature controlled ionic liquid dispersive liquid phase micro extraction combined with high performance liquid phase chromatography. *Mikrochim. Acta,* **2011**, *173*, 477-483.
[http://dx.doi.org/10.1007/s00604-011-0587-2]

[177] Zhou, Q.; Gao, Y.; Xie, G. Determination of bisphenol A, 4-n-nonylphenol, and 4-tert-octylphenol by temperature-controlled ionic liquid dispersive liquid-phase microextraction combined with high performance liquid chromatography-fluorescence detector. *Talanta,* **2011**, *85*(3), 1598-1602.
[http://dx.doi.org/10.1016/j.talanta.2011.06.050] [PMID: 21807227]

[178] Zhao, R.S.; Wang, X.; Yuan, J.P.; Wang, S.S.; Cheng, C.G. Trace determination of hexabromocyclododecane diastereomers in water samples with temperature controlled ionic liquid dispersive liquid phase microextraction. *Chin. Chem. Lett.,* **2011**, *22*, 97-100.
[http://dx.doi.org/10.1016/j.cclet.2010.07.010]

[179] Zeeb, M.; Ganjali, M.R.; Norouzi, P. Modified ionic liquid cold-induced aggregation dispersive liquid-liquid microextraction combined with spectrofluorimetry for trace determination of ofloxacin in pharmaceutical and biological samples. *Daru,* **2011**, *19*(6), 446-454.
[PMID: 23008691]

[180] Baghdadi, M.; Shemirani, F. Cold-induced aggregation microextraction: a novel sample preparation technique based on ionic liquids. *Anal. Chim. Acta,* **2008**, *613*(1), 56-63.
[http://dx.doi.org/10.1016/j.aca.2008.02.057] [PMID: 18374702]

[181] Bai, H.; Zhou, Q.; Xie, G.; Xiao, J. Temperature-controlled ionic liquid-liquid-phase microextraction for the pre-concentration of lead from environmental samples prior to flame atomic absorption spectrometry. *Talanta,* **2010**, *80*(5), 1638-1642.
[http://dx.doi.org/10.1016/j.talanta.2009.09.059] [PMID: 20152389]

[182] Berton, P.; Martinis, E.M.; Martinez, L.D.; Wuilloud, R.G. Room temperature ionic liquid-based microextraction for vanadium species separation and determination in water samples by electrothermal atomic absorption spectrometry. *Anal. Chim. Acta,* **2009**, *640*(1-2), 40-46.
[http://dx.doi.org/10.1016/j.aca.2009.03.028] [PMID: 19362617]

[183] Zeeb, M.; Sadeghi, M. Modified ionic liquid cold-induced aggregation dispersive liquid-liquid microextraction followed by atomic absorption spectrometry for trace determination of zinc in water and food samples. *Mikrochim. Acta,* **2011**, *175*, 159-165.
[http://dx.doi.org/10.1007/s00604-011-0653-9]

[184] Absalan, G.; Akhond, M.; Sheikhian, L.; Goltz, D.M. Temperature-controlled ionic liquid-based dispersive liquid-phase micro extraction, preconcentration and quantification of nano-amounts of silver ion by using disulfiram as complexing agent. *Anal. Methods,* **2011**, *3*, 2354-2359.
[http://dx.doi.org/10.1039/c1ay05226h]

[185] Zhou, Q.; Zhang, X.; Xiao, J. Ultrasound-assisted ionic liquid dispersive liquid-phase micro-extraction: a novel approach for the sensitive determination of aromatic amines in water samples. *J. Chromatogr. A,* **2009**, *1216*(20), 4361-4365.
[http://dx.doi.org/10.1016/j.chroma.2009.03.046] [PMID: 19329122]

[186] Mao, T.; Hao, B.; He, J.; Li, W.; Li, S.; Yu, Z. Ultrasound assisted ionic liquid dispersive liquid phase extraction of lovastatin and simvastatin: a new pretreatment procedure. *J. Sep. Sci.,* **2009**, *32*(17), 3029-3033.
[http://dx.doi.org/10.1002/jssc.200900337] [PMID: 19662643]

[187] Han, D.; Yan, H.; Row, K.H. Ionic liquid-based dispersive liquid-liquid microextraction for sensitive determination of aromatic amines in environmental water. *J. Sep. Sci.,* **2011**, *34*(10), 1184-1189.
[http://dx.doi.org/10.1002/jssc.201000912] [PMID: 21462339]

[188] Arvand, M.; Bozorgzadeh, E.; Shariati, S.; Zanjanchia, M.A. Trace determination of linear alkylbenzene sulfonates using ionic liquid based ultrasound-assisted dispersive liquid-liquid micro extraction and response surface methodology. *Anal. Methods,* **2012**, *4*, 2272-2277.
[http://dx.doi.org/10.1039/c2ay25302j]

[189] Zhang, Y.; Lee, H.K. Ionic liquid-based ultrasound-assisted dispersive liquid-liquid microextraction followed high-performance liquid chromatography for the determination of ultraviolet filters in environmental water samples. *Anal. Chim. Acta,* **2012**, *750*, 120-126.
[http://dx.doi.org/10.1016/j.aca.2012.04.014] [PMID: 23062433]

[190] Wang, S.; Liu, C.; Liu, F.; Ren, L. IL-USA-DLLME method to simultaneously extract and determine four phenylurea herbicides in water samples. *Curr. Anal. Chem.,* **2011**, *7*, 357-364.
[http://dx.doi.org/10.2174/1573411111797183128]

[191] Eisapour, M.; Shemirani, F.; Majidi, B.; Baghdadi, M. Ultrasound assisted cold induced aggregation: an improved method for trace determination of volatile phenol. *Mikrochim. Acta,* **2012**, *177*, 349-355.
[http://dx.doi.org/10.1007/s00604-012-0783-8]

[192] Ge, D.; Lee, H.K. Ionic liquid based dispersive liquid-liquid microextraction coupled with micro-solid phase extraction of antidepressant drugs from environmental water samples. *J. Chromatogr. A,* **2013**, *1317*, 217-222.
[http://dx.doi.org/10.1016/j.chroma.2013.04.014] [PMID: 23639124]

[193] Zhang, J.; Gao, H.; Peng, B.; Li, S.; Zhou, Z. Comparison of the performance of conventional, temperature-controlled, and ultrasound-assisted ionic liquid dispersive liquid-liquid microextraction combined with high-performance liquid chromatography in analyzing pyrethroid pesticides in honey samples. *J. Chromatogr. A,* **2011**, *1218*(38), 6621-6629.
[http://dx.doi.org/10.1016/j.chroma.2011.07.102] [PMID: 21862027]

[194] Wang, S.; Ren, L.; Xu, Y.; Liu, F. Application of ultrasound-assisted ionic liquid dispersive liquid-phase micro extraction followed high-performance liquid chromatography for the determination of fungicides in red wine. *Mikrochim. Acta,* **2011**, *173*, 453-457.

[http://dx.doi.org/10.1007/s00604-011-0577-4]

[195] Sun, J.N.; Shi, Y.P.; Chen, J. Ultrasound-assisted ionic liquid dispersive liquid-liquid microextraction coupled with high performance liquid chromatography for sensitive determination of trace celastrol in urine. *J. Chromatogr. B Analyt. Technol. Biomed. Life Sci.,* **2011**, *879*(30), 3429-3433. [http://dx.doi.org/10.1016/j.jchromb.2011.09.019] [PMID: 21963272]

[196] Li, S.; Cai, S.; Hu, W.; Chen, H.; Liu, H. Ionic liquid-based ultrasound-assisted dispersive liquid-liquid micro extraction combined with electrothermal atomic absorption spectrometry for a sensitive determination of cadmium in water samples. Spectroc. Acta Pt. B-Atom. *Spectr.,* **2009**, *64*, 666-671.

[197] Molaakbari, E.; Mostafavi, A.; Afzali, D. Ionic liquid ultrasound assisted dispersive liquid-liquid microextraction method for preconcentration of trace amounts of rhodium prior to flame atomic absorption spectrometry determination. *J. Hazard. Mater.,* **2011**, *185*(2-3), 647-652. [http://dx.doi.org/10.1016/j.jhazmat.2010.09.067] [PMID: 20971554]

[198] Tuzen, M.; Pekiner, O.Z. Ultrasound-assisted ionic liquid dispersive liquid-liquid microextraction combined with graphite furnace atomic absorption spectrometric for selenium speciation in foods and beverages. *Food Chem.,* **2015**, *188*, 619-624. [http://dx.doi.org/10.1016/j.foodchem.2015.05.055] [PMID: 26041239]

[199] Ye, L.; Liu, J.; Yang, X.; Peng, Y.; Xu, L. Orthogonal array design for the optimization of ionic liquid-based dispersive liquid-liquid microextraction of benzophenone-type UV filters. *J. Sep. Sci.,* **2011**, *34*(6), 700-706. [http://dx.doi.org/10.1002/jssc.201000552] [PMID: 21290603]

[200] Vázquez, M.M.; Vázquez, P.P.; Galera, M.M.; García, M.D. Determination of eight fluoroquinolones in groundwater samples with ultrasound-assisted ionic liquid dispersive liquid-liquid microextraction prior to high-performance liquid chromatography and fluorescence detection. *Anal. Chim. Acta,* **2012**, *748*, 20-27. [http://dx.doi.org/10.1016/j.aca.2012.08.042] [PMID: 23021803]

[201] Qin, H.; Li, B.; Liu, M.S.; Yang, Y.L. Separation and pre-concentration of glucocorticoids in water samples by ionic liquid supported vortex-assisted synergic microextraction and HPLC determination. *J. Sep. Sci.,* **2013**, *36*(8), 1463-1469. [http://dx.doi.org/10.1002/jssc.201200989] [PMID: 23418157]

[202] Asensio-Ramos, M.; Hernández-Borges, J.; Borges-Miquel, T.M.; Rodríguez-Delgado, M.A. Ionic liquid-dispersive liquid-liquid microextraction for the simultaneous determination of pesticides and metabolites in soils using high-performance liquid chromatography and fluorescence detection. *J. Chromatogr. A,* **2011**, *1218*(30), 4808-4816. [http://dx.doi.org/10.1016/j.chroma.2010.11.030] [PMID: 21145553]

[203] Zhang, L.; Chen, F.; Liu, S.; Chen, B.; Pan, C. Ionic liquid-based vortex-assisted dispersive liquid-liquid microextraction of organophosphorus pesticides in apple and pear. *J. Sep. Sci.,* **2012**, *35*(18), 2514-2519. [http://dx.doi.org/10.1002/jssc.201101060] [PMID: 22997036]

[204] Shirkhanloo, H.; Rouhollahi, A.; Mousavi, H.Z. Ultra-trace arsenic determination in urine and whole blood samples by flow injection-hydride generation atomic absorption spectrometry after preconcentration and speciation based on dispersive liquid-liquid micro extraction. *Bull. Korean Chem. Soc.,* **2011**, *32*, 3923-3927. [http://dx.doi.org/10.5012/bkcs.2011.32.11.3923]

[205] Berton, P.; Wuilloud, R.G. An online ionic liquid-based micro extraction system coupled to electrothermal atomic absorption spectrometry for cobalt determination in environmental samples and pharmaceutical formulations. *Anal. Methods,* **2011**, *3*, 664-672. [http://dx.doi.org/10.1039/c0ay00616e]

[206] Xu, X.; Su, R.; Zhao, X.; Liu, Z.; Li, D.; Li, X.; Zhang, H.; Wang, Z. Determination of formaldehyde in beverages using microwave-assisted derivatization and ionic liquid-based dispersive liquid-liquid

microextraction followed by high-performance liquid chromatography. *Talanta,* **2011**, *85*(5), 2632-2638.
[http://dx.doi.org/10.1016/j.talanta.2011.08.037] [PMID: 21962694]

[207]   Xu, X.; Liu, Z.; Zhao, X.; Su, R.; Zhang, Y.; Shi, J.; Zhao, Y.; Wu, L.; Ma, Q.; Zhou, X.; Zhang, H.; Wang, Z. Ionic liquid-based microwave-assisted surfactant-improved dispersive liquid-liquid microextraction and derivatization of aminoglycosides in milk samples. *J. Sep. Sci.,* **2013**, *36*(3), 585-592.
[http://dx.doi.org/10.1002/jssc.201200801] [PMID: 23303586]

[208]   Zhong, Q.; Su, P.; Zhang, Y.; Wang, R.; Yang, Y. *In situ* ionic liquid-based microwave-assisted dispersive liquid-liquid micro extraction of triazine herbicides. *Mikrochim. Acta,* **2012**, *178*, 341-347.
[http://dx.doi.org/10.1007/s00604-012-0847-9]

[209]   Li, S.; Gao, H.; Zhang, J.; Li, Y.; Peng, B.; Zhou, Z. Determination of insecticides in water using *in situ* halide exchange reaction-assisted ionic liquid dispersive liquid-liquid microextraction followed by high-performance liquid chromatography. *J. Sep. Sci.,* **2011**, *34*(22), 3178-3185.
[http://dx.doi.org/10.1002/jssc.201100577] [PMID: 22012623]

[210]   Bi, W.; Tian, M.; Row, K.H. Ultrasonication-assisted extraction and preconcentration of medicinal products from herb by ionic liquids. *Talanta,* **2011**, *85*(1), 701-706.
[http://dx.doi.org/10.1016/j.talanta.2011.04.054] [PMID: 21645761]

[211]   Germán-Hernández, M.; Pino, V.; Anderson, J.L.; Afonso, A.M. A novel *in situ* preconcentration method with ionic liquid-based surfactants resulting in enhanced sensitivity for the extraction of polycyclic aromatic hydrocarbons from toasted cereals. *J. Chromatogr. A,* **2012**, *1227*, 29-37.
[http://dx.doi.org/10.1016/j.chroma.2011.12.097] [PMID: 22265775]

[212]   Fan, C.; Li, N.; Cao, X. Determination of chlorophenols in honey samples using *In situ* ionic liquid-dispersive liquid-liquid microextraction as a pretreatment method followed by high-performance liquid chromatography. *Food Chem.,* **2015**, *174*, 446-451.
[http://dx.doi.org/10.1016/j.foodchem.2014.11.050] [PMID: 25529704]

[213]   Shu, Y.; Gao, M.; Wang, X.; Song, R.; Lu, J.; Chen, X. Separation of curcuminoids using ionic liquid based aqueous two-phase system coupled with *in situ* dispersive liquid-liquid microextraction. *Talanta,* **2016**, *149*, 6-12.
[http://dx.doi.org/10.1016/j.talanta.2015.11.009] [PMID: 26717808]

[214]   Ayala-Cabrera, J.F.; Trujillo-Rodríguez, M.J.; Pino, V.; Hernández-Torres, Ó.M.; Afonso, A.M.; Sirieix-Plénet, J. Ionic liquids *versus* ionic liquid-based surfactants in dispersive liquid-liquid microextraction for determining copper in water by flame atomic absorption spectrometry. *Int. J. Environ. Anal. Chem.,* **2016**, *96*(2), 101-118.
[http://dx.doi.org/10.1080/03067319.2015.1128538]

[215]   Mahpishanian, S.; Shemirani, F. Preconcentration procedure using *in situ* solvent formation microextraction in the presence of ionic liquid for cadmium determination in saline samples by flame atomic absorption spectrometry. *Talanta,* **2010**, *82*(2), 471-476.
[http://dx.doi.org/10.1016/j.talanta.2010.04.060] [PMID: 20602922]

[216]   Meysam vaezzadeh, ; Shemirani, F.; Majidi, B. Microextraction technique based on ionic liquid for preconcentration and determination of palladium in food additive, sea water, tea and biological samples. *Food Chem. Toxicol.,* **2010**, *48*(6), 1455-1460.
[http://dx.doi.org/10.1016/j.fct.2010.03.005] [PMID: 20226224]

[217]   Vaezzadeh, M.; Shemirani, F.; Majidi, B. Determination of silver in real samples using homogeneous liquid-liquid microextraction based on ionic liquid. *J. Anal. Chem.,* **2012**, *67*, 28-34.
[http://dx.doi.org/10.1134/S1061934812010170]

[218]   Wang, Y.; Sun, Y.; Xu, B.; Li, X.; Jin, R.; Zhang, H.; Song, D. Rapid and sensitive analysis of polychlorinated biphenyls and acrylamide in food samples using ionic liquid-based *In situ* dispersive liquid-liquid microextraction coupled to headspace gas chromatography. *J. Chromatogr. A,* **2014**,

1373, 9-16.
[http://dx.doi.org/10.1016/j.chroma.2014.11.009] [PMID: 25464995]

[219]   Beiraghi, A.; Shokri, M.; Seidi, S.; Godajdar, B.M. Magnetomotive room temperature dicationic ionic liquid: a new concept toward centrifuge-less dispersive liquid-liquid microextraction. *J. Chromatogr. A*, **2015**, *1376*, 1-8.
[http://dx.doi.org/10.1016/j.chroma.2014.12.004] [PMID: 25528072]

[220]   Gao, S.; Yang, X.; Yu, W.; Liu, Z.; Zhang, H. Ultrasound-assisted ionic liquid/ionic liquid-dispersive liquid-liquid microextraction for the determination of sulfonamides in infant formula milk powder using high-performance liquid chromatography. *Talanta,* **2012**, *99*, 875-882.
[http://dx.doi.org/10.1016/j.talanta.2012.07.050] [PMID: 22967637]

[221]   Li, L.; Huang, M.; Shao, J.; Lin, B.; Shen, Q. Rapid determination of alkaloids in Macleaya cordata using ionic liquid extraction followed by multiple reaction monitoring UPLC-MS/MS analysis. *J. Pharm. Biomed. Anal.,* **2017**, *135*, 61-66.
[http://dx.doi.org/10.1016/j.jpba.2016.12.016] [PMID: 28011444]

[222]   Vaghar-Lahijani, G.; Aberoomand-Azar, P.; Saber-Tehrani, M.; Soleimani, M. Application of ionic liquid-based ultrasonic-assisted microextraction coupled with HPLC for determination of citalopram and nortriptyline in human plasma. *J. Liq. Chromatogr. Relat. Technol.,* **2017**, *40*(1), 1-7.
[http://dx.doi.org/10.1080/10826076.2016.1274999]

[223]   Jiang, N.; Chang, X.; Zheng, H.; He, Q.; Hu, Z. Selective solid-phase extraction of nickel(II) using a surface-imprinted silica gel sorbent. *Anal. Chim. Acta,* **2006**, *577*(2), 225-231.
[http://dx.doi.org/10.1016/j.aca.2006.06.049] [PMID: 17723676]

[224]   Memon, S.Q.; Bhanger, M.I.; Khuhawar, M.Y. Preconcentration and separation of Cr(III) and Cr(VI) using sawdust as a sorbent. *Anal. Bioanal. Chem.,* **2005**, *383*(4), 619-624.
[http://dx.doi.org/10.1007/s00216-005-3391-1] [PMID: 16184363]

[225]   Tian, M.; Yan, H.; Row, K.H. Solid-phase extraction of tanshinones from Salvia Miltiorrhiza Bunge using ionic liquid-modified silica sorbents. *J. Chromatogr. B Analyt. Technol. Biomed. Life Sci.,* **2009**, *877*(8-9), 738-742.
[http://dx.doi.org/10.1016/j.jchromb.2009.02.012] [PMID: 19237325]

[226]   Fontanals, N.; Ronka, S.; Borrull, F.; Trochimczuk, A.W.; Marcé, R.M. Supported imidazolium ionic liquid phases: a new material for solid-phase extraction. *Talanta,* **2009**, *80*(1), 250-256.
[http://dx.doi.org/10.1016/j.talanta.2009.06.068] [PMID: 19782223]

[227]   Dai, X.; Wang, D.; Li, H.; Chen, Y.; Gong, Z.; Xiang, H.; Shi, S.; Chen, X. Hollow porous ionic liquids composite polymers based solid phase extraction coupled online with high performance liquid chromatography for selective analysis of hydrophilic hydroxybenzoic acids from complex samples. *J. Chromatogr. A,* **2017**, *1484*, 7-13.
[http://dx.doi.org/10.1016/j.chroma.2017.01.022] [PMID: 28088360]

[228]   Liu, S.; Wang, C.; He, S.; Bai, L.; Liu, H. On-line SPE using ionic liquid-based monolithic column for the determination of antihypertensives in human plasma. *Chromatographia,* **2016**, *79*, 441-449.
[http://dx.doi.org/10.1007/s10337-016-3045-9]

[229]   Li, J.; Cai, Y.; Shi, Y.; Mou, S.; Jiang, G. Analysis of phthalates *via* HPLC-UV in environmental water samples after concentration by solid-phase extraction using ionic liquid mixed hemimicelles. *Talanta,* **2008**, *74*(4), 498-504.
[http://dx.doi.org/10.1016/j.talanta.2007.06.008] [PMID: 18371667]

[230]   Fontanals, N.; Ronka, S.; Borrull, F.; Trochimczuk, A.W.; Marcé, R.M. Supported imidazolium ionic liquid phases: a new material for solid-phase extraction. *Talanta,* **2009**, *80*(1), 250-256.
[http://dx.doi.org/10.1016/j.talanta.2009.06.068] [PMID: 19782223]

[231]   Tian, M.; Yan, H.; Row, K.H. Solid-phase extraction of tanshinones from Salvia Miltiorrhiza Bunge using ionic liquid-modified silica sorbents. *J. Chromatogr. B Analyt. Technol. Biomed. Life Sci.,* **2009**, *877*(8-9), 738-742.

[http://dx.doi.org/10.1016/j.jchromb.2009.02.012] [PMID: 19237325]

[232] Tian, M.; Bi, W.; Row, K.H. Solid-phase extraction of liquiritin and glycyrrhizic acid from licorice using ionic liquid-based silica sorbent. *J. Sep. Sci.,* **2009**, *32*(23-24), 4033-4039.
[http://dx.doi.org/10.1002/jssc.200900497] [PMID: 19882630]

[233] Li, M.; Pham, P.J.; Wang, T.; Pittman, C.U. Jr., Li, T. Selective extraction and enrichment of polyunsaturated fatty acid methyl esters from fish oil by novel π-complexing sorbents. *Separ. Purif. Tech.,* **2009**, *66*, 1-8.
[http://dx.doi.org/10.1016/j.seppur.2008.12.009]

[234] Li, M.; Pham, P.J.; Wang, T.; Pittman, C.U., Jr; Li, T. Solid phase extraction and enrichment of essential fatty acid methyl esters from soy-derived biodiesel by novel π-complexing sorbents. *Bioresour. Technol.,* **2009**, *100*(24), 6385-6390.
[http://dx.doi.org/10.1016/j.biortech.2009.07.054] [PMID: 19692237]

[235] Liang, P.; Peng, L. Ionic liquid-modified silica as sorbent for preconcentration of cadmium prior to its determination by flame atomic absorption spectrometry in water samples. *Talanta,* **2010**, *81*(1-2), 673-677.
[http://dx.doi.org/10.1016/j.talanta.2009.12.056] [PMID: 20188980]

[236] Chen, M.L.; Zhao, Y.N.; Zhang, D.W.; Tian, Y.; Wang, J.H. The immobilization of hydrophilic ionic liquid for Cr(VI) retention and chromium speciation. *J. Anal. At. Spectrom.,* **2010**, *25*, 1688-1694.
[http://dx.doi.org/10.1039/c0ja00026d]

[237] Tian, M.; Yan, H.; Row, H.K. Solid-Phase Extraction of Caffeine and Theophylline from Green Tea by a New Ionic Liquid-Modified Functional Polymer Sorbent. *Anal. Lett.,* **2010**, *43*, 110-118.
[http://dx.doi.org/10.1080/00032710903276554]

[238] Bi, W.; Tian, M.; Row, K.H. Solid-phase extraction of matrine and oxymatrine from Sophora Flavescens Ait using amino-imidazolium polymer. *J. Sep. Sci.,* **2010**, *33*(12), 1739-1745.
[http://dx.doi.org/10.1002/jssc.200900835] [PMID: 20437409]

[239] Zhu, T.; Row, K. Extraction and Determination of β-Sitosterol from Salicornia herbacea L. Using Monolithic Cartridge. *Chromatogr.,* **2010**, *71*, 981-985.
[http://dx.doi.org/10.1365/s10337-010-1574-1]

[240] Bi, W.; Tian, M.; Row, K.H. Solid-phase extraction of liquiritin and glycyrrhizin from licorice using porous alkyl-pyridinium polymer sorbent. *Phytochem. Anal.,* **2010**, *21*(5), 496-501.
[http://dx.doi.org/10.1002/pca.1227] [PMID: 20931626]

[241] Fang, G.; Chen, J.; Wang, J.; He, J.; Wang, S. N-methylimidazolium ionic liquid-functionalized silica as a sorbent for selective solid-phase extraction of 12 sulfonylurea herbicides in environmental water and soil samples. *J. Chromatogr. A,* **2010**, *1217*(10), 1567-1574.
[http://dx.doi.org/10.1016/j.chroma.2010.01.010] [PMID: 20116796]

[242] Yang, F.; Shen, R.; Long, Y.; Sun, X.; Tang, F.; Cai, Q.; Yao, S. Magnetic microsphere confined ionic liquid as a novel sorbent for the determination of chlorophenols in environmental water samples by liquid chromatography. *J. Environ. Monit.,* **2011**, *13*(2), 440-445.
[http://dx.doi.org/10.1039/C0EM00389A] [PMID: 21157608]

[243] Guo, L.; Deng, Q.; Fang, G.; Gao, W.; Wang, S. Preparation and evaluation of molecularly imprinted ionic liquids polymer as sorbent for on-line solid-phase extraction of chlorsulfuron in environmental water samples. *J. Chromatogr. A,* **2011**, *1218*(37), 6271-6277.
[http://dx.doi.org/10.1016/j.chroma.2011.07.016] [PMID: 21807367]

[244] Ayata, S.; Bozkurt, S.S.; Ocakoglu, K. Separation and preconcentration of Pb(II) using ionic liquid-modified silica and its determination by flame atomic absorption spectrometry. *Talanta,* **2011**, *84*(1), 212-215.
[http://dx.doi.org/10.1016/j.talanta.2011.01.006] [PMID: 21315922]

[245] Tian, M.; Row, K. SPE of Tanshinones from Salvia miltiorrhiza Bunge by using Imprinted

Functionalized Ionic Liquid-Modified Silic. *Chromatogr.,* **2011**, *73*, 25-31.
[http://dx.doi.org/10.1007/s10337-010-1836-y]

[246]   Tian, M.; Bi, W.; Row, K.H. Molecular imprinting in ionic liquid-modified porous polymer for recognitive separation of three tanshinones from Salvia miltiorrhiza Bunge. *Anal. Bioanal. Chem.,* **2011**, *399*(7), 2495-2502.
[http://dx.doi.org/10.1007/s00216-010-4641-4] [PMID: 21221534]

[247]   Bi, W.; Zhou, J.; Row, K.H. Solid phase extraction of lactic acid from fermentation broth by anion-exchangeable silica confined ionic liquids. *Talanta,* **2011**, *83*(3), 974-979.
[http://dx.doi.org/10.1016/j.talanta.2010.11.006] [PMID: 21147346]

[248]   Vidal, L.; Parshintsev, J.; Hartonen, K.; Canals, A.; Riekkola, M-L. Ionic liquid-functionalized silica for selective solid-phase extraction of organic acids, amines and aldehydes. *J. Chromatogr. A,* **2012**, *1226*, 2-10.
[http://dx.doi.org/10.1016/j.chroma.2011.08.075] [PMID: 21925663]

[249]   Bratkowska, D.; Fontanals, N.; Ronka, S.; Trochimczuk, A.W.; Borrull, F.; Marcé, R.M. Comparison of different imidazolium supported ionic liquid polymeric phases with strong anion-exchange character for the extraction of acidic pharmaceuticals from complex environmental samples. *J. Sep. Sci.,* **2012**, *35*(15), 1953-1958.
[http://dx.doi.org/10.1002/jssc.201100923] [PMID: 22865758]

[250]   Bi, W.; Tian, M.; Row, K.H. Selective extraction and separation of oxymatrine from Sophora flavescens Ait. extract by silica-confined ionic liquid. *J. Chromatogr. B Analyt. Technol. Biomed. Life Sci.,* **2012**, *880*(1), 108-113.
[http://dx.doi.org/10.1016/j.jchromb.2011.11.025] [PMID: 22138590]

[251]   Bi, W.; Tian, M.; Row, K.H. Separation of phenolic acids from natural plant extracts using molecularly imprinted anion-exchange polymer confined ionic liquids. *J. Chromatogr. A,* **2012**, *1232*, 37-42.
[http://dx.doi.org/10.1016/j.chroma.2011.08.054] [PMID: 21903215]

[252]   Wang, X.; Xing, L.; Shu, Y.; Chen, X.; Wang, J. Novel polymeric ionic liquid microspheres with high exchange capacity for fast extraction of plasmid DNA. *Anal. Chim. Acta,* **2014**, *837*, 64-69.
[http://dx.doi.org/10.1016/j.aca.2014.06.002] [PMID: 25000859]

[253]   Liu, S.; Wang, C.; He, S.; Bai, L.; Liu, H. On-line SPE using ionic liquid-based monolithic column for the determination of antihypertensives in human plasma. *Chromatogr.,* **2016**, *79*, 441-449.
[http://dx.doi.org/10.1007/s10337-016-3045-9]

[254]   Wen, Q.; Wang, Y.; Xu, K.; Li, N.; Zhang, H.; Yang, Q.; Zhou, Y. Magnetic solid-phase extraction of protein by ionic liquid-coated Fe@graphene oxide. *Talanta,* **2016**, *160*, 481-488.
[http://dx.doi.org/10.1016/j.talanta.2016.07.031] [PMID: 27591642]

[255]   Huang, Y.; Wang, Y.; Wang, Y.; Pan, Q.; Ding, X.; Xu, K.; Li, N.; Wen, Q. Ionic liquid-coated Fe3O4/APTES/graphene oxide nanocomposites: synthesis, characterization and evaluation in protein extraction processes. *RSC Advances,* **2016**, *6*, 5718-5728.
[http://dx.doi.org/10.1039/C5RA22013K]

[256]   Yang, X.; Qiao, K.; Ye, Y.; Yang, M.; Li, J.; Gao, H.; Zhang, S.; Zhou, W.; Lu, R. Facile synthesis of multifunctional attapulgite/Fe3O4/polyaniline nanocomposites for magnetic dispersive solid phase extraction of benzoylurea insecticides in environmental water samples. *Anal. Chim. Acta,* **2016**, *934*, 114-121.
[http://dx.doi.org/10.1016/j.aca.2016.06.027] [PMID: 27506351]

[257]   Dang, M.; Deng, Q.; Fang, G.; Zhang, D-D.; Liu, J-M.; Wang, S. Preparation of novel anionic polymeric ionic liquids materials and their potential in proteins adsorption. *J. Mater. Chem. B Mater. Biol. Med.,* **2017**, *5*, 6339-6347.
[http://dx.doi.org/10.1039/C7TB01234A]

[258]   Yang, Q.; Wang, Y.; Zhang, H.; Xu, K.; Wei, X.; Xu, P.; Zhou, Y. A novel dianionic amino acid ionic

liquid-coated PEG 4000 modified Fe_3O_4 nanocomposite for the magnetic solid-phase extraction of trypsin. *Talanta*, **2017**, *174*, 139-147.
[http://dx.doi.org/10.1016/j.talanta.2017.06.011] [PMID: 28738559]

[259] Liu, C.; Liao, Y.; Huang, X. Fabrication of polymeric ionic liquid-modified magnetic adsorbent for extraction of apolar and polar pollutants in complicated samples. *Talanta*, **2017**, *172*, 23-30.
[http://dx.doi.org/10.1016/j.talanta.2017.05.030] [PMID: 28602299]

[260] Shahriman, M.S.; Ramachandran, M.R.; Zain, N.N.M.; Mohamad, S.; Manan, N.S.A.; Yaman, S.M. Polyaniline-dicationic ionic liquid coated with magnetic nanoparticles composite for magnetic solid phase extraction of polycyclic aromatic hydrocarbons in environmental samples. *Talanta*, **2018**, *178*, 211-221.
[http://dx.doi.org/10.1016/j.talanta.2017.09.023] [PMID: 29136814]

[261] Arthur, C.L.; Pawliszyn, J. Solid phase microextraction with thermal desorption using fused silica optical fibers. *Anal. Chem.*, **1990**, *62*, 2145-2148.
[http://dx.doi.org/10.1021/ac00218a019]

[262] Liu, J.F.; Li, N.; Jiang, G.B.; Liu, J.M.; Jönsson, J.A.; Wen, M.J. Disposable ionic liquid coating for headspace solid-phase microextraction of benzene, toluene, ethylbenzene, and xylenes in paints followed by gas chromatography-flame ionization detection. *J. Chromatogr. A*, **2005**, *1066*(1-2), 27-32.
[http://dx.doi.org/10.1016/j.chroma.2005.01.024] [PMID: 15794551]

[263] Ho, T.D.; Canestraro, A.J.; Anderson, J.L. Ionic liquids in solid-phase microextraction: a review. *Anal. Chim. Acta*, **2011**, *695*(1-2), 18-43.
[http://dx.doi.org/10.1016/j.aca.2011.03.034] [PMID: 21601027]

[264] Amini, R.; Rouhollahi, A.; Adibi, M.; Mehdinia, A. A new disposable ionic liquid based coating for headspace solid-phase microextraction of methyl tert-butyl ether in a gasoline sample followed by gas chromatography-flame ionization detection. *Talanta*, **2011**, *84*(1), 1-6.
[http://dx.doi.org/10.1016/j.talanta.2010.10.043] [PMID: 21315889]

[265] Hsieh, Y.N.; Huang, P.C.; Sun, I.W.; Whang, T.J.; Hsu, C.Y.; Huang, H.H.; Kuei, C.H. Nafion membrane-supported ionic liquid-solid phase microextraction for analyzing ultra-trace PAHs in water samples. *Anal. Chim. Acta*, **2006**, *557*, 321-328.
[http://dx.doi.org/10.1016/j.aca.2005.10.019]

[266] Zhao, F.; Meng, Y.; Anderson, J.L. Polymeric ionic liquids as selective coatings for the extraction of esters using solid-phase microextraction. *J. Chromatogr. A*, **2008**, *1208*(1-2), 1-9.
[http://dx.doi.org/10.1016/j.chroma.2008.08.071] [PMID: 18805539]

[267] Meng, Y.; Anderson, J.L. Tuning the selectivity of polymeric ionic liquid sorbent coatings for the extraction of polycyclic aromatic hydrocarbons using solid-phase microextraction. *J. Chromatogr. A*, **2010**, *1217*(40), 6143-6152.
[http://dx.doi.org/10.1016/j.chroma.2010.08.007] [PMID: 20800235]

[268] Shearrow, A.M.; Harris, G.A.; Fang, L.; Sekhar, P.K.; Nguyen, L.T.; Turner, E.B.; Bhansali, S.; Malik, A. Ionic liquid-mediated sol-gel coatings for capillary microextraction. *J. Chromatogr. A*, **2009**, *1216*(29), 5449-5458.
[http://dx.doi.org/10.1016/j.chroma.2009.04.093] [PMID: 19515375]

[269] Yang, P.; Lau, C.; Liu, X.; Lu, J. Direct solid-support sample loading for fast cataluminescence determination of acetone in human plasma. *Anal. Chem.*, **2007**, *79*(22), 8476-8485.
[http://dx.doi.org/10.1021/ac0702488] [PMID: 17939643]

[270] Huang, K.P.; Wang, G.R.; Huang, B.Y.; Liu, C.Y. Preparation and application of ionic liquid-coated fused-silica capillary fibers for solid-phase microextraction. *Anal. Chim. Acta*, **2009**, *645*(1-2), 42-47.
[http://dx.doi.org/10.1016/j.aca.2009.04.037] [PMID: 19481629]

[271] Meng, Y.; Pino, V.; Anderson, J.L. Exploiting the versatility of ionic liquids in separation science: determination of low-volatility aliphatic hydrocarbons and fatty acid methyl esters using headspace

solid-phase microextraction coupled to gas chromatography. *Anal. Chem.,* **2009**, *81*(16), 7107-7112.
[http://dx.doi.org/10.1021/ac901377w] [PMID: 19637849]

[272] He, Y.; Pohl, J.; Engel, R.; Rothman, L.; Thomas, M. Preparation of ionic liquid based solid-phase
microextraction fiber and its application to forensic determination of methamphetamine and
amphetamine in human urine. *J. Chromatogr. A,* **2009**, *1216*(24), 4824-4830.
[http://dx.doi.org/10.1016/j.chroma.2009.04.028] [PMID: 19426983]

[273] Martín-Calero, A.; Ayala, J.H.; González, V.; Afonso, A.M. Ionic liquids as desorption solvents and
memory effect suppressors in heterocyclic aromatic amines determination by SPME-HPLC
fluorescence. *Anal. Bioanal. Chem.,* **2009**, *394*(4), 937-946.
[http://dx.doi.org/10.1007/s00216-008-2568-9] [PMID: 19104777]

[274] Shearrow, A.M.; Bhansali, S.; Malik, A. Ionic liquid-mediated bis[(3-methyldimethoxysilyl)propyl]
polypropylene oxide-based polar sol-gel coatings for capillary microextraction. *J. Chromatogr. A,*
2009, *1216*(36), 6349-6355.
[http://dx.doi.org/10.1016/j.chroma.2009.07.028] [PMID: 19643422]

[275] López-Darias, J.; Pino, V.; Meng, Y.; Anderson, J.L.; Afonso, A.M. Utilization of a benzyl
functionalized polymeric ionic liquid for the sensitive determination of polycyclic aromatic
hydrocarbons; parabens and alkylphenols in waters using solid-phase microextraction coupled to gas
chromatography-flame ionization detection. *J. Chromatogr. A,* **2010**, *1217*(46), 7189-7197.
[http://dx.doi.org/10.1016/j.chroma.2010.09.016] [PMID: 20933234]

[276] Zhao, Q.; Wajert, J.C.; Anderson, J.L. Polymeric ionic liquids as CO(2) selective sorbent coatings for
solid-phase microextraction. *Anal. Chem.,* **2010**, *82*(2), 707-713.
[http://dx.doi.org/10.1021/ac902438k] [PMID: 20038114]

[277] Liu, M.; Zhou, X.; Chen, Y.; Liu, H.; Feng, X.; Qiu, G.; Liu, F.; Zeng, Z. Innovative chemically
bonded ionic liquids-based sol-gel coatings as highly porous, stable and selective stationary phases for
solid phase microextraction. *Anal. Chim. Acta,* **2010**, *683*(1), 96-106.
[http://dx.doi.org/10.1016/j.aca.2010.10.004] [PMID: 21094387]

[278] Carda-Broch, S.; Ruiz-Angel, M.; Armstrong, D.; Berthod, A. Ionic liquid based headspace solid
phase microextraction-gas chromatography for the determination of volatile polar organic compounds.
Sep. Sci. Technol., **2010**, *45*, 2322-2328.
[http://dx.doi.org/10.1080/01496395.2010.497526]

[279] Meng, Y.; Pino, V.; Anderson, J.L. Role of counteranions in polymeric ionic liquid-based solid-phase
microextraction coatings for the selective extraction of polar compounds. *Anal. Chim. Acta,* **2011**,
687(2), 141-149.
[http://dx.doi.org/10.1016/j.aca.2010.11.046] [PMID: 21277416]

[280] Graham, C.M.; Meng, Y.; Ho, T.; Anderson, J.L. Sorbent coatings for solid-phase microextraction
based on mixtures of polymeric ionic liquids. *J. Sep. Sci.,* **2011**, *34*(3), 340-346.
[http://dx.doi.org/10.1002/jssc.201000367] [PMID: 21268258]

[281] Amini, R.; Rouhollahi, A.; Adibi, M.; Mehdinia, A. A novel reusable ionic liquid chemically bonded
fused-silica fiber for headspace solid-phase microextraction/gas chromatography-flame ionization
detection of methyl tert-butyl ether in a gasoline sample. *J. Chromatogr. A,* **2011**, *1218*(1), 130-136.
[http://dx.doi.org/10.1016/j.chroma.2010.10.114] [PMID: 21130999]

[282] Ebrahimi, M.; Es'haghi, Z.; Samadi, F.; Hosseini, M.S. Ionic liquid mediated sol-gel sorbents for
hollow fiber solid-phase microextraction of pesticide residues in water and hair samples. *J.
Chromatogr. A,* **2011**, *1218*(46), 8313-8321.
[http://dx.doi.org/10.1016/j.chroma.2011.09.058] [PMID: 21993517]

[283] Gao, Z.; Li, W.; Liu, B.; Liang, F.; He, H.; Yang, S.; Sun, C. Nano-structured polyaniline-ionic liquid
composite film coated steel wire for headspace solid-phase microextraction of organochlorine
pesticides in water. *J. Chromatogr. A,* **2011**, *1218*(37), 6285-6291.
[http://dx.doi.org/10.1016/j.chroma.2011.07.041] [PMID: 21821255]

[284]   Zhao, F.; Wang, M.; Ma, Y.; Zeng, B. Electrochemical preparation of polyaniline-ionic liquid based solid phase microextraction fiber and its application in the determination of benzene derivatives. *J. Chromatogr. A,* **2011**, *1218*(3), 387-391.
[http://dx.doi.org/10.1016/j.chroma.2010.12.017] [PMID: 21185028]

[285]   Zhou, X.; Xie, P.F.; Wang, J.; Zhang, B.B.; Liu, M.M.; Liu, H.L.; Feng, X.H. Preparation and characterization of novel crown ether functionalized ionic liquid-based solid-phase microextraction coatings by sol-gel technology. *J. Chromatogr. A,* **2011**, *1218*(23), 3571-3580.
[http://dx.doi.org/10.1016/j.chroma.2011.03.048] [PMID: 21531419]

[286]   Ebrahimi, M.; Es'haghi, Z.; Samadi, F.; Bamoharram, F.F.; Hosseini, M.S. Rational design of heteropolyacid-based nanosorbent for hollow fiber solid phase microextraction of organophosphorus residues in hair samples. *J. Chromatogr. A,* **2012**, *1225*, 37-44.
[http://dx.doi.org/10.1016/j.chroma.2011.12.077] [PMID: 22236566]

[287]   Ho, T.D.; Yu, H.; Cole, W.T.; Anderson, J.L. Ultraviolet photoinitiated on-fiber copolymerization of ionic liquid sorbent coatings for headspace and direct immersion solid-phase microextraction. *Anal. Chem.,* **2012**, *84*(21), 9520-9528.
[http://dx.doi.org/10.1021/ac302316c] [PMID: 22991947]

[288]   Zhang, Y.; Wang, X.; Lin, C.; Fang, G.; Wang, S. A novel SPME fiber chemically linked with 1-vinyl-3- hexadecylimidazolium hexafluorophosphate ionic liquid coupled with GC for the simultaneous determination of pyrethroids in vegetables. *Chromatogr.,* **2012**, *75*, 789-797.
[http://dx.doi.org/10.1007/s10337-012-2244-2]

[289]   Pang, L.; Liu, J.F. Development of a solid-phase microextraction fiber by chemical binding of polymeric ionic liquid on a silica coated stainless steel wire. *J. Chromatogr. A,* **2012**, *1230*, 8-14.
[http://dx.doi.org/10.1016/j.chroma.2012.01.052] [PMID: 22340892]

[290]   Feng, J.; Sun, M.; Xu, L.; Wang, S.; Liu, X.; Jiang, S. Novel double-confined polymeric ionic liquids as sorbents for solid-phase microextraction with enhanced stability and durability in high-ioni--strength solution. *J. Chromatogr. A,* **2012**, *1268*, 16-21.
[http://dx.doi.org/10.1016/j.chroma.2012.10.037] [PMID: 23127811]

[291]   Feng, J.; Sun, M.; Li, J.; Liu, X.; Jiang, S. A novel aromatically functional polymeric ionic liquid as sorbent material for solid-phase microextraction. *J. Chromatogr. A,* **2012**, *1227*, 54-59.
[http://dx.doi.org/10.1016/j.chroma.2012.01.010] [PMID: 22265783]

[292]   Zhou, X.; Shao, X.; Shu, J.J.; Liu, M.M.; Liu, H.L.; Feng, X.H.; Liu, F. Thermally stable ionic liquid-based sol-gel coating for ultrasonic extraction-solid-phase microextraction-gas chromatography determination of phthalate esters in agricultural plastic films. *Talanta,* **2012**, *89*, 129-135.
[http://dx.doi.org/10.1016/j.talanta.2011.12.001] [PMID: 22284470]

[293]   Shu, J.; Li, C.; Liu, M.; Feng, X.; Tan, W.; Liu, F. Role of counter anions in sol-gel-derived alkoxyl functionalized ionic-liquid-based organic-inorganic hybrid coatings for SPME. *Chromatogr.,* **2012**, *75*, 1421-1433.
[http://dx.doi.org/10.1007/s10337-012-2323-4]

[294]   Ho, T.T.; Chen, C.Y.; Li, Z.G.; Yang, T.C.C.; Lee, M.R. Determination of chlorophenols in landfill leachate using headspace sampling with ionic liquid-coated solid-phase microextraction fibers combined with gas chromatography-mass spectrometry. *Anal. Chim. Acta,* **2012**, *712*, 72-77.
[http://dx.doi.org/10.1016/j.aca.2011.11.025] [PMID: 22177067]

[295]   Sarafraz-Yazdi, A.; Vatani, H. A solid phase microextraction coating based on ionic liquid sol-gel technique for determination of benzene, toluene, ethylbenzene and o-xylene in water samples using gas chromatography flame ionization detector. *J. Chromatogr. A,* **2013**, *1300*, 104-111.
[http://dx.doi.org/10.1016/j.chroma.2013.03.039] [PMID: 23582769]

[296]   Xu, L.; Jia, J.; Feng, J.; Liu, J.; Jiang, S. Polymeric ionic liquid modified stainless steel wire as a novel fiber for solid-phase microextraction. *J. Sep. Sci.,* **2013**, *36*(2), 369-375.
[http://dx.doi.org/10.1002/jssc.201200644] [PMID: 23335459]

[297] Gao, Z.; Deng, Y.; Hu, X.; Yang, S.; Sun, C.; He, H. Determination of organophosphate esters in water samples using an ionic liquid-based sol-gel fiber for headspace solid-phase microextraction coupled to gas chromatography-flame photometric detector. *J. Chromatogr. A,* **2013**, *1300*, 141-150.
[http://dx.doi.org/10.1016/j.chroma.2013.02.089] [PMID: 23541656]

[298] Ge, D.; Lee, H.K. Ionic liquid based dispersive liquid-liquid microextraction coupled with micro-solid phase extraction of antidepressant drugs from environmental water samples. *J. Chromatogr. A,* **2013**, *1317*, 217-222.
[http://dx.doi.org/10.1016/j.chroma.2013.04.014] [PMID: 23639124]

[299] Joshi, M.D.; Ho, T.D.; Cole, W.T.S.; Anderson, J.L. Determination of polychlorinated biphenyls in ocean water and bovine milk using crosslinked polymeric ionic liquid sorbent coatings by solid-phase microextraction. *Talanta,* **2014**, *118*, 172-179.
[http://dx.doi.org/10.1016/j.talanta.2013.10.014] [PMID: 24274285]

[300] Vatani, H.; Yazdi, A.S. Preparation of an ionic liquid–mediated carbon nanotube–poly(dimethylsiloxane) fiber by sol-gel technique for determination of polycyclic aromatic hydrocarbons in urine samples using headspace solid–phase microextraction coupled with gas chromatography. *J. Iran Chem. Soc.,* **2014**, *11*, 969-977.
[http://dx.doi.org/10.1007/s13738-013-0363-9]

[301] Shu, J.; Xie, P.; Lin, D.; Chen, R.; Wang, J.; Zhang, B.; Liu, M.; Liu, H.; Liu, F. Two highly stable and selective solid phase microextraction fibers coated with crown ether functionalized ionic liquids by different sol-gel reaction approaches. *Anal. Chim. Acta,* **2014**, *806*, 152-164.
[http://dx.doi.org/10.1016/j.aca.2013.11.006] [PMID: 24331051]

[302] Khan, S.; Kazi, T.G.; Soylak, M. Rapid ionic liquid-based ultrasound assisted dual magnetic microextraction to preconcentrate and separate cadmium-4-(2-thiazolylazo)-resorcinol complex from environmental and biological samples. *Spectrochim. Acta A Mol. Biomol. Spectrosc.,* **2014**, *123*, 194-199.
[http://dx.doi.org/10.1016/j.saa.2013.12.065] [PMID: 24398463]

[303] Wu, L.; Song, Y.; Hu, M.; Xu, X.; Zhang, H.; Yu, A.; Ma, Q.; Wang, Z. Determination of sulfonamides in butter samples by ionic liquid magnetic bar liquid-phase microextraction high-performance liquid chromatography. *Anal. Bioanal. Chem.,* **2015**, *407*(2), 569-580.
[http://dx.doi.org/10.1007/s00216-014-8288-4] [PMID: 25384336]

[304] Wang, Y.; Sun, Y.; Xu, B.; Li, X.; Jin, R.; Zhang, H.; Song, D. Magnetic ionic liquid-based dispersive liquid-liquid microextraction for the determination of triazine herbicides in vegetable oils by liquid chromatography. *J. Chromatogr. A,* **2014**, *1373*, 9-16.
[http://dx.doi.org/10.1016/j.chroma.2014.11.009] [PMID: 25464995]

[305] Wang, T.T.; Chen, Y.H.; Ma, J.F.; Hu, M.J.; Li, Y.; Fang, J.H.; Gao, H.Q. A novel ionic liquid-modified organic-polymer monolith as the sorbent for in-tube solid-phase microextraction of acidic food additives. *Anal. Bioanal. Chem.,* **2014**, *406*(20), 4955-4963.
[http://dx.doi.org/10.1007/s00216-014-7923-4] [PMID: 24939131]

[306] Feng, J.; Sun, M.; Li, L.; Wang, X.; Duan, H.; Luo, C. Multiwalled carbon nanotubes-doped polymeric ionic liquids coating for multiple headspace solid-phase microextraction. *Talanta,* **2014**, *123*, 18-24.
[http://dx.doi.org/10.1016/j.talanta.2014.01.030] [PMID: 24725859]

[307] Naeemullah, K.; Kazi, T.G.; Tuzen, M. Magnetic stirrer induced dispersive ionic-liquid microextraction for the determination of vanadium in water and food samples prior to graphite furnace atomic absorption spectrometry. *Food Chem.,* **2015**, *172*, 161-165.
[http://dx.doi.org/10.1016/j.foodchem.2014.09.053] [PMID: 25442538]

[308] Soylak, M.; Yilmaz, E. Determination of cadmium in fruit and vegetables by ionic liquid magnetic microextraction and flame atomic absorption spectrometry. *Anal. Lett.,* **2015**, *48*, 464-476.
[http://dx.doi.org/10.1080/00032719.2014.949732]

[309] Nacham, O.; Clark, K.D.; Anderson, J.L. Analysis of bacterial plasmid DNA by solid-phase

microextraction. *Anal. Methods,* **2015**, *7*, 7202-7207.
[http://dx.doi.org/10.1039/C5AY00532A]

[310] Cagliero, C.; Ho, T.D.; Zhang, C.; Bicchi, C.; Anderson, J.L. Determination of acrylamide in brewed coffee and coffee powder using polymeric ionic liquid-based sorbent coatings in solid-phase microextraction coupled to gas chromatography-mass spectrometry. *J. Chromatogr. A,* **2016**, *1449*, 2-7.
[http://dx.doi.org/10.1016/j.chroma.2016.04.034] [PMID: 27157428]

[311] Feng, J.; Sun, M.; Bu, Y.; Luo, C. Hollow fiber membrane-coated functionalized polymeric ionic liquid capsules for direct analysis of estrogens in milk samples. *Anal. Bioanal. Chem.,* **2016**, *408*(6), 1679-1685.
[http://dx.doi.org/10.1007/s00216-015-9279-9] [PMID: 26753984]

[312] Liao, K.; Mei, M.; Li, H.; Huang, X.; Wu, C. Multiple monolithic fiber solid-phase microextraction based on a polymeric ionic liquid with high-performance liquid chromatography for the determination of steroid sex hormones in water and urine. *J. Sep. Sci.,* **2016**, *39*(3), 566-575.
[http://dx.doi.org/10.1002/jssc.201501156] [PMID: 26608868]

[313] Nacham, O.; Clark, K.D.; Anderson, J.L. Extraction and purification of DNA from complex biological sample matrices using solid phase microextraction coupled with real-time PCR. *Anal. Chem.,* **2016**, *88*(15), 7813-7820.
[http://dx.doi.org/10.1021/acs.analchem.6b01861] [PMID: 27373463]

[314] Pei, M.; Zhang, Z.; Huang, X.; Wu, Y. Fabrication of a polymeric ionic liquid-based adsorbent for multiple monolithic fiber solid-phase microextraction of endocrine disrupting chemicals in complicated samples. *Talanta,* **2017**, *165*, 152-160.
[http://dx.doi.org/10.1016/j.talanta.2016.12.043] [PMID: 28153235]

[315] Nacham, O.; Clark, K.D.; Varona, M.; Anderson, J.L. Selective and Efficient RNA Analysis by Solid-Phase Microextraction. *Anal. Chem.,* **2017**, *89*(20), 10661-10666.
[http://dx.doi.org/10.1021/acs.analchem.7b02733] [PMID: 28872298]

[316] Pelit, F.O.; Pelit, L.; Dizdaş, T.N.; Aftafa, C.; Ertaş, H.; Yalçınkaya, E.E.; Türkmen, H.; Ertaş, F.N. A novel polythiophene - ionic liquid modified clay composite solid phase microextraction fiber: Preparation, characterization and application to pesticide analysis. *Anal. Chim. Acta,* **2015**, *859*, 37-45.
[http://dx.doi.org/10.1016/j.aca.2014.12.043] [PMID: 25622604]

134   *Recent Advances in Analytical Techniques*, 2019, *Vol. 3*, 134-178

CHAPTER 4

Electrospun Nanofibers: Functional and Attractive Materials for the Sensing and Separation Approaches in Analytical Chemistry

Betul Unal, Fatma Ozturk Kirbay, Irem Yezer, Dilek Odaci Demirkol* and Suna Timur

Faculty of Science, Biochemistry Department, Ege University, 35100 Bornova-Izmir, Turkey

Abstract: Electrospinning was patented in 1902. This technology is attracting considerable research attention as a result of increasing demand to nanoscience and nanotechnology. Nanofiber structures (from 10 nm to 100 nm or bigger) have been obtained from synthetic and natural polymers *via* electrospinning techniques. The resulted electrospun nanofibers are used in various areas such as public health, biotechnology, environmental engineering, textile, defense industry, and energy deposition. Among them, analytical applications have a great potential as a sorbent for chromatographic separations and a modifer of surfaces in electrochemistry or spectroscopy. In this chapter, electrospun nanofibers based separation and detection systems will be explained in detail.

Keywords: Affinity membrane, Biosensor, Chromatographic sorbent bed, Electrospinning, Electrospun nanofiber, Filtration membranes, Immobilization matrix, Nanofiber, Nanomaterial, Nanotechnology.

INTRODUCTION TO ELECTROSPUN NANOFIBERS

Electrospinning is an ancient and interesting technique, first observed by Rayleigh in 1897 and first patented by John Francis Cooley in 1900s and then, Antonin Formhals patented electrospinning of cellulose acetate in 1934 [1]. This technique is a widely-used fabrication technique for producing nanofibers. A variety of materials such as natural and synthetic polymer, polymer alloys, ceramics, composites, chromophores, as well as active metallic agents can be used to prepare nanofibers [2 - 9]. The fundamental requirements of an electrospinning set-up consist of a needle, a high voltage power supply (HVPS), and a collector (Fig. **1**).

---

* **Corresponding author Dilek Odaci Demirkol:** Faculty of Science, Biochemistry Department, Ege University, 35100 Bornova-Izmir, Turkey; Tel: +90 232 311 1711; Fax: +90 232 311 5485;
E-mails: dilek.odaci.demirkol@ege.edu.tr and dilek.demirkol@yahoo.com

Atta-ur-Rahman & Sibel A. Ozkan (Eds.)

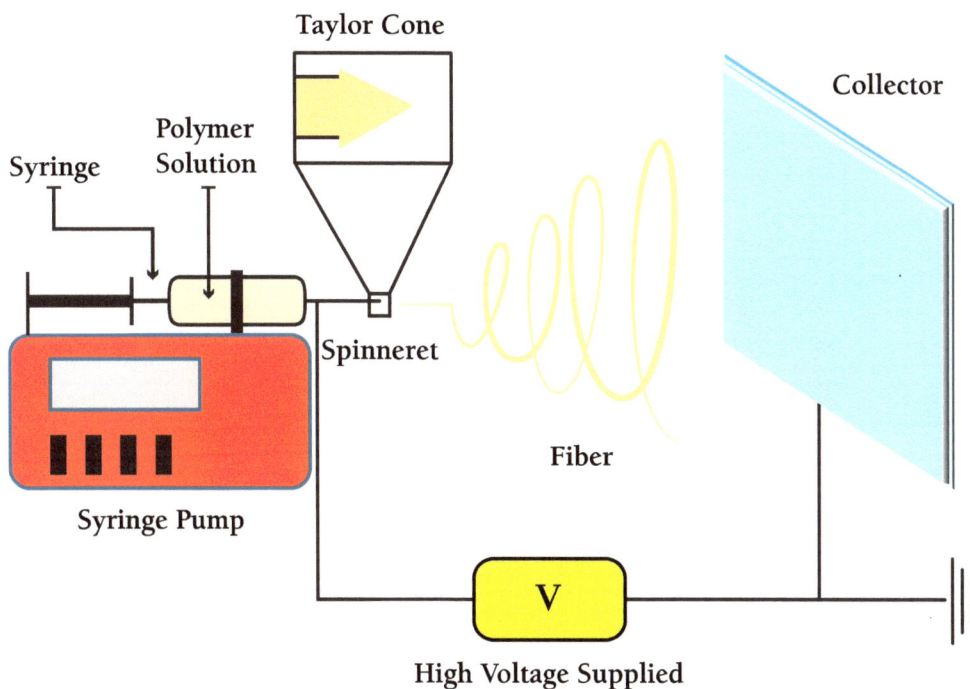

Fig. (1). Schematic of horizontal electrospinning system.

The electrospinning process can be explained as an applied electric field between a tip nozzle and the collector. During electrospinning, a polymer solution or a polymer melt is loaded into the syringe tipped with a needle. A high voltage is turned onto the charged surface of a polymer solution at the tip of the needle. The polymer solution is sprayed on the collector which is controlled by a syringe pump system. The surface tension of the feedstock solution is overcome by the force of the electric field, and fiber is formed in a conical shape called the Taylor cone [10]; on the surface of the pendant drop. The nanofiber extends and is deposited on the collector. Most of the polymer solution evaporates from the nanofiber film line and solidifies the nanofibers before it is collected on the metal screen and thus is ready for the next processing. The diameter and morphology of generated nanofibers depend on four main parameters: the solution properties, the processing, the operating parameters, and the equipment design [11 - 15]. Table **1** shows the main parameters that affect nanofiber morphology during the electrospinning process.

Table 1. Main parameters, which effect electrospinning process.

		Ref
	Solvent Select	[16]
	Molecular weight of Polymer	[11]
	Polymer concentration	[2, 11]
Material Parameters	Conductivity	[2, 11]
	Viscosity	[11]
	Dielectric constant	[16]
	Dipole moment	[17]
	Surface tension	[2]
Processing Parameters	Applied voltage	[11]
	Distance between needle and collector	[18]
	Flow rate	[11]
Operating Parameters	Temperature	[2]
	Humidity	[2]
	Atmospheric pressure	[11]
	Air velocity	[19]
Equipment Design	Needle design	[18]
	Collector geometry	[16]

The Effect of Material Parameters on the Electrospinning Process

Solution Properties

In the process, the first step is to choose an appropriate solvent which can be used for dissolving of a proper polymer and converting it into fiber structures through electrospinning. The solution parameters can be explained as solvent type, solution viscosity, surface tension, dipole moment, dielectric constant, and conductivity. These parameters are usually related to each other. Table **2** shows some of the common solvents and their features.

Table 2. Common solvents in electrospinning process and their rheological properties.

Solvent	Boiling Point (°C)	Density (g.cm^{-3})	Dipole Moment (D)	Dielectric Constant	Surface Tension (mN^2.m^2)	Viscosity (cP at 25°C)	Ref
Acetic acid	118.0	1.05	1.68	6.15	26.9	1.12	[20, 21]
Acetone	78.0	1.39	2.88	27.0	21.4	0.32	[20]

(Table 2) contd.....

Solvent	Boiling Point (°C)	Density (g.cm^{-3})	Dipole Moment (D)	Dielectric Constant	Surface Tension (mN^2.m^2)	Viscosity (cP at 25°C)	Ref
Chloroform	61.6	1.50	1.15	4.8	26.5	0.53	[20, 21]
Carbon disulfide	46.2	1.263	0.0	2.6	35.3	0.36	[22]
Cyclohexane	81	0.778	0.0	2.02	24.65	0.98	[22]
"Dichloromethane	40.0	1.33	1.60	8.93	28.1	0.41	[20]
Dimethylacetamide	165.0	0.94	3.72	37.8	36.7	1.96	[20]
Dimethylformamide	153.0	0.99	36.70	38.3	37.1	0.80	[20, 21]
Dimethylsulfoxide	189.0	1.10	3.90	46.7	43.0	2.00	[20]
Ethylene glycol	197.0	1.11	2.20	37.7	47.0	16.13	[20]
Formamide	211.0	1.13	3.37	110.0	59.1	3.30	[20]
Formic acid	101.0	1.22	1.41	57.9	37.6	1.57	[20]
Glycerol	290.0	1.26	2.62	42.5	64.0	950	[20]
Hexafluoro isopropanol	58.2	1.60	1.85	16.7	16.1	1.02	[20, 21]
Methanol	65.0	0.79	1.70	33.0	22.7	0.54	[20]
Tetrahydrofurane	66.0	0.89	1.75	7.52	26.4	0.46	[20, 21]
Toluene	111.0	0.87	0.36	2.4	28.52	0.56	[22]
Trifluoroacetic acid	72.4	1.489	2.28	8.55	13,63	0.93	[?1]
Triflouroethanol	74.0	1.38	2.52	8.55	43.3	1.24	[20]
Water	100.0	1.0	-	21.0	25.2	1.0	[20, 21]

Two major criteria in solvent selection are the solubility of the polymer and its volatility. Solvent volatility affects fiber porosity as well as morphology. In the literature, the nanofibers of polystyrene (PS) in 100% tetrahydrofuran (THF) (more volatile) are porous while the same polymer in 100% dimethylformamide (DMF) exhibits a smooth morphology [22]. Although volatile solvents are preferred for the nanofiber formation, they have disadvantages such as their low boiling point, which causes clogging at the tip of the needle [23, 24].

Polymer Properties

Electrospun nanofibers have been formed from various synthetic, natural, and melt polymers and blends including nanocomposite [25], nucleic acid [26], proteins [27], and polysaccharides [28]. Long ultrafine fibers have been obtained by using over 200 polymer solutions for a wide range of applications such as tissue engineering, drug delivery, filtration membranes, biosensor systems, biomedical materials, *etc* [29, 30]. The concentration of the polymer solution affects on the viscosity and the surface tension which influences spinnability. Moreover, using lower polymer concentrations causes micro/nanoparticle formation instead of electrospun nanofiber [14]. Fig. (**2**) shows the effect of polymer concentration on nanofiber morphology. When concentration-dependent viscosity is slightly higher, the bead structure changes from spherical to spindle-like nanofibers. Additionally, if the polymer concentration is selected properly, resulted fibers are obtained with a smooth morphology [31 - 33]. On the other hand, if the higher polymer concentration is used, the fibers are observed in microscale size [34]. The molecular weight of polymer, similar to viscosity, is the key parameter for the morphology of the fibers. If the molecular weight is too high, the fiber diameters increase.

Increasing Viscosity

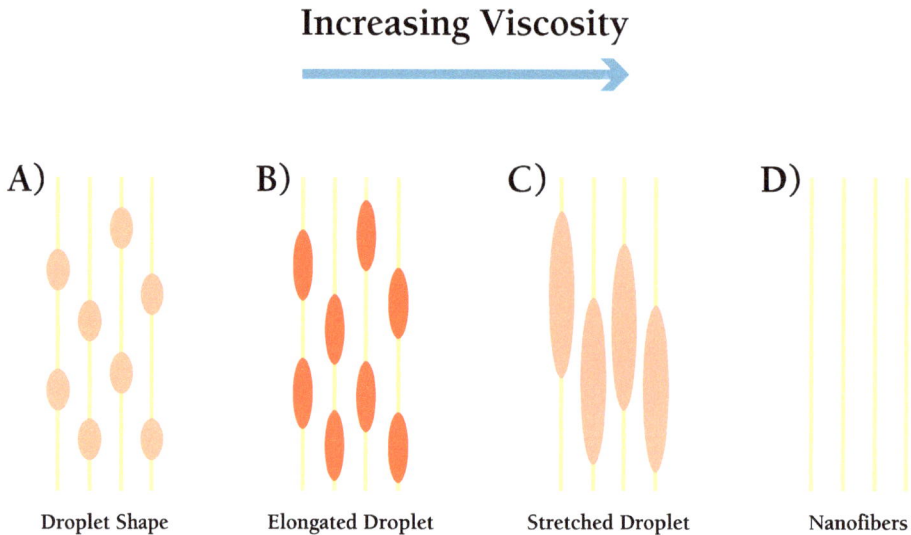

Fig. (2). Shape of beads as a result of viscosity.

Processing Parameters

The Effect of Distance between the Needle and the Collector

The distance between the tip of needle and collector also affects the fiber diameter and morphology [35]. If this distance is too short, polymer fiber reaches the collector without drying and leads to the formation of thicker fibers. In addition, when the distance between them is too long, beaded fibers are occurred [36, 37]. The solvent properties used in nanofiber formation should be well-controlled during the electrospinning process because the solvent evaporation rate from the fibers is very important in adjusting the correct distance between the needle and collector [38].

Applied Voltage

Generally, in the process, electrical field is applied at 5.0-40 kV to the needle in laboratory conditions [39]. The voltage of the electrical field is a critical factor because of the occurring of the Taylor cone so that enough voltage should be applied to the solution to overcome surface tension. The applied voltage affects the formation of beads and beaded nanofibers as well as the diameter of the fibers. This is related to the size of the Taylor cone [40]. For instance, Ren *et al.* produced poly (dimethyl siloxane)/poly (methyl methacrylate) (PDMS/PMMA) electrospun membranes by applying various voltage values [41]. That study showed that the increased voltage leads to smaller fiber diameter, bead formation, and larger pore size. Hence, the applied voltage effects on fiber diameters and morphologies. Clearly, polymers, their concentrations and the distance between needle and collector have a significant impact on the fibers' characteristics.

Flow Rate

Flow rate is regarded as another significant parameter in fiber diameter and its distribution, initiating droplet shape, trajectory of the jet, maintenance of the Taylor cone, and deposition area [42]. When applied flow rate is too high, the nanofibers are unable to dry completely prior to reaching the deposition area and consequently, higher bead sizes can be occurred [43]. Also, higher flow rate has a negative effect on the fibers' quality. It causes the beaded fiber formation and increment in fiber diameters [15]. On the other hand, application of a lower flow rate leads to clogging (Fig. **3D**) at the tip of the needle due to rapid evaporation of the solution.

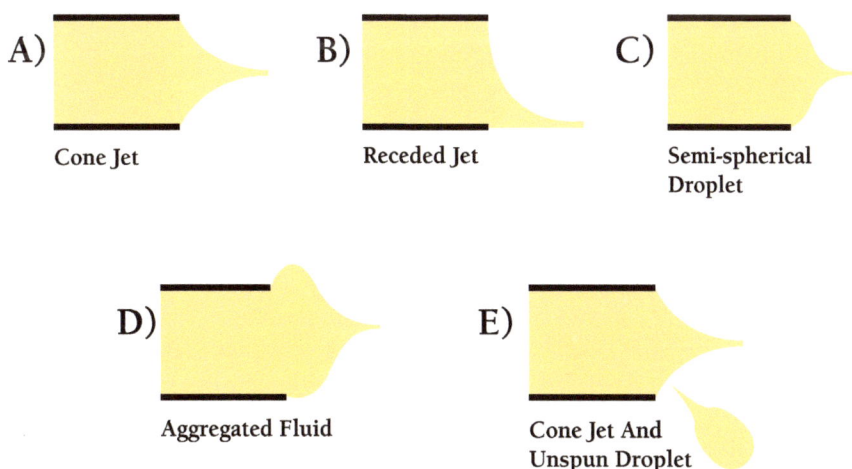

Fig. (3). Schematic illustration of various formation jet modes in electrospinning process.

Operating Parameters

Operating parameters such as humidity, temperature, air velocity and atmospheric pressure also affect on fiber diameter as well as morphology. In a recent work, Pelipenkol *et al.* [44] examined the effect of relative humidity (RH) on fiber morphology and mechanical properties. At a lower RH, thicker fibers are formed because of rapid evaporation of the polymer solvent under these conditions, but applying of higher RH values lead to slow evaporation of the solvent, resulting in formation of thinner nanofibers. A higher temperature causes an increase in fiber diameter because of rapid solvent evaporation. Moreover, the operation parameters are critical in giving enough time for nanofiber elongation before the solidification of the polymer melt during the electrospinning process [45].

Equipment Design

Needle Design and Collector Design

The proper design of needle is the most critical parameter of fiber morphology. A needle with a higher diameter causes increment in the volume of released solution to the target surface, hence, nanofibers with enlarged fiber diameters are produced [46]. When a thinner needle is used, the solution containing a polymer with other ingredients (nanoparticles, biological materials, *etc.*) causes clogging of the nozzle. Various types of needles can be utilized to produce the core-sheath and hollow fibers [15, 47]. Different fiber patterns can be produced by variation of collectors such as a flat plate, parallel plates/strips, a rotating drum, a rotating disk (flat edge), a rotating disk (sharp edge), and a conveyor [48]. Randomly oriented fibers are obtained by a stationary collector which is particularly preferred in most laboratories. However, the speed of the rotating drums provides distinct mechanical properties to the fibers which result in alignment fibers and a smaller fiber diameter [49].

ELECTROSPUN NANOFIBERS IN DETECTION

Sensors and Biosensors

Owing to the requirements in detection of various compounds from metals to mammalian cells in the fields of medical diagnosis, the food industry, bioprocess monitoring, and environmental screening with highly sensitive and selective detection systems have a great attention. Numerous smart architectures and nanomaterials such as quantum dots, gold nanoparticles, dendrimers, magnetic nanoparticles, and nanowires, *etc.* have been used to prepare a novel analysis systems [50 - 55]. Among them, nanofibers have been applied in a wide range due to their large surface area, availability of controlling diameter and thickness, flexibility, and porous structure [56 - 59]. Both chemical and bio-chemical sensing systems have been successfully applied for analysis of wide range of chemicals, biomolecules *etc*.

The applications of electrospun nanofibers as sensing component in sensors and biosensors design are common for a wide range of samples in various areas. Some examples of sensor systems that are designed using electrospun nanofibers are summarized in Table **3**.

Table 3. Electrospun nanofiber based non-enzymatic biosensors.

Polymer	Solution	Nanofiber Content	Analyte	Detection Principle	Linear Range	Ref
PEI	Water	AuNPs/PPyNFs	Glc	EC	0.2-13 mM	[60]
PAN/ NiAA	DMF	NiCF nanocomposite	Glc	AMP	0.002-2.5 mM	[61]
PMMA/ P3HT	CF	PMMA-P3HT/ AuNPs/FTO	-	Optical and EC	-	[62]
PVP	DMF	ELP-OPH/BSA/ TiO_2NFs/AuNPs	OPs	EC	0.1-5.4 µM	[63]
PA6/ Chit	FA	PA6-Chit/ AuNPs/FTO	Cd^{+2}	EC	25-75 µg/L	[64]
PSMA	DMF/ DCM	PSMA-PhB+TPE/Pro	Hpr/ Try	RM	0.4-0.8 U/mL (Hpr) and up to 8 µg/mL(Try)	[65]
PAN	DMF	PAN-TPDI/ FNFM	Hg^{+2}	FCM	0.001-1.0 mM	[66]
PVDF-HFP	DMF	PVDF-HFP/ Ni-CoNPs	Urea	EC	0.020-2.0 mM	[67]
PCL	HFIP/ FA/AA	RGO/PDDA/ PCL-PPy/GCE	Dopamine	EC	0.004-0.69 mM	[68]
PVP	DMF/ EtOH	In_2O_3/PANI composite	Ammonia	EC	0.1-1.0 M	[69]
PVP	DMF	RGO/α-Fe_2O_3NFs	AC	EC	8.9-100 mg/L	[70]
PPy/ PAN	DMF	NCNPFs/GCE	H_2O_2	AMP	0.005-27 mM	[71]
PAN	DMF	NCNFs-GCE	HQ CC RC	EC	1 - 400 µM (HQ) 1 - 400 µM (CC) 2 - 500 µM (RC)	[72]
PVP	HTAB	RuO_2NWs-TiO_2NFs/GCE	ASC	AMP	0.01-1.5 mM	[73]
PVP/ PMMA	DMF	ZnONFs/GCE	Cd^{+2}	EC	$4.8*10^{-3}$-1.3 µM	[74]

(Table 3) contd.....

Polymer	Solution	Nanofiber Content	Analyte	Detection Principle	Linear Range	Ref
PAA/ PVA	Water	PAA/PVA HydrogelNF-LAPS	*E. coli*	PTM	10^3-10^7 CFU/ml	[75]
PVP	AA/ EtOH	TiO$_2$NFs/ GONs/SPCE	Adenine	CAMP	0.1-1.0 μM	[76]
PVP	DMF/ AA	TiO$_2$/Fe$_2$O$_3$/PPy	H$_2$O$_2$	CLR	2.0-50 μM	[77]
PEI/ PVA	Water	AgNPs/PEI/PVA	pH	SP	2.56-11.20	[78]
PAN	DMF	CuO/NCNFs/GCE	Glc	AMP	0.25-2.0 mM	[79]
PSU	DMAc	AuNC-FNFM	H$_2$O$_2$	CLR	-	[80]
N6-DMG/ PVA	FA/ Water	N6-DMG/ Glass slide	Ni^{+2}	CLR	5.0-100 μg/mL	[81]
PVP	AA/ EtOH	TiO$_2$-AuHNF/FTO	GSH	PEC	0.022-0.7 mM	[82]
PAN	DMF	QDs/mesoSiO$_2$/ PAN NFM	Fe^{+3}	Fluorescent	0.0-99.1 μM	[83]
CA	AC/ DMAc	CC-CA NFs	Pb^{+2}	CLR	0.010-1.0 mM	[84]
PANI	HCL/ Water	GCE/PANI-NiO	Ph	EC	7.6-14 pM	[85]
PVP	EtOH/ DMF	PdNPs-CoFe$_2$O$_4$NTs	H$_2$O$_2$	CLR	0.01-0.1 mM	[86]
PAN	DMF	CNF-PtNPs/GCE	H$_2$O$_2$	AMP	0.01-9.38 mM and 9.38-74.38 mM	[87]
PAN	DMF	PdCo-CNF/CPE	H$_2$O$_2$ and Nitrite	EC	0.2 μM-23.5 mM (H$_2$O$_2$); 0.4-30 μM and 30-400 μM (nitrite)	[88]
PMMA	DCM	PMAH/CNTs-NFs/GCE	FM	EC	0.001-10 mM	[89]
PVP	DMF	LSMCO/CPE	H$_2$O$_2$	AMP	0.5-1000 μM	[90]

EC: Electrochemical; PEC: Photoelectrochemical; FCM: Fluorescent chemosensor; AMP: Amperometric; PTM: Potentiometric; RM: Ratiometric; SP: Spectroscopic; CAMP: Chronoamperometric; CLR: Colorimetric; GCE: Glassy carbon electrode; FTO: Fluorine-doped tin oxide glass; RGO: Reduced graphene oxide; SPCE: Screen printed carbon electrode; CPE: Carbon paste electrode; PEI: Poly (ethylene imine); PAN: Poly (acrylonitrile); PCL: Poly (caprolactone); PMMA: Poly (methyl methacrylate); PMAH: Poly (methacryloylhydrazide); PVP: Poly (vinyl pyrrolidone); PAA: Poly (acrylic acid); PA6: Polyamide 6; PVA: Poly (vinyl alcohol); PSU: Polysulfone; PANI: Polyaniline; PVDF-HFP: Poly (vinylidene) fluoride-*co*-hexafluoropropylene; PSMA: Polystyrene-*co*-maleic anhydride; P3HT: Poly (3-hexyl thiophene-2,5-diyl); PDDA: Poly (dimethyl diallyl ammonium chloride); PPy: Polypyrrole; N6: Poly (caprolactam); CA: Cellulose acetate; DMF: Dimethylformamide; DMG: Dimethylglyoxime; DCM: Dichloromethane; HFIP: Hexafluoroisopropanol; DMAc: Dimethylacetylamide; FA: Formic acid; AA: Acetic acid; EtOH: Ethyl alcohol; AC: Acetone; FM: Formaldehyde; HCL: Hydrochloric acid; CF: Chloroform; HTAB: Hexadecyltrimethyl ammonium bromide; AuNC: Gold nanoclusters; AuNPs: Gold nanoparticles; Ni-CoNPs: Nickel-cobalt nanoparticles; AgNPs: Silver nanoparticles; PtNPs: Platinum nanoparticles; PPyNFs: Polypyrrole nanofibers; NCNFs: Nitrogen-doped carbon nanofibers; NCNPFs: Nitrogen-doped carbon nanoparticles-embedded carbon nanofibers; NiCF: Nickel nanoparticle-loaded carbon nanofiber; TiO$_2$NFs: Titanium dioxide nanofibers; TiO$_2$-AuHNF: Titanium dioxide-gold hybrid nanofibers; PdCo-CNF: Palladium-Cobalt carbon nanofibers; In$_2$O$_3$/PANI: Indium oxide/polyaniline nanofibers; ZnONFs: Zinc oxide nanofibers; PANI-NiO: Nickel oxide-doped polyaniline nanofibers; FNFM: Fluorescent nanofibrous membrane; RuO$_2$NWs: Ruthenium dioxide nanowires; GONs: Graphene oxide nanosheets; PdNPs: Palladium nanoparticles; CNTs: Carbon nanotubes; CoFe$_2$O$_4$NTs: Cobalt ferrite nanotubes; NiAA: Ni-acetylacetonate; ELP-OPH: Elastin-like polypeptide-organophosphate hydrolase; OPs: Organophosphate pesticides; TiO$_2$: Titanium dioxide; Fe$_2$O$_3$: Iron (III) oxide; α-Fe$_2$O$_3$: Hematite; Chit: Chitosan; PhB: Phloxine B; TPE: Tetraphenylethene; Pro: Protamine; DTPDI: Dithioacetal-modified perylenediimide; H$_2$O$_2$: Hydrogen peroxide; CuO: Copper oxide; BSA: Bovine serum albumin; Glc: Glucose; Try: Trypsin; Hpr: Heparin; Ph: Phenanthrene; GSH: Glutathione; ASC: Ascorbic acid; HQ: Hydroquinone; CC: Catechol; RC: Resorcinol; QDs: Carbon quantum dots; mesoSiO$_2$: Mesoporous silica; CC: Curcumin; LSMCO: La$_{0.7}$Sr$_{0.3}$Mn$_{0.75}$Co$_{0.25}$O$_3$ nanofiber; LAPS: Light addressable potentiometric sensor.

Biosensors consist of biological molecules and transducers which convert a biological signal to an analytical response. The obtained signal is proportional to the concentration of analyte or group of analytes. The biological part of a biosensor can be biomaterials whether it has catalytically features or not. Members of a catalytically active group are enzymes, microorganisms, and tissues, non-catalytics are antibodies, receptors, and nucleic acids. The transducer, a part of the biosensors, consists of electrochemical (amperometric, potentiometric, conductometric), optical, piezoelectric and magnetic systems.

Recently, various types of biosensors have been developed and used to analyze species in clinical, medical and environmental applications. In parallel to microsensor technology and biotechnology areas, research and development in biosensor technology has increased. The most important advantage of biosensors is their specificity to the compounds that are analyzed [91]. When traditional analytical techniques are used, generally pre-processing of the sample is required. It is known that the sensitivity of traditional techniques is high, whereas miniaturization of the apparatus is difficult. Electronic or electrochemical detection has an advantage when compared to other technologies due to the ease of miniaturization [92]. Biosensors are suitable for real-time analysis; thus, they can be applied to the analysis of various targets in different areas such as the pharmaceutics industry, medical diagnostics, the food and beverage industry, environmental screening, and process monitoring.

Enzyme Sensors

Enzyme-modified electrodes are the main parts of amperometric biosensors as well as biofuel cells. Enzyme electrodes are miniaturized chemical transducers which combine an electrochemical process with the catalytic activity of an immobilized enzyme [93]. Enzyme sensors are highly selective to substrate molecules because of the presence of enzymes as a biological element. Thus, the selectivity of the enzymes provides advantages to enzyme sensors. Typically, the kinetic aspect of the enzymatic reaction is based on the velocity of the product formation or the consumption of the substrate. If the substrate or the products are electrochemically active, these species can be followed using amperometric detection directly. A huge number of electrochemical biosensors are based on oxidoreductase enzymes, which take or produce electrons during the catalytic reaction. These electrochemical reactions are realized on the surfaces of solid electrodes and used to generate current. The produced electrons are transferred to the anode. Here, they are transported to high electron acceptors such as oxygen.

Enzymatic biosensors are divided into three main groups according to their oxidation mechanism. In first generation biosensors, electrons, which are formed by enzymatic oxidation, are transferred to molecular oxygen and the consumption of oxygen or produced hydrogen peroxide is measured. In second generation biosensors, artificial molecules such as ferrocene, osmium polymers, and hexacyanoferrate are used to transport electrons to the electrode for the generation of an enzyme active site. They are also known as mediated biosensors because of the presence of mediator molecules in the reaction mechanism. In third generation biosensors, electrons are transported from enzyme to electrode directly without any molecules such as oxygen or another mediator. In other words, direct electron transfer (DET) occurs.

The main parts of enzyme biosensors are enzymes and electrodes. The most important parameters in the choice of the electrode material are electrical conductivity and the rigidity of the material. Thus, gold (foil and rod) or carbon (paper, rod, paste, metal doped carbon, glassy carbon, carbon fiber, nanotube film) electrodes are generally preferred. Enzymes, as a biorecognition element are chosen according to reaction type of the analyte. Redox enzymes (oxidoreductases) are the most frequently used enzymes. Moreover, lipases and esterases are combined with oxidoreductases to prepare bi-enzymatic biosensors. On the other hand, the interaction between the biological system and the transducer is then most important parameter for the preparation of biosensors. Therefore, immobilization technology has a critical role. Immobilization matrices provide a connection between the biomaterial and transducer. Additionally, they assist reproducible and repeatable biosensor design with higher stabilization. Biomaterials can be immobilized on to the transducer surface *via* physical adsorption, covalent binding, cross linking and entrapment techniques. For the immobilization of biological material on the electrode surface, conducting polymers, functionalized polymers, clays, calixarenes, composite materials, sol-gel, and natural polymers such as gelatin, chitosan and nanomaterials can be used.

Among these materials, electrospun nanofibers are a useful alternative as an immobilization matrix. Nanofibers are used in public health, environmental engineering, textile, defense and security, and energy deposition and production. These materials can be produced on the scale of 10-100 nm or larger by the electrospinning process. The potential of these nanofibers is promising for human health, for example, tissue/organ regeneration, a vector to transport drugs and other therapeutics, biocompatible and biodegradable medical implants, medical diagnostics, fabrics against infection agents in hospitals, and even cosmetics and dental applications [94 - 96]. The uses of these matrices in biomedical applications has a great advantage because of mechanical strength and ease of modification of nanofibers in, for instance, surface morphology and fiber orientation [97]. Recently, nanofibers have been a novel tool for forming biopolymer sheets for tissue engineering [98]. The basic principle of tissue engineering is the development of *in vitro* cell culture systems like *in vivo* cell functions. Biomimetic matrices, which were obtained using electrospinning, provide cell attachment, cell proliferation, and cell differentiation. Collagen and elastin are two key proteins in the extracellular matrix of a lot of tissues and natural materials for electrospinning [98]. Furthermore, DNA, silk fibroin, fibrinogens, dextran, and some viruses are used to prepare electrospun nanofibers. However, most of the natural polymers have a low mechanical stability, and their hydrophilicity and viscosity is limited to producing electrospun nanofibers. Synthetic polymers are optimized more easily than natural polymers, but their features are not sufficient for use in biological applications, and they do not provide biological conditions. Thus, multifunctional electrospun nanofibers are prepared using one or more polymers in the production. After forming fibers, they can be subjected to additional processes.

Surface modification after the forming of electrospun nanofibers is one of the options to create functional materials. This could be *via* physical adsorption of various compounds. Their surfaces can also be covered with various types of polymers to form graft-polymers. Furthermore, (bio-) molecules can be bound covalently on the surface of nanofibers. Different techniques have been used to vary the chemical composition of nanofibers' surfaces. Plasma treatment, radiation (UV or gamma rays, or chemical methods (treatment with acids or bases), surface modifications can be used for this purpose. Thus, hydroxyl, carboxyl or amine groups can be inserted to the main structure. In another application, surface characteristics can be changed after adsorption of biomolecules. Biological structures are adsorbed to the surface *via* electrostatic interactions. The most popular techniques in the immobilization of biomolecules are covalent binding. Alternatively, biomolecules can be immobilized *via* covalent bonds after the addition of functional groups on the surface of nanofibers as mentioned before [99].The designed biomolecule-modified or functional electrospun nanofibers have been successfully utilized in biosensor construction. In the literature, there are many examples that use electrospun nanofibers in biosensor fabrication. In Table **4**, the comparison of these biosensors is summarized.

Table 4. Summary of enzymatic biosensors that uses electrospun nanofibers to fabricate electrochemical and optical biosensors.

Polymer	Solution	Enzyme	Immobilization Technique	Analyte	Detection Principle	Linear Range	Ref
PAN	DMF	Lac	Lac-Nafion-ECNFs/GCE	CC	CAMP	0.001-1.31 mM	[100]
PAN	DMF	Lac	Cu/CNFs/Lac/Nafion/GCE	CC	AMP	9.95-9760 µM	[101]
PAN	DMF	FDH	PAN/HAuCl$_4$/FDH	Frc	EC	0.1-3.0 mM	[102]
PU	DMAc	GOx-HRP	Co-encapsulation	Glc	CLR and Optical	0.01-20 mM (CLR) and 0.5-50 mM (optical)	[103]
PVA	PBS	GOx	PVA/GOx	Glc	AMP	1.0-10 mM	[104]
PVA	Water	Uricase	Uricase/Chit-CNTsNF/AgNPs/Au	UA	AMP	0.001-0.4 mM	[105]
Nylon 6,6	FA	GOx	Nylon6,6/4MWCNT/PBIBA/GOx	Glc	AMP	0.01-2.0 mM	[106]
PVP	DMF	GOx	Mn$_2$O$_3$-Ag-GOx/GCE	Glc	AMP	up to 1.1 mM	[107]
PVP	IPA/AA	GOx	Chit/GOx/TiO$_2$/Pt	Glc	AMP	0.01-6.98 mM	[108]
PVA/Chit	Water	GOx	Chit-PVA/PB/ITO	Glc	AMP	3.3*10^{-3}-55.6 mM	[109]
PVA-SbQ	Water	GOx	PVA-SbQ/MWCNT-COOH/GOx	Glc	EC	up to 4.0 mM	[110]
PVA	Water	GOx	PVA/GOx/Graphene	Glc	CAMP	0-10 mM	[111]
PVA/Chit	AA	GOx	Nafion/PVA/Chit/GOx/GONs/Pt	Glc	EC	0.005-3.5 mM	[112]
PAN	DMF	Lip/GLDH	Lip-GLDH/AgCNF/ITO	Trg	EC	0.5-4.0 mg/mL	[113]
PVA/PEI	Water	GOx	4-ATP/(PVA/PEI/GOxNFs)/AuNPs	Glc	EC	0.01-0.2 mM	[114]
PVP	FeTS	NiR	NiR/PEDOTNFs/GONs	Nitrate ions	EC	0.44-442 mg/L	[115]
PAN	DMF	ChlOx	ChOx/CuO/ECNFs	Chl	EC	0.12-12.93 mM	[116]
PVA/Chit	Water/AA	AChE	AChE/Chit-PVA NFM/SPE	ACh	AMP	0.1-8.0 nM	[117]

(Table 4) contd.....

Polymer	Solution	Enzyme	Immobilization Technique	Analyte	Detection Principle	Linear Range	Ref
PVAc	DMF	GOx	AuNPs-P(ANA-*co*-CNTA--PVAc/GOx/ITO	Glc	EC	0.02-0.1 mM	[118]
PAN	DMF	Lac	Nafion/TiO$_2$-CuCNFs/Lac/GCE	HQ	AMP	1.0-89.8 μM	[119]
PVA	Water	HRP	PVA/BSA/HRP	H$_2$O$_2$	AMP	0.01-6.4 mM	[120]
PS/PSMA	DMF	MMPs	MMP-9-specific FITC-Peptides/PS/PSMA	MMP-9	FL	0-100 pM	[121]
PAN-PU/P3ANA	DMF	Tyr	PAN-PU/P3ANA/Tyr	-	EC	-	[122]
PAN/PPyNSs	DMF	GOx	GOx/NCNSs-CNFs/GCE	Glc	CAMP	0.012-1.0 mM	[123]
CA	AC/DMAc	Lac	Lac/AgNPs-CMC-CA/GCE	CC	AMP	4.98-3650 μM	[124]
PAN	DMF	Lac	Lac/NiCuCNF-Nafion/GCE	HQ	EC	0.4-2.37 μM	[125]

EC: Electrochemical; AMP: Amperometric; CAMP: Chronoamperometric; FL: Fluorometric; CLR: Colorimetric; GCE: Glassy carbon electrode; Pt: Platinum electrode; Au: Gold electrode; SPE: Screen-printed electrode; ITO: Indium tin oxide-coated glass; GOx: Glucose oxidase; Lac: Laccase; FDH: Fructose dehydrogenase; HRP: Horseradish peroxidase; Lip: Lipase; NiR: Nitrate reductase; ChlOx: Cholesterol oxidase; AChE: Acetylcholinesterase; MMPs: Matrix metalloproteinases; Tyr: Tyrosinase; GLDH: Glycerol dehydrogenase; PAN: Poly (acrylonitrile); PVA: Poly (vinyl alcohol); PVAc: Poly (vinyl acetate); PVA-SbQ: PVA with styrylpyridinium pendent groups; PEI: Poly (ethylene imine); PU: Polyurethane; PVP: Poly (vinyl pyrrolidone); P3ANA: Poly (m-anthranilic acid); P(ANA-*co*-CNTA): Poly (anthranilic acid-co-3-carbo-y-N-(2-thenylidene) aniline); PBIBA: Poly-4-(4,7-di(thiophen-2-yl)-1H-benzo[d]imidazol-2-yl)-benzaldehyde; PS: Polystyrene; PSMA: Polystyrene-*co*-maleic anhydride; CA: Cellulose acetate; PPyNSs: Polypyrrole nanospheres; FeTS: Ferric p-toluenesulfonate; DMF: Dimethylformamide; DMAc: Dimethylacetylamide; FA: Formic acid; AA: Acetic acid; AC: Acetone; PBS: Phosphate-buffered saline; IPA: Isopropyl alcohol; CNFs: Carbon nanofibers; ECNFs: Electrospun CNFs; CNTsNF: Carbon nanotubes nanofiber; MWCNTs: Multiwalled carbon nanotubes; MWCNT-COOHs: Carboxylated MWCNTs; Mn$_2$O$_3$-Ag: Manganese oxide-silver nanofibers; AgCNF: Silver-carbon composite nanofiber; PEDOTNFs: Poly (3,4-ethylene dioxythiophene) nanofibers; TiO$_2$NFs: Titanium dioxide nanofibers; TiO$_2$/CuCNFs: TiO$_2$ loaded copper and carbon composite nanofibers; NiCuCNFs: Nickel-Copper alloy nanoparticle-loaded carbon nanofibers; NCNSs-CNFs: Nitrogen-doped carbon nanospheres-carbon nanofibers; AgNPs: Silver nanoparticles; AuNPs: Gold nanoparticles; NFM: Nanofibrous membrane; GONs: Graphene oxide nanosheets; Cu: Copper; CuO: Copper oxide; HAuCl$_4$: Chloroauric acid; PB: Prussian blue film; 4-ATP: 4-aminothiophenol; BSA: Bovine serum albumin; Chit: Chitosan; CC: Catechol; Glc: Glucose; Frc: Fructose; HQ: Hydroquinone; H$_2$O$_2$: Hydrogen peroxide; UA: Uric acid; Trg: Triglyceride; ACh: Acetylthiocholine; Chl: Cholesterol; FITC: Fluorescein isocyanate; CMC: Carboxymethyl cellulose.

In the fabrication of enzymatic biosensors, glucose oxidase (GOx) and laccase (Lac) are the most used enzymes. GOx was described as the "ideal enzyme" for glucose oxidation in a review by Wilson and Turner [126]. GOx is a flavoprotein and composed of two identical subunits. It is produced by fungi, particularly the *Aspergillus* and *Penicillium* species. GOx (Glucose-1-oxidase; β-D-glucose: oxygen 1-oxidoreductase) is a member of the oxidoreductases and catalyzes the oxidation of β-D-glucose to D-glucono-1,5-lactone in the presence of oxygen as a co-substrate of the reaction and produces hydrogen peroxide [127]:

$$\beta\text{-D-glucose} + O_2 \rightarrow \text{D-glucono-1,5-lactone} + H_2O_2$$

Following of the consumed oxygen or the produced hydrogen peroxide plays a key role in the monitoring of sensor responses. In a amperometric detection mode, a consumed oxygen or produced hydrogen peroxidase are followed by a potentiostat at -0.7 V or +0.7 V, respectively [128]. GOx has been immobilized on the surface or inside of electrospun nanofibers by different researchers. Ren *et al.* used electrospun poly (vinyl alcohol) (PVA) nanofibers as an immobilization matrix for GOx [104]. In that study, PVA and GOx was dissolved in a phosphate buffer saline (PBS) and this mixture was placed in a syringe. The electrospun PVA/GOx nanofibers were formed on the surface of a gold electrode. Then, a crosslinking process was carried out using glutaraldehyde. The diameter of resulted 'bead-free' nanofibers was varied from 70 to 250 nm whereas the addition of GOx into the PVA solution caused bead formation. In another study, nylon 6,6 electrospun nanofibers or multiwalled carbon nanotubes (MWCNTs) incorporating nylon 6,6 nanofibers were formed on the surface of a glassy carbon electrode. Then, the conducting polymer namely (poly-4-(4,7-di(thiophen--yl)-1H-benzo[d]imidazol-2-yl) benzaldehyde) (PBIBA) was electropolymerized on the electrospun nanofiber coated electroactive surface. Finally, GOx was immobilized *via* covalent binding. Resulted Naylon 6,6/MWCNT/PBIBA/GOx biosensor showed high I_{max} and low LOD values for glucose [106].

Lac is the other enzyme that is used to construct electrospun nanofiber based biosensors. Lac (Polyphenol oxidase; EC 1.10.3.2) is a member of the copper-containing oxidoreductases (blue multi-copper-oxidase family) and oxidases various compounds such as ortho- and para-substituted phenol, aminophenols, polyphenols, methoxy-phenols, aromatic amines, ascorbate, benzenethiols, and a series of other oxidizable compounds with the concomitant reduction of oxygen to water [129, 130]:

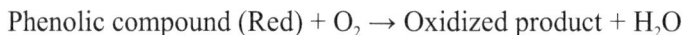

$$\text{Phenolic compound (Red)} + O_2 \rightarrow \text{Oxidized product} + H_2O$$

Lac was produced in a variety of organisms such as bacteria, fungi, plants, and insects as an extracellular enzyme. On the other hand, some species produce it as an intracellular enzyme [131]. Lac is widely used in phenol biosensors. Various types of materials have been utilized for Lac immobilization in biosensor fabrication. Electrospun nanofibers also are a good alternative to other immobilization platforms. Liu *et al.* immobilized *Trametes versicolor* Lac (TvL) in PVA, PEO-PPO-PEO (F108) and gold nanoparticles (AuNPs) containing nanofibers [132]. A PVA/F108/AuNPs/Lac biosensor was applied to analyze chlorophenols.

Another enzyme used to fabricate biosensors in electrospun nanofibers, is fructose dehydrogenase (FD, from *Gluconobacter* species; EC 1.1.99.11) [133]. FDH is a member of the membrane-bound oxidoreductases and does not need oxygen for the regeneration of active sites. Pyrroloquinoline/quinine (PQQ) redox sites and a few heme c sites are found in the active site of FDH. It catalyzes the oxidation of fructose and mediators such as ferrocene and hexacyanoferrate, *etc.* are used in electrochemical sensing.

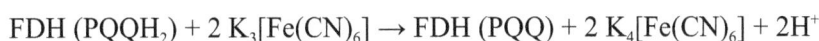

$$\text{D-fructose} + \text{FDH (PQQ)} \rightarrow \text{5-keto-D-fructose} + \text{FDH (PQQH}_2)$$

$$\text{FDH (PQQH}_2) + 2\ K_3[\text{Fe(CN)}_6] \rightarrow \text{FDH (PQQ)} + 2\ K_4[\text{Fe(CN)}_6] + 2H^+$$

$$2\ K_4[\text{Fe(CN)}_6] \rightarrow 2\ K_3[\text{Fe(CN)}_6] + 2e^-$$

In a previous work, poly (acrylonitrile) (PAN) and $HAuCl_4$ were dissolved in dimethylformamide (DMF). Electrospinning was applied using the PAN/$HAuCl_4$ mixture, and then the obtained nanofibers were immersed in a $NaBH_4$ solution. After that, cysteamine and glutaraldehyde immersion was performed, and incubation with FDH was carried out. After the optimization of the system, the fructose was analyzed electrochemically [102].

Microbial Sensors

Microorganisms are also used in biosensor preparation as a biological part. Microbial cells are immobilized on the surface of the transducer using the similar immobilization procedures as with enzymes. Transducers convert biochemical signals to appropriate physical signals such as electrochemical or optical, *etc* [91, 134, 135]. The signal is proportional to the concentration of the detected compound. Microbial sensors are divided in two groups according to their measurements principle: respiration (metabolic) activity detection and electrochemical active metabolite detection. In the detection of metabolic activity, microbial sensors are based on the oxygen consumption by aerobic microorganisms. After substrate injection, respiration activity increases in this configuration. According to the substrate concentration, the oxygen concentration decreases [91]. In metabolite detection based microbial sensors, the electrochemical active products such as H_2, CO_2, NH_3 and organic acids that are released from microorganism to medium are detected. In this system, it is not necessary to use aerobic microorganisms as a biological component in the preparation of microbial sensors. In the literature, metabolic activity detection-based microbial sensors are more common. Also, mediated microbial sensors can be designed [136].

Electrospun nanofibers have been used to immobilize various microbial cells. For instance, *Escherichia coli* (*Staphylococcus albus*) and bacterial viruses (T7, T4, λ) were encapsulated in PVA based electrospun nanofibers and the prepared network tested for cell viability [137]. In another study, Klein *et al.* immobilized whole microbial cells and used them for the bioremediation of pollutants in water [138]. Furthermore, San *et al.* immobilized *Aeromonas eucrenophila, Clavibacter michiganensis* and *Pseudomonas aeruginosa* in cellulose acetate nanofibers and used decolorization of methylene blue [139]. To our best knowledge, microbial cells have been not yet immobilized in electrospun nanofibers for the design of microbial sensors.

Immunoassays

The interaction between antigen and antibody is based on molecular recognition, and it is very selective. This principle is taken as the basis for the construction of immunoassays. Immunoassay technology is especially important in clinical laboratories, and research in this field continues to gain attention. Moreover, immunoassays play key roles in the food industry, and environmental screening to detect trace amounts of compounds. The needs in these areas increase the attention for continuous analysis-based immunosensors [140]. The measurement principles in immunosensors can be direct (un-labelled) or indirect (labelled). In direct immunosensors, after the forming of an immune complex, a physical change is detected. In indirect immunosensors, signal-producing labels are necessary. Various labels such as enzymes (peroxidase, phosphatase, catalase, and luciferase), ferrocene, In^{2+} salts, fluorophores (rhodamine, fluorescein, ruthenium complexes, profirin dyes), and nanomaterials (quantum dots) can be used in the design of indirect immunosensors [140].

Electrospun nanofibers have been successfully used as a platform for immunosensor fabrication based on colorimetric, electrochemical, or optical detection strategies. Table **5** summarizes the electrospun-based immunosensors.

Table 5. Electrospun nanofiber-based immunosensors.

Polymer	Solution	Immobilization Technique	Analyte	Detection Principle	Linear Range	Ref
PLA	DMF	PLA/PLA-*b*-PEG/Biotin	Biotin	CLR	-	[141]
PA6	Cresol/FA	MWCNTs-PA6-PPy	p53 tumor	EC	0.001-0.1 nM	[142]
PAN	DMF	BSA/Anti-CA125/ MZnONF/GCE	CA125	EC	0.001-1000 U/mL	[143]
PAN	DMF	GCE/MNF/Probe	DENV	EC	0.001-1 μM	[144]

(Table 5) contd.....

Polymer	Solution	Immobilization Technique	Analyte	Detection Principle	Linear Range	Ref
Chit	TFA/DCM	MI-ESNFs-BiOINFs/FTO	TPhP	PEC	0.01-500 ng/mL	[145]
PMMA	DMF	CNHs/Fe$_3$O$_4$-NFs/MGCE	AFB1	CLM	0.05-200 ng/mL	[146]
PMMA/ PAN	DMF	BSA/Anti-AFB1/ Microchannel/ Carbon-Pt	AFB1	EC	10^{-12}-10^{-7} g/mL	[147]
PAN	DMF	CuZnONFs/ Anti-HRP2/GCE	HRP2	EC	10 ag/ml-10 µg/ml	[148]
SU-8	-	Anti(Myo/cTn I/CK-MB)/MWCNTs-SU-8/GCE	Myo cTn I CK-MB	EC	1.0-50 ng/ml (Myo) 0.1-10 ng/mL (CTn I) 0.01-10 µg/mL (CK-MB)	[149]
PAN	DMF	ITO-plastic/ ZnONW/Anti-BSA	BSA	EC	100 fg/mL-100 µg/mL	[150]
PANI/ PEG	-	GCE/PANI/PEG/C1	BRCA1	EC	0.01 pM-1.0 nM	[151]
PCL/PPy	THF/DMF	PCL/PPy/ssDNA/ ITO-PET	-	EC	2.0 0 µM	[152]
PAN	DMF	NiCNF/GCE	Adenine-Guanine	EC	0.05-2.0 µM	[153]

EC: Electrochemical; PEC: Photoelectrochemical; CLR: Colorimetric; CLM: Chemiluminescent; GCE: Glassy carbon electrode; MGCE: Magnetic glassy carbon electrode; FTO: Fluorine-doped tin oxide glass; ITO: Indium tin oxide-coated glass; ITO-PET: Indium tin oxide coated polyester film; PLA: Poly (lactic acid); PA6: Polyamide 6; PAN: Poly (acrylonitrile); PPy: Polypyrrole; PMMA: Poly (methyl methacrylate); SU-8: Epoxy-based negative photoresist; PANI: Polyaniline; PCL: Poly (caprolactone); PEG: Poly (ethylene glycol); PLA-*b*-PEG: Poly (lactic acid)-block-poly (ethylene glycol); DMF: Dimethylformamide; TFA: Trifluoroacetic acid; DCM: Dichloromethane; THF: Tetrahydrofurane; FA: Formic acid; MWCNTs: Multiwalled carbon nanotubes; MZnONF: Multiwalled carbon nanotubes embedded zinc oxide nanofibers; Fe$_3$O$_4$-NFs: Magnetite nanofibers; CuZnONFs: Copper-doped zinc oxide nanofibers; BiOINFs: BiOI nanoflake arrays; CNHs: Carbon nanohorns; ZnONW: Zinc oxide nanowires; Carbon-Pt: Carbon electrodes decorated with platinum nanoparticles; MNF: Manganese (III) oxide nanofibers; MI-ESNFs: Molecularly imprinted electrospun nanofibers; NiCNF: Nickel loaded porous carbon nanofibers; BSA: Bovine serum albumin; Chit: Chitosan; CA125: Carcinoma antigen-125; TPhP: Triphenyl phosphate; AFB1: Aflatoxin B1; HRP2: Histidine-rich protein-2; Myo: Myoglobin; cTn I: Troponin I; CK-MB: Creatine Kinase-MB; C1: Capture DNA probe; BRCA1: Breast cancer susceptibility gene; ssDNA: Single-stranded DNA; Consensus DENV: Consensus Dengue virus.

DNA Sensors

Deoxyribonucleic acid (DNA) is a key macromolecule in the identification of genetic characteristics and the transfer of knowledge for replication. Because of the importance of the chemical and biophysical features of DNA, the analysis of its sequence is vital to the exploration of the genetic code and sequencing of the genome [154]. DNA sensors are constructed by the association of the receptor and the transducer. The most important point of the DNA sensors is the immobilization of a single chain probe in the correct orientation. After the hybridization of DNA with a probe chain, a signal is obtained. Like immunosensors, DNA sensors can be prepared in a direct or indirect format [155]. DNA sensors can be used in the detection of toxic agents, metals, genetically modified foods and organisms, DNA drug interaction mechanisms, bioterrorism, diagnostics of diseases, DNA base damage, *etc* [155].

Electrospun nanofibers have also been used in DNA sensors. Wang *et al.* prepared cellulose acetate membranes, and DNA-functionalized gold nanoparticles on cellulose acetate electrospun nanofiber were assembled on the surfaces. An 18-nucleotide BRCA I gene fragment related to breast cancer13 was detected as the model for investigating the performance of the fabricated sensor [156]. In another study, Tripathy *et al.* prepared electrospun semi-conducting manganese (III) oxide (Mn_2O_3) nanofibers to design electrochemical DNA sensors. Zeptomolar detection of Dengue consensus primer in spiked serum samples was achieved using a label-free electrochemical detection strategy [157]. There is no additional study for nanofibers. Usage of electrospun nanofibers in DNA sensors is an open area for novel designs and approaches.

ELECTROSPUN NANOFIBERS IN PURIFICATION AND SEPARATION

Electrospun Nanofibers as a Filtration Membrane

Air Filtration Applications

Electrospun nanofiber membranes are used for air filtration applications. In the early 1980s, the Donaldson Company (US) began to manufacture electrospun nanofiber air filters, and the application of nanofibers in the fabrication of air filters has increased steadily over the last 35 years [158]. With industrialization, air pollution has become one of the major environmental problems. Particulate matter (PM) contains microscopic solids or liquid droplets found in the air that are extremely small and can be inhaled. Both PM and gaseous pollutants cause serious health problems (asthma, lung cancer, heart failure, *etc*.). From an environmental point of view, they contribute to acid rain, depleting the nutrients in the soil, reducing the diversity of ecosystems, *etc* [159]. Thus, to reduce the health risks and environmental impact of air pollution, stricter limits for the emission of air pollutants are set. Fiberglass, activated carbon, spun-bonded fibers, and melt-blown fibers are commercially used matrices in air filtration application. Although these commercial filtration materials perform well in their filtering performance for micrometer-sized particles, stricter emission limits require higher quality filtration membranes for sub-micrometer sized PM and microbial filtration [160, 161].

Today, electrospun nanofibers are used in various air filtration applications to achieve the desired filtration performance. These approaches include vesicle cabin air filters, indoor air filters, industrial gas cleaning, protective suits and face masks, *etc.* When they are compared to conventional filtration membranes, electrospun nanofiber mats have a smaller pore size, finer diameters, and a higher surface area to volume ratio that boosts filtration efficiency due to inertial impaction and interception. At the same pressure drop of both commercial materials and electrospun nanofibers during filtration, nanofibers are more desirable because they provide higher efficiency as a result of increased inertial impaction and interception. In addition, controllable morphology, and functionalization of electrospun nanofibrous membranes provide an opportunity to constitute desirable nanofibrous mats for air filtration processes from various synthetic and natural polymers. To date, diverse polymers and composite materials have been used for air filtration processes. These polymers include poly (acrylonitrile) (PAN), poly (vinyl alcohol) (PVA), poly (ethylene terephthalate) (PET), polyamide (PA 6 and PA 6/6), polystyrene (PS), polyurethane (PU), poly (ethylene oxide) (PEO), polysulfone (PSU), poly (vinyl acetate) (PVAc), cellulose acetate (CA), poly (vinyl chloride) (PVC), keratin, chitosan, poly (lactic acid) (PLA), polypropylene (PP), and their composites [162].

Liu *et al.* investigated different polymer mats for PM capture to evaluate their filtration efficiency for indoor air protection applications. In this study, PM was generated by burning incense which contained pollutant gases such as CO, CO_2, NO_2, SO_2, and volatile compounds like benzene, aromatic hydrocarbons, aldehydes, *etc.* Commercial metal-coated window screen mesh was used for the collector in this electrospinning process. Thus PAN, PVP, PS, and PVA nanofibers could be formed as a transparent air filters thanks to the window screen. Since PP, copper, and carbon are commonly used in commercial air filtration media, these materials were used for performance comparison with electrospun nanofiber filters. As-made transparent filters had the same transmittance (70%) and similar morphology, with approximately 200 nm fiber sizes. The efficiency comparison showed that PAN had the best performance for the removal of both $PM_{2.5}$ (98.11±1.41%) and $PM_{10-2.5}$ (99.31±0.08%) followed by PVP, PVA, PS, PP, copper, and carbon at the same filter transmittance.

Controlling the surface chemistry of the PAN electrospun nanofiber enabled

>95% removal of $PM_{2.5}$ with ~90% transparency under extremely hazardous air quality conditions [163].

Volatile Organic Compounds (VOC) are toxic chemicals that have a high vapor pressure at room temperature. The sources of VOCs are industrial waste gases, pesticides, paints, aerosol sprays, smoking, heating, *etc.* Conventional activated carbon is commonly used for adsorption of VOCs because of its high surface area [164]. Due to its high surface area, electrospun nanofibers are a new candidate for the removal of VOCs from the air. Scholten *et al.* reported that the adsorption of volatile compounds by PU-based nanofibers was found to be similar to adsorption by conventional activated carbon although activated carbon has a higher surface area than PU-based nanofibers. Furthermore, while regeneration of the activated carbon was impossible, PU-based nanofibers had reversible absorption and desorption. In this study, PU-based nanofibers were found to have a high affinity towards toluene and chloroform [165]. Uyar *et al.* reported that poly (methyl methacrylate) (PMMA) nanofibers which contained beta-cyclodextrin (β-CD) had the ability to entrap aniline, styrene, and toluene. β-CD molecules, which were present at the surface of PMMA, could trap organic vapors with the help of inclusion complexation. This study showed that electrospun nanofibers that were functionalized with cyclodextrin are a potential candidate for air filtration applications [166].

Silver ion itself and silver compounds are well known as antimicrobial agents. Silver nanoparticles that incorporated three nanofibers (CA, PAN, PVC) were tested against *E. coli* and *P. aeruginosa* by Neeta *et al.* This study demonstrated that PAN nanofiber in DMF that contained 5.0 wt% $AgNO_3$ and were UV-irradiated for 30 min gave the best result. In conclusion, these nanofibers could be used as a filter for protective suits to protect individuals from bacterial contamination [167]. In 2007, Jeong and Youk discovered that electrospun nanofiber mats that consisted of Polyurethane Cationomers (PUCs) contained different amounts of quaternary ammonium groups exhibited strong antimicrobial activity against *Staphylococcus aureus* and *E. coli*. Thus, this study demonstrated that PUCs containing quaternary ammonium groups could be used as antimicrobial air nanofilters [168]. The summary of air filtration application of nanofibers is given in Table **6**.

Table 6. Summary of air filtration applications based on electrospun nanofibers.

Polymer	Solvent	Pressure Drop (Pa)	Filtration Efficiency (%)	Area of Application	Ref
Nylon 6	-	42	>99.97	PM removal	[169]

(Table 6) contd.....

Polymer	Solvent	Pressure Drop (Pa)	Filtration Efficiency (%)	Area of Application	Ref
PI	DMF	45	99.97 (at 60% transmittance)	PM removal	[170]
PVA	Water	65	99.1	PM removal	[171]
PLA	DCM/DMAc	165.3	99.99	PM removal	[172]
Nylon-6/ Ag NPs	FA/AA	-	-	Antimicrobial filter	[173]
PSU/MgO	DMF	-	-	Protective clothing	[174]
HE incorporated PVP	EtOH	46.6	99.99	Antimicrobial filter	[175]
PSU/PAN/ PA 6	DMF/FA	118	99.99	PM removal	[176]
PVA/TiO$_2$	DMF/FA	-	90	VOC (ethanol) and PM removal	[177]
PLA/PHB/IL	CCl$_4$	140-320	-	Antimicrobial filter and PM removal	[178]
Chit/TiO$_2$ or AgNPs	AA	-	>99	PM removal	[179]
PVDF/ PTFE NPs	DMF	57	99.97	PM removal	[180]
Nylon 6,6/CD	FA	-	-	VOC (toluene) removal	[181]
PU/FAP	DMF/MEK	-	-	VOC (Chloroform, benzene, toluene o-xylene and styrene) removal	[182]

PI: Polyimide; PVA: Poly (vinyl alcohol); PLA: Poly (lactic acid); PSU: Polysulfone; PVP: Poly (vinyl pyrrolidone); PAN: Poly (acrylonitrile); PA 6: Polyamide 6; PHB: Poly (hydroxybutyrate); PVDF: Poly (vinylidene fluoride); PTFE: Poly (tetrafluoro ethylene); PU: Polyurethane; PM: Particulate matter; DMF: Dimethylformamide; DCM: Dichloromethane; DMAc: Dimethylacetamide; FA: Formic acid; AA: Acetic acid; EtOH: Ethanol; MgO: Magnesium oxide; HE: Herbal extract TiO$_2$: Titanium dioxide; VOC: Volatile organic compound; IL: Ionic liquid; AgNPs: Silver Nanoparticules; CD: Cyclodextrin; FAP: Fly ash particles; MEK: Methyl ethyl ketone; Chit: Chitosan.

Liquid Filtration Applications

Thermally Driven Membrane Process: Membrane Distillation

Membrane Distillation (MD) is a thermally driven membrane separation process which is usually used for the desalination of water, the concentration of wastewater, the production of pure water, and the removing of VOCs from wastewater. MD membranes have begun to attract a great deal of interest in membrane systems prepared by the electrospinning technique for the past few years. In the MD process, only vapor molecules can migrate through a porous barrier of hydrophobic/superhydrophobic membranes from a high pressure site to a lower one. Sweeping-gas MD (air stripping), vacuum MD, air gap MD, and direct contact MD are four different configurations of MD. The differences between these configurations are the way they create condensation of vapor on the permeated side. Direct contact MD is the most commonly used configuration due to its simple operational conditions, and over 60% of MD processes are carried out with this MD system. However, MD technology remains in the developmental phase and has not been utilized on an industrial scale [183].

High hydrophobicity, small microporous structure, and breathability of the membrane (allowing moisture and vapor to be transmitted through the membrane) are key factors for the MD process. A good MD performance is generally evaluated by hindering any liquid water penetration into the pores of the nanofiber which refers to high liquid entry pressure (LEP). High liquid entry pressure provides a higher permeated production rate and also prevents the pore wetting of membranes [184, 185]. In recent years, electrospinning techniques have gained more attention in terms of providing electrospun nanofibers with high surface hydrophobicity which is used directly as MD membranes [186]. Commercial membranes used for MD process are generally made of polypropylene (PP), poly (vinylidene fluoride) (PVDF), and poly (tetrafluoroethylene) polymers, which have hydrophobic characteristics [187].

Lee *et al.* have studied electrospun nanofiber membranes that incorporated fluorosilane functionalized TiO_2 for direct contact membrane distillation. 10%, 15%, and 20% of poly (vinylidene fluoride-*co*-hexafluoropropylene) (PVDF-HFP) as the dope solution mixed with various amounts of functionalized TiO_2 (1.0%, 5.0%, 10%) were used to construct nanofiber mats. Electrospun nanofibers that were generated by 20% PH and 10% TiO_2 showed the highest surface hydrophobicity with a high contact angle (149°) and the highest liquid entry pressure (194.5 kPa). Moreover, electrospun nanofibers containing 10% TiO_2 exposed more stable salt rejection than commercial membranes and electrospun nanofiber membranes without TiO_2. Severe wetting of nanofiber was not observed even with the high salt concentration (7.0 wt.% NaCl) of feed over one week of operation [188].

Woo *et al.* have studied dual-layer nonwoven membranes for the air gap distillation process. The top layer consisted of PH (Hydrophobic part), and the other side was made of nylon-6, PVA, or PAN nanofibers. A dual-layer was produced by heat-press post-treatment. The dual layer membrane exhibited a higher water permeation flux using 3.5 wt % NaCl solution as a feed when it was compared with the commercial PVDF membrane. The PH and N6 dual-layer nanofiber membrane showed high salt rejection (99.2%) and flux (15.5 LMH) which demonstrated that this dual-layer membrane has valuable features for air gap membrane distillation [189].

Feng at al. discovered that electrospun poly (vinylidene fluoride) (PVDF) nanofiber for the gas stripping process can be used for removing VOCs. Chloroform was used as VOC, and nitrogen was used as the sweeping gas in this system. The overall mass transfer coefficient of chloroform was determined at room temperature. The overall mass transfer coefficient through the nanofiber membrane was found to be higher than in the hollow fiber air-stripping system. The activation energy of the overall mass transfer coefficient was determined as almost equal to that of the Henry constant for chloroform absorption in water [190].

Pressure Driven Membrane Processes: Microfiltration, Ultrafiltration and Nanofiltration

With increasing population and urbanization, water/wastewater treatment has become one of the most important issues in recent years. Therefore, membrane science has focused on the development of novel effective membranes to improve the removal of impurities from liquid samples. While hydrophobicity is a key factor in MD processes, higher hydrophilicity and low surface roughness are important in pressure-driven membrane processes. In pressure-driven membrane processes, external pressure is applied to the system and then molecules can flow to a higher concentration from the lower concentration area. The main types of pressure-driven membrane processes are microfiltration (MF), ultrafiltration (UF), nanofiltration (NF), and reverse osmosis (RO) [191]. Microfiltration membranes are widely used to remove large particles and bacteria from feed streams. MF membranes have pore sizes ranging from 0.1 to 1 μm. Ultrafiltration membranes have smaller pore sizes which range from 0.01 to 0.1 μm. UF membranes are used for protein concentration application, virus removal, and wastewater treatment [192]. Nanofiltration membranes contain thin-film composite layers with small pore sizes from 1 to 10 nanometers. Reverse osmosis membranes are tighter than nanofiltration membranes which are able to reject even monovalent ions and allow the passage of water molecules. However, higher operational pressure is required for both RO and NF applications which decreases overall yield [193].

Electrospun nanofiber membranes have attracted a great deal of attention in the literature for pressure-driven membrane processes because they provide high internal surface, porosity and 3D interconnected pore structure [194 - 197]. Electrospun nanofibrous membranes have promising features, not only for purification but also for disinfection of water since they allow surface modifications such as heat and chemical treatment, coating and grafting which can increase filtration performance. Many research groups have showed great interest in electrospun filtration membranes for liquid filtration with the aim of generating filtrations on an industrial scale. In the following part, studies about electrospun nanofiber membranes which are widely used in MF, UF, and NF applications will be mentioned along with a summary of the pressure driven membrane processes based on electrospun nanofibers as shown in Table **7** [198].

Kaur *et al.* have demonstrated that electrospun nanofiber membranes have a water flux higher than the commercial MF membrane. The same polymeric materials and the same applied pressure were used to generate electrospun nanofiber membranes. To reduce the pore size of electrospun nanofibers, graft copolymerization was carried out with methacrylic acid. The grafted electrospun nanofiber membrane exhibited 150-200% higher water flux than commercial membranes [199]. Wang *et al.* have generated PVA based UF membrane by an electrospinning technique. 96% hydrolyzed PVA was found to have a good mechanical performance. To decrease the fouling problem in the ultrafiltration process, PVA based electrospun nanofibrous mats were coated with PVA hydrogel. According to results, hydrophilic composite membrane showed a higher flux rate than commercial UF membranes. However, filtration efficiency with the rejection rate of >99.5% was found to be same as commercial membranes [200]. Uyar at al. have studied electrospun PS/CD nanofibers to evaluate the removal efficiency of membrane for organic compounds (phenolphthalein). The dependence between β-CD content and the trapping efficiency of phenolphthalein was examined. X-ray photoelectron spectroscopy (XPS) was used to determine the amount of β-CD on the fiber surface. Results demonstrated that electrospun PS/CD membrane has a potential as nanofiltration membrane for the purification/separation of various organic molecules [201].

Table 7. Summary of pressure driven membrane processes based on electrospun nanofibers.

Polymer	Solvent	Fiber Diameter (nm)	Type of Process	Area of Application	Ref
PES	NMP mixture	600	Microfiltration	Water treatment	[202]
RC grafted PHEMA/PAAS	AC/DMAc	500	Ultrafiltration	Water filtration	[203]
PAN	DMF	530	Nanofiltration	Water filtration	[204]
Nafion/PVDF	DMF	240	Microfiltration	Water treatment	[205]
GO/PAN	DMF	300	Nanofiltration	Water treatment	[206]
TFNC (PVA/PAN/PET)	H₂O	170	Ultrafiltration	Oil/water emulsions	[207]

(Table 7) contd.....

Polymer	Solvent	Fiber Diameter (nm)	Type of Process	Area of Application	Ref
Chit/PVA	DMF	1000	Nanofiltration, Ultrafiltration	Water treatment	[208]
PAN	DMF	100	Microfiltration	Water filtration	[209]
PVA	Water	100	Microfiltration	Water treatment	[210]
PVA/TiO$_2$	Citric acid (10%)	524.5	Microfiltration	Oil/water filtration	[211]
GO/APAN	DMF	500-900	Microfiltration	Oil/water emulsion	[212]
Cellulose/PEI	NMMO	64-110	Nanofiltration	Water purification	[213]
PAN/PVA	DMF/H$_2$O	250	Ultrafiltration	Oil/water emulsion	[214]
Zeolite/PVDF	DMAC/AC	580	Ultrafiltration	Waster treatment	[215]
PVDF/PMMA	DMF/AC	250	Microfiltration	Waste water treatment	[216]
PVDF/TiO$_2$	DMAc	300-500	Ultrafiltration	Water treatment	[217]
Polysaccharide (cellulose-chitin)	Water (contained NaBr and TEMPO)	5-10	Ultrafiltration	Water purification	[218]

PES: Poly (ether sulphone); PHEMA: Poly (hydroxyethyl methacrylate); PAAS: Poly (sodium acrylate); PAN: Poly (acrylonitrile); APAN: Aminated PAN; PVDF: Poly (vinylidene fluoride); PVA: Poly (vinyl alcohol); PET: Poly (ethylene terephthalate); PEI: Poly (ethylene imine); PMMA: Poly (methyl methacrylate); NMP: *N*-Methyl-2-pyrrolidone; RC: Regenerated cellulose; DMAc: Dimethylacetamide; DMF: Dimethylformamide; AC: Acetone; GO: Graphene oxide; TFNC: Thin film nanocomposite; TiO$_2$: Titanium dioxide; NMMO: N-methylmorpholine-N-oxide; NaBr: Sodium bromide; Chit: Chitosan; TEMPO: (2,2,6,6-Tetramethylpiperidin-1-yl)oxyl.

Electrospun Nanofibers as an Affinity Membrane

Microfiltration, ultrafiltration and reverse osmosis are pressure-driven types of processes which do not have high selectivity to produce highly purified biomolecules from a complex mixture. The purification of biomolecules such as proteins is a crucial step in many biochemical processes. For instance, the purification of antibodies is a critical process in the application of immunodiagnostics and immunotherapy. Traditional packed-bed columns have some disadvantages in terms of lower flow rate and higher pressure drop in comparison with affinity membrane systems which cause insufficient throughput [219]. Affinity membranes are usually formed by attaching specific, selective ligands to microfiltration membranes [220]. By introducing specific affinity ligands into electrospun nanofibers, effective purification of the target molecules from the filtrate can be achieved. In addition, due to the high surface area of electrospun nanofibers, the performance of the purification process can be increased. Ma *et al.* have studied surface modified polysulphone (PSU) electrospun nanofibers to develop affinity membranes. The surface modification was carried out by introducing carboxyl groups onto a fiber surface which was generated by graft co-polymerization of methacrylic acid (MAA). Toluidine Blue O (TBO) is a dye that was selected as model target in this study. TBO dye can interact with carboxyl groups on modified PSU nanofibers. Modified PSU nanofibers showed lower pressure drops (0.7-1.5 psi) and a higher flux rate when compared with conventional MF membranes. Lower pressure drops and a higher flux rate meet the demand for affinity membranes; also, these properties are the most important advantages of membrane chromatography compared to column chromatography [221]. The same research group demonstrated that Cibacron blue F3GA immobilized PSU (MAA modified) electrospun nanofibers were successfully designed for albumin adsorption. CB is a reactive dye which is one of the most commonly used dye-ligands for protein purification. Amino groups were obtained by the reaction of carboxylic acid with diamino-dipropylamine (DADPA). The capturing capacity of CB immobilized PSU membranes was evaluated to be 22 mg/g for BSA. Until this study, the dynamic binding efficiency of affinity membranes which were prepared by electrospinning was not determined in a quantitative way. The dynamic binding efficiency was determined with the help of frontal analysis [222]. Ma and Ramakrishna (2008) have constructed regenerated cellulose (RC) nanofiber for IgG purification. Regenerated cellulose nanofiber mats were obtained by alkaline treatment of cellulose acetate nanofiber mats. Protein A/G was covalently bound to the RC mats to produce affinity membranes which have specificity towards to IgG. XPS and ATR-FTIR were used for characterization of membranes during surface modification. Bovine serum albumin was used as a model impurity to evaluate the purification ability of the RC membrane. The capturing capacity of the modified RC affinity membranes was determined to be 18 µg/mg. Current commercial products have a lower capturing capacity than these modified RC membranes. Therefore, this simply produced membrane is a candidate for antibody purification on a small scale [223].

Electrospun Nanofibers as a Chromatographic Sorbent Bed

Electrospun nanofibers as stationary phase for ultra-thin layer chromatography (UTLC) have been an attractive topic in analytical chemistry in recent years. UTLC is a planar chromatographic technique that has a monolithic layer unlike thin layer chromatography (TLC) and high performance thin layer chromatography (HPTLC) which are particle based techniques. For selective separation, planar chromatography requires new advanced stationary phases. However, development of new stationary phases for UTLC application is a current research area in separation science. As compared with TLC and HPTLC, UTLC provides higher separation efficiency, shorter development times, and very thin layer membranes (~5-30 μm). Furthermore, it requires lower volumes of solvent and sample [224].

Poly (acrylonitrile) polymer was used to generate an electrospun UTLC device by Clark and Olesik (2009). 10% PAN polymer was utilized, and the fiber diameter of the mat was determined to be 400 nm. To evaluate the effect of thickness on retention behavior, different operation times of electrospinning were carried out, and the effects of various mobile phase compositions were tested. A mixture of laser dyes and steroidal compounds were used to illustrate the separation ability of PAN based UTLC devices. In comparison with typical TLC methods, a shorter development time and a lower solvent requirement after complete analysis demonstrated that electrospun PAN nanofibers with ~25 μm thickness are a new candidate for UTLC methods [225]. They were also studied on electrospun glassy carbon UTLC devices. Glassy carbon (GC), also called vitreous carbon, is a commonly used sorbent in chromatography. However, the preparation of porous glassy carbon (PGC) as a widely used type of GC is very expensive. In this study, 75% (v/v) of SU-8 2100 were prepared in cyclopentanone for the electrospinning procedure. ~200-350 nm of fiber diameter and electrospun glassy carbon with a thickness of ~15 μm was acquired after the pyrolysis of SU-8 2100 photoresist. FITC-labeled threonine, phenylalanine and lysine, and a set of six laser dyes were tested to illustrate the performance of the UTLC device. According to the results, three essential amino acids and triple mixtures of laser dyes could be resolved completely on this device [226]. Moheman *et al.* have investigated performance of PAN nanofibers for the UTLC method. UTLC plates were fabricated by different concentrations of PAN (8.0, 10 and 12 wt %) at different flow rates. The effect of the thickness of the nanofibers and the mobile phase composition on chromatographic performance was also tested. The uses of n-butanol-ethylene glycol-ethyl acetate (5:3:2 v/v) as a mobile phase and 10% (w/v) PAN solution in dimethylformamide (1 h operation time at 0.2 mL h^{-1} of flow rate) was determined have the most efficient impart differential migration among the amino acids. Moreover, PAN derived UTLC plates were also compared with silica gel TLC plates and ion exchange resin (Na$^+$ form) for amino acids sensitivity. These electrospun UTLC plates were found more useful with regards to sensitivity, rapidity, and low solvent requirements. PAN derived UTLC could also successfully identify lysine and methionine amino acids in the commercial drug samples [227].

CONCLUSION

In this chapter, electrospun nanofiber based separation systems and detection systems have been explained and illustrated. Nanomaterials were chosen for the fabrication of functional surfaces for analytical applications. Among them, nanofibers were also found to be useful materials because of their unique properties such as their large surface area, availability of controlling diameter and thickness, flexibility and porous structure. The produced electrospun nanofibers are used in various areas such as public health, biotechnology, environmental engineering, textiles, the defense industry, and energy deposition. Special applications of electrospun nanofibers in analytical chemistry have great potential as a sorbent for chromatographic separations and modifier of surfaces in electrochemistry or spectroscopy. In recent years, biosensors based on electrospun nanofibers have attracted increasing attention. PVA, PCL, PAA, and nafion, *etc.* were used to modify transducer surface to construct platforms for the immobilization of biological molecules. Also, the incorporation of gold nanoparticles and carbon nanotubes into electrospun nanofibers increased the sensor performance parameters such as low detection limit, wide linear range, increased stability, and higher selectivity. The major drawbacks in the fabrication of electrospun nanofibers based analytical devices are biocompatibility and toxicity of the prepared nanofibers. Using natural polymers such as gelatin, chitosan, *etc.* or non-toxic and biocompatible polymers such as PCL, *etc.* can decrease the risk elements. Therefore, studies can focus on the generation of novel nanofibers which are suitable for preparing sensitive, biocompatible, and reproducible analytical systems with higher stability.

CONSENT FOR PUBLICATION

Not applicable.

CONFLICT OF INTEREST

The author confirms that he has no conflict of interest to declare for this publication.

ACKNOWLEDGEMENTS

Declared none.

REFERENCES

[1]   Bhattarai, P.; Thapa, K.B.; Basnet, R.B.; Sharma, S. Electrospinning: How to Produce Nanofibers Using Most Inexpensive Technique? An Insight into the Real Challenges of Electrospinning Such Nanofibers and Its Application Areas. *Int. J. Biol. Adv. Res.,* **2014**, *5*, 401-405.
       [http://dx.doi.org/10.7439/ijbar.v5i9.854]

[2]   Greiner, A.; Wendorff, J.H. Electrospinning: a fascinating method for the preparation of ultrathin fibers. *Angew. Chem. Int. Ed. Engl.,* **2007**, *46*(30), 5670-5703.
       [http://dx.doi.org/10.1002/anie.200604646] [PMID: 17585397]

[3]   Zhao, J.; Sun, Z.; Shao, Z.; Xu, L. Effect of surface-active agent on morphology and properties of electrospun PVA nanofibres. *Fibers Polym.,* **2016**, *17*, 896-901.
[http://dx.doi.org/10.1007/s12221-016-6163-y]

[4]   Wu, H.; Pan, W.; Lin, D.; Li, H. Electrospinning of ceramic nanofibers: Fabrication, assembly and applications. *J. Adv. Ceram.,* **2012**, *1*, 2-23.
[http://dx.doi.org/10.1007/s40145-012-0002-4]

[5]   Khattab, T.A.; Abdelmoez, S.; Klapötke, T.M. Electrospun Nanofibers from a Tricyanofuran-Based Molecular Switch for Colorimetric Recognition of Ammonia Gas. *Chemistry,* **2016**, *22*(12), 4157-4163.
[http://dx.doi.org/10.1002/chem.201504448] [PMID: 26864701]

[6]   Sahay, R.; Kumar, P.S.; Sridhar, R.; Sundaramurthy, J.; Venugopal, J.; Mhaisalkar, S.G.; Ramakrishna, S. Electrospun composite nanofibers and their multifaceted applications. *J. Mater. Chem.,* **2012**, *22*, 12953.
[http://dx.doi.org/10.1039/c2jm30966a]

[7]   Khalil, A.; Singh Lalia, B.; Hashaikeh, R.; Khraisheh, M. Electrospun metallic nanowires: Synthesis, characterization, and applications. *J. Appl. Phys.,* **2013**, *114*, 171301.
[http://dx.doi.org/10.1063/1.4822482]

[8]   Goyal, R.; Vega, M.E.; Pastino, A.K.; Singh, S.; Guvendiren, M.; Kohn, J.; Murthy, N.S.; Schwarzbauer, J.E. Development of hybrid scaffolds with natural extracellular matrix deposited within synthetic polymeric fibers. *J. Biomed. Mater. Res. A,* **2017**, *105*(8), 2162-2170.
[http://dx.doi.org/10.1002/jbm.a.36078] [PMID: 28371271]

[9]   Soujanya, G.K.; Hanas, T.; Chakrapani, V.Y.; Sunil, B.R.; Kumar, T.S. Electrospun Nanofibrous Polymer Coated Magnesium Alloy for Biodegradable Implant Applications. *Procedia Mater. Sci.,* **2014**, *5*, 817-823.
[http://dx.doi.org/10.1016/j.mspro.2014.07.333]

[10]   Reneker, D.H.; Yarin, A.L.; Fong, H.; Koombhongse, S. Bending instability of electrically charged liquid jets of polymer solutions in electrospinning. *J. Appl. Phys.,* **2000**, *87*, 4531-4547.
[http://dx.doi.org/10.1063/1.373532]

[11]   Tan, S.H.; Inai, R.; Kotaki, M.; Ramakrishna, S. Systematic parameter study for ultra-fine fiber fabrication *via* electrospinning process. *Polymer (Guildf.),* **2005**, *46*, 6128-6134.
[http://dx.doi.org/10.1016/j.polymer.2005.05.068]

[12]   Topuz, F.; Uyar, T. Electrospinning of gelatin with tunable fiber morphology from round to flat/ribbon. *Mater. Sci. Eng. C,* **2017**, *80*, 371-378.
[http://dx.doi.org/10.1016/j.msec.2017.06.001] [PMID: 28866176]

[13]   Li, D.; Wang, Y.; Xia, Y.; Li, D.; Wang, Y.; Wang, Y.; Xia, Y.; Xia, Y. Electrospinning Nanofibers as Uniaxially Aligned Arrays and Layer-by-Layer Stacked Films. *Adv. Mater.,* **2004**, *16*, 361-366.
[http://dx.doi.org/10.1002/adma.200306226]

[14]   Deitzel, J. Kleinmeyer, J.; Harris, D.; Beck Tan, N. The effect of processing variables on the morphology of electrospun nanofibers and textiles. *Polymer (Guildf.),* **2001**, *42*, 261-272.
[http://dx.doi.org/10.1016/S0032-3861(00)00250-0]

[15]   Sas, I.; Gorga, R.E.; Joines, J.A.; Thoney, K.A. Literature review on superhydrophobic self-cleaning surfaces produced by electrospinning. *J. Polym. Sci., B, Polym. Phys.,* **2012**, *50*, 824-845.
[http://dx.doi.org/10.1002/polb.23070]

[16]   Bhardwaj, N.; Kundu, S. C. *Electrospinning : A fascinating fiber fabrication technique.,* **2010**.
[http://dx.doi.org/10.1016/j.biotechadv.2010.01.004]

[17]   Wannatong, L.; Sirivat, A.; Supaphol, P. *Effects of solvents on electrospun polymeric fibers : preliminary study on polystyrene.,* **2004**.
[http://dx.doi.org/10.1002/pi.1599]

[18]   Sill, T. J.; Recum, H. A. *Von Electrospinning: Applications in drug delivery and tissue engineering*; , **2008**, 29, pp. 1989-2006.
       [http://dx.doi.org/10.1016/j.biomaterials.2008.01.011]

[19]   Si, Y.; Tang, X.; Yu, J.; Ding, B. *Electrospun Nanofibers for Energy and Environmental Applications*; Nanostructure Sci. Technol, **2014**, p. 525.

[20]   Mokhena, T.C.; Jacobs, V.; Luyt, A.S. A review on electrospun bio-based polymers for water treatment. *Express Polym. Lett.,* **2015**, 9, 839-880.
       [http://dx.doi.org/10.3144/expresspolymlett.2015.79]

[21]   Bhardwaj, N.; Kundu, S.C. Electrospinning: a fascinating fiber fabrication technique. *Biotechnol. Adv.,* **2010**, 28(3), 325-347.
       [http://dx.doi.org/10.1016/j.biotechadv.2010.01.004] [PMID: 20100560]

[22]   Megelski, S.; Stephens, J.S.; Bruce Chase, D.; Rabolt, J.F. Micro- and nanostructured surface morphology on electrospun polymer fibers. *Macromolecules,* **2002**, 35, 8456-8466.
       [http://dx.doi.org/10.1021/ma020444a]

[23]   Guo, C.; Zhou, L.; Lv, J. Effects of expandable graphite and modified ammonium polyphosphate on the flame-retardant and mechanical properties of wood flour-polypropylene composites. *Polym. Polymer Compos.,* **2013**, 21, 449-456.

[24]   Kanjanapongkul, K.; Wongsasulak, S.; Yoovidhya, T. Prediction of clogging time during electrospinning of zein solution: Scaling analysis and experimental verification. *Chem. Eng. Sci.,* **2010**, 65, 5217-5225.
       [http://dx.doi.org/10.1016/j.ces.2010.06.018]

[25]   Homaeigohar, S.; Elbahri, M. Nanocomposite electrospun nanofiber membranes for environmental remediation. *Materials (Basel),* **2014**, 7(2), 1017-1045.
       [http://dx.doi.org/10.3390/ma7021017] [PMID: 28788497]

[26]   Fang, X.; Reneker, D.H. DNA fibers by electrospinning. *J. Macromol. Sci. Part B,* **1997**, 36, 169-173.
       [http://dx.doi.org/10.1080/00222349708220422]

[27]   Nuansing, W.; Frauchiger, D.; Huth, F.; Rebollo, A.; Hillenbrand, R.; Bittner, A.M. Electrospinning of peptide and protein fibres: approaching the molecular scale. *Faraday Discuss.,* **2013**, 166, 208-221.
       [http://dx.doi.org/10.1039/c3fd00069a] [PMID: 24611278]

[28]   Lee, K.Y.; Jeong, L.; Kang, Y.O.; Lee, S.J.; Park, W.H. Electrospinning of polysaccharides for regenerative medicine. *Adv. Drug Deliv. Rev.,* **2009**, 61(12), 1020-1032.
       [http://dx.doi.org/10.1016/j.addr.2009.07.006] [PMID: 19643155]

[29]   Ramakrishna, S. *An Introduction to Electrospinning and Nanofibers*; World Scientific, **2005**.
       [http://dx.doi.org/10.1142/5894]

[30]   Masilela, N.; Kleyi, P.; Tshentu, Z.; Priniotakis, G.; Westbroek, P.; Nyokong, T. Photodynamic inactivation of Staphylococcus aureus using low symmetrically substituted phthalocyanines supported on a polystyrene polymer fiber. *Dyes Pigments,* **2013**, 96, 500-508.
       [http://dx.doi.org/10.1016/j.dyepig.2012.10.001]

[31]   Eda, G.; Shivkumar, S. Bead-to-fiber transition in electrospun polystyrene. *J. Appl. Polym. Sci.,* **2007**, 106, 475-487.
       [http://dx.doi.org/10.1002/app.25907]

[32]   Fong, H.; Chun, I.; Reneker, D.H. Beaded nanofibers formed during electrospinning. *Polymer (Guildf.),* **1999**, 40, 4585-4592.
       [http://dx.doi.org/10.1016/S0032-3861(99)00068-3]

[33]   Lee, K.H.; Kim, H.Y.; Bang, H.J.; Jung, Y.H.; Lee, S.G. The change of bead morphology formed on electrospun polystyrene fibers. *Polymer (Guildf.),* **2003**, 44, 4029-4034.
       [http://dx.doi.org/10.1016/S0032-3861(03)00345-8]

[34]   Chuangchote, S.; Sagawa, T.; Yoshikawa, S. Electrospinning of poly(vinyl pyrrolidone): Effects of solvents on electrospinnability for the fabrication of poly(p-phenylene vinylene) and TiO$_2$ nanofibers. *J. Appl. Polym. Sci.,* **2009**, *114*, 2777-2791.
[http://dx.doi.org/10.1002/app.30637]

[35]   Ki, C.S.; Baek, D.H.; Gang, K.D.; Lee, K.H.; Um, I.C.; Park, Y.H. Characterization of gelatin nanofiber prepared from gelatin-formic acid solution. *Polymer (Guildf.),* **2005**, *46*, 5094-5102.
[http://dx.doi.org/10.1016/j.polymer.2005.04.040]

[36]   Lee, J.S.; Choi, K.H.; Do Ghim, H.; Kim, S.S.; Chun, D.H.; Kim, H.Y.; Lyoo, W.S. Role of molecular weight of atactic poly (vinyl alcohol) (PVA) in the structure and properties of PVA nanofabric prepared by electrospinning. *J. Appl. Polym. Sci.,* **2004**, *93*, 1638-1646.
[http://dx.doi.org/10.1002/app.20602]

[37]   Geng, X.; Kwon, O.H.; Jang, J. Electrospinning of chitosan dissolved in concentrated acetic acid solution. *Biomaterials,* **2005**, *26*(27), 5427-5432.
[http://dx.doi.org/10.1016/j.biomaterials.2005.01.066] [PMID: 15860199]

[38]   Yuan, X.Y.; Zhang, Y.Y.; Dong, C.; Sheng, J. Morphology of ultrafine polysulfone fibers prepared by electrospinning. *Polym. Int.,* **2004**, *53*, 1704-1710.
[http://dx.doi.org/10.1002/pi.1538]

[39]   Subbiah, T.; Bhat, G.S.; Tock, R.W.; Parameswaran, S.; Ramkumar, S.S. Electrospinning of nanofibers. *J. Appl. Polym. Sci.,* **2005**, *96*, 557-569.
[http://dx.doi.org/10.1002/app.21481]

[40]   Haider, A.; Haider, S.; Kang, I.K. A comprehensive review summarizing the effect of electrospinning parameters and potential applications of nanofibers in biomedical and biotechnology. *Arab. J. Chem.,* **2015**.
[http://dx.doi.org/10.1016/j.arabjc.2015.11.015]

[41]   Ren, L.F.; Xia, F.; Shao, J.; Zhang, X.; Li, J. Experimental investigation of the effect of electrospinning parameters on properties of superhydrophobic PDMS/PMMA membrane and its application in membrane distillation. *Desalination,* **2017**, *404*, 155-165.
[http://dx.doi.org/10.1016/j.desal.2016.11.023]

[42]   Zargham, S.; Bazgir, S.; Tavakoli, A.; Rashidi, A.S.; Damerchely, R. The Effect of Flow Rate on Morphology and Deposition Area of Electrospun Nylon 6 Nanofiber. *J. Eng. Fibers Fabrics,* **2012**, *7*, 42-49.

[43]   Pillay, V.; Dott, C.; Choonara, Y. E.; Tyagi, C.; Tomar, L.; Kumar, P.; Du Toit, L. C.; Ndesendo, V. M. K. *A review of the effect of processing variables on the fabrication of electrospun nanofibers for drug delivery applications.,* **2013**.
[http://dx.doi.org/10.1155/2013/789289]

[44]   Pelipenko, J.; Kristl, J.; Janković, B.; Baumgartner, S.; Kocbek, P. The impact of relative humidity during electrospinning on the morphology and mechanical properties of nanofibers. *Int. J. Pharm.,* **2013**, *456*(1), 125-134.
[http://dx.doi.org/10.1016/j.ijpharm.2013.07.078] [PMID: 23939535]

[45]   Hutmacher, D.W.; Dalton, P.D. Melt electrospinning. *Chem. Asian J.,* **2011**, *6*(1), 44-56.
[http://dx.doi.org/10.1002/asia.201000436] [PMID: 21080400]

[46]   Persano, L.; Camposeo, A.; Tekmen, C.; Pisignano, D. Industrial upscaling of electrospinning and applications of polymer nanofibers: A review. *Macromol. Mater. Eng.,* **2013**, *298*, 504-520.
[http://dx.doi.org/10.1002/mame.201200290]

[47]   Muthuraman, T.N. *Development and Optimization of an Alternative Electrospinning Process for High Throughput*; North Carolina State University, **2012**.

[48]   Teo, W.E.; Ramakrishna, S. A review on electrospinning design and nanofibre assemblies. *Nanotechnology,* **2006**, *17*(14), R89-R106.

[http://dx.doi.org/10.1088/0957-4484/17/14/R01] [PMID: 19661572]

[49]   Alfaro De Prá, M.A.; Ribeiro-do-Valle, R.M.; Maraschin, M.; Veleirinho, B. Effect of collector design on the morphological properties of polycaprolactone electrospun fibers. *Mater. Lett.,* **2017**, *193*, 154-157.
[http://dx.doi.org/10.1016/j.matlet.2017.01.102]

[50]   Wang, J. Nanomaterial-based electrochemical biosensors. *Analyst (Lond.),* **2005**, *130*(4), 421-426.
[http://dx.doi.org/10.1039/b414248a] [PMID: 15846872]

[51]   Kerman, K.; Saito, M.; Tamiya, E.; Yamamura, S.; Takamura, Y. Nanomaterial-based electrochemical biosensors for medical applications. *TrAC -. Trends Analyt. Chem.,* **2008**, *27*, 585-592.
[http://dx.doi.org/10.1016/j.trac.2008.05.004]

[52]   Putzbach, W.; Ronkainen, N.J. Immobilization techniques in the fabrication of nanomaterial-based electrochemical biosensors: a review. *Sensors (Basel),* **2013**, *13*(4), 4811-4840.
[http://dx.doi.org/10.3390/s130404811] [PMID: 23580051]

[53]   Silvestre, C.; Duraccio, D.; Cimmino, S. Food packaging based on polymer nanomaterials. *Prog. Polym. Sci.,* **2011**, *36*, 1766-1782.
[http://dx.doi.org/10.1016/j.progpolymsci.2011.02.003]

[54]   Marín, S.; Merkoçi, A. Nanomaterials Based Electrochemical Sensing Applications for Safety and Security. *Electroanalysis,* **2012**, *24*, 459-469.
[http://dx.doi.org/10.1002/elan.201100576]

[55]   Parolo, C.; Merkoçi, A. Paper-based nanobiosensors for diagnostics. *Chem. Soc. Rev.,* **2013**, *42*(2), 450-457.
[http://dx.doi.org/10.1039/C2CS35255A] [PMID: 23032871]

[56]   Wang, Z.G.; Wan, L.S.; Liu, Z.M.; Huang, X.J.; Xu, Z.K. Enzyme immobilization on electrospun polymer nanofibers: An overview. *J. Mol. Catal., B Enzym.,* **2009**, *56*, 189-195.
[http://dx.doi.org/10.1016/j.molcatb.2008.05.005]

[57]   Matlock-Colangelo, L.; Baeumner, A.J. Recent progress in the design of nanofiber-based biosensing devices. *Lab Chip,* **2012**, *12*(15), 2612-2620.
[http://dx.doi.org/10.1039/c2lc21240d] [PMID: 22596104]

[58]   Ding, B.; Wang, M.; Yu, J.; Sun, G. Gas sensors based on electrospun nanofibers. *Sensors (Basel),* **2009**, *9*(3), 1609-1624.
[http://dx.doi.org/10.3390/s90301609] [PMID: 22573976]

[59]   Senthamizhan, A.; Balusamy, B.; Uyar, T. Glucose sensors based on electrospun nanofibers: a review. *Anal. Bioanal. Chem.,* **2016**, *408*(5), 1285-1306.
[http://dx.doi.org/10.1007/s00216-015-9152-x] [PMID: 26573168]

[60]   Li, C.; Su, Y.; Lv, X.; Xia, H.; Shi, H.; Yang, X.; Zhang, J.; Wang, Y. Controllable anchoring of gold nanoparticles to polypyrrole nanofibers by hydrogen bonding and their application in nonenzymatic glucose sensors. *Biosens. Bioelectron.,* **2012**, *38*(1), 402-406.
[http://dx.doi.org/10.1016/j.bios.2012.04.049] [PMID: 22727516]

[61]   Liu, Y.; Teng, H.; Hou, H.; You, T. Nonenzymatic glucose sensor based on renewable electrospun Ni nanoparticle-loaded carbon nanofiber paste electrode. *Biosens. Bioelectron.,* **2009**, *24*(11), 3329-3334.
[http://dx.doi.org/10.1016/j.bios.2009.04.032] [PMID: 19450966]

[62]   Sanfelice, R.C.; Mercante, L.A.; Pavinatto, A.; Tomazio, N.B.; Mendonça, C.R.; Ribeiro, S.J.; Mattoso, L.H.; Correa, D.S. Hybrid composite material based on polythiophene derivative nanofibers modified with gold nanoparticles for optoelectronics applications. *J. Mater. Sci.,* **2017**, *52*, 1919-1929.
[http://dx.doi.org/10.1007/s10853-016-0481-8]

[63]   Bao, J.; Hou, C.; Huo, D.; Dong, Q.; Ma, X.; Sun, X.; Yang, M.; El Galil, K.H.; Chen, W.; Lei, Y. Sensitive and Selective Electrochemical Biosensor Based on ELP-OPH/BSA/TiO$_2$NFs/AuNPs for Determination of Organophosphate Pesticides with *p*-Nitrophenyl Substituent. *J. Electrochem. Soc.,*

2017, *164*, G17-G22.
[http://dx.doi.org/10.1149/2.0311702jes]

[64]   Migliorini, F. L.; Sanfelice, R. C.; Pavinatto, A.; Steffens, J.; Steffens, C.; Correa, D. S. Voltammetric
        cadmium(II) sensor based on a fluorine doped tin oxide electrode modified with polyamide 6/chitosan
        electrospun nanofibers and gold nanoparticles. *Microchim. Acta,* **2017**, 2-9.

[65]   Zhao, L.; Wang, T.; Wu, Q.; Liu, Y.; Chen, Z.; Li, X. Fluorescent Strips of Electrospun Fibers for
        Ratiometric Sensing of Serum Heparin and Urine Trypsin. *ACS Appl. Mater. Interfaces,* **2017**, *9*(4),
        3400-3410.
        [http://dx.doi.org/10.1021/acsami.6b14118] [PMID: 28067489]

[66]   Ma, L.; Liu, K.; Yin, M.; Chang, J.; Geng, Y.; Pan, K. Fluorescent nanofibrous membrane (FNFM) for
        the detection of mercuric ion (II) with high sensitivity and selectivity. *Sens. Actuators B Chem.,* **2017**,
        238, 120-127.
        [http://dx.doi.org/10.1016/j.snb.2016.07.049]

[67]   Babu, K.J.; Senthilkumar, N.; Kim, A.R. kumar, G. G. Freestanding and binder free PVdF-HFP/Ni-Co
        nanofiber membrane as a versatile platform for the electrocatalytic oxidation and non-enzymatic
        detection of urea. *Sens. Actuators B Chem.,* **2017**, *241*, 541-551.
        [http://dx.doi.org/10.1016/j.snb.2016.10.069]

[68]   Wang, Z.; Ying, Y.; Li, L.; Xu, T.; Wu, Y.; Guo, X.; Wang, F.; Shen, H.; Wen, Y.; Yang, H. Stretched
        graphene tented by polycaprolactone and polypyrrole net–bracket for neurotransmitter detection. *Appl.
        Surf. Sci.,* **2017**, *396*, 832-840.
        [http://dx.doi.org/10.1016/j.apsusc.2016.11.038]

[69]   Pang, Z.; Nie, Q.; Wei, A.; Yang, J.; Huang, F.; Wei, Q. Effect of In_2O_3 nanofiber structure on the
        ammonia sensing performances of In_2O_3/PANI composite nanofibers. *J. Mater. Sci.,* **2017**, *52*, 686-
        695.
        [http://dx.doi.org/10.1007/s10853-016-0362-1]

[70]   Guo, L.; Kou, X.; Ding, M.; Wang, C.; Dong, L.; Zhang, H.; Feng, C.; Sun, Y.; Gao, Y.; Sun, P.; Lu,
        G. Reduced graphene oxide/α-Fe_2O_3 composite nanofibers for application in gas sensors. *Sens.
        Actuators B Chem.,* **2016**, *244*, 233-242.
        [http://dx.doi.org/10.1016/j.snb.2016.12.137]

[71]   Zhang, X.; Liu, D.; Yu, B.; You, T. A novel nonenzymatic hydrogen peroxide sensor based on
        electrospun nitrogen-doped carbon nanoparticles-embedded carbon nanofibers film. *Sens. Actuators B
        Chem.,* **2016**, *224*, 103-109.
        [http://dx.doi.org/10.1016/j.snb.2015.10.033]

[72]   Huang, J.; Zhang, X.; Zhou, L.; You, T. Simultaneous electrochemical determination of
        dihydroxybenzene isomers using electrospun nitrogen-doped carbon nanofiber film electrode. *Sens.
        Actuators B Chem.,* **2015**, *224*, 568-576.
        [http://dx.doi.org/10.1016/j.snb.2015.10.102]

[73]   Kim, S.J.; Cho, Y.K.; Lee, C.; Kim, M.H.; Lee, Y. Real-time direct electrochemical sensing of
        ascorbic acid over rat liver tissues using RuO_2 nanowires on electrospun TiO_2 nanofibers. *Biosens.
        Bioelectron.,* **2016**, *77*, 1144-1152.
        [http://dx.doi.org/10.1016/j.bios.2015.11.012] [PMID: 26569445]

[74]   Liu, J.; Zhu, G.; Chen, M.; Ma, X.; Yang, J. Fabrication of electrospun ZnO nanofiber-modified
        electrode for the determination of trace Cd(II). *Sens. Actuators B Chem.,* **2016**, *234*, 84-91.
        [http://dx.doi.org/10.1016/j.snb.2016.04.073]

[75]   Shaibani, P.M.; Jiang, K.; Haghighat, G.; Hassanpourfard, M.; Etayash, H.; Naicker, S.; Thundat, T.
        The detection of *Escherichia coli* (E. coli) with the pH sensitive hydrogel nanofiber-light addressable
        potentiometric sensor (NF-LAPS). *Sens. Actuators B Chem.,* **2016**, *226*, 176-183.
        [http://dx.doi.org/10.1016/j.snb.2015.11.135]

[76]   Arvand, M.; Ghodsi, N.; Zanjanchi, M.A. A new microplatform based on titanium dioxide

nanofibers/graphene oxide nanosheets nanocomposite modified screen printed carbon electrode for electrochemical determination of adenine in the presence of guanine. *Biosens. Bioelectron.,* **2016**, *77,* 837-844.
[http://dx.doi.org/10.1016/j.bios.2015.10.055] [PMID: 26556182]

[77]   Jiang, Y.; Nie, G.; Chi, M.; Yang, Z.; Zhang, Z.; Wang, C.; Lu, X. Synergistic effect of ternary electrospun TiO₂/Fe₂O₃/PPy composite nanofibers on peroxidase-like mimics with enhanced catalytic performance. *RSC Advances,* **2016**, *6,* 31107-31113.
[http://dx.doi.org/10.1039/C5RA26706D]

[78]   Yang, T.; Ma, J.; Zhen, S.J.; Huang, C.Z. Electrostatic Assemblies of Well-Dispersed AgNPs on the Surface of Electrospun Nanofibers as Highly Active SERS Substrates for Wide-Range pH Sensing. *ACS Appl. Mater. Interfaces,* **2016**, *8*(23), 14802-14811.
[http://dx.doi.org/10.1021/acsami.6b03720] [PMID: 27214514]

[79]   Lu, N.; Shao, C.; Li, X.; Miao, F.; Wang, K.; Liu, Y. CuO nanoparticles/nitrogen-doped carbon nanofibers modified glassy carbon electrodes for non-enzymatic glucose sensors with improved sensitivity. *Ceram. Int.,* **2016**, *42,* 11285-11293.
[http://dx.doi.org/10.1016/j.ceramint.2016.04.046]

[80]   Senthamizhan, A.; Balusamy, B.; Aytac, Z.; Uyar, T. Ultrasensitive electrospun fluorescent nanofibrous membrane for rapid visual colorimetric detection of H2O2. *Anal. Bioanal. Chem.,* **2016**, *408*(5), 1347-1355.
[http://dx.doi.org/10.1007/s00216-015-9149-5] [PMID: 26637215]

[81]   Najarzadekan, H.; Sereshti, H. Development of a colorimetric sensor for nickel ion based on transparent electrospun composite nanofibers of polycaprolactam-dimethylglyoxime/polyvinyl alcohol. *J. Mater. Sci.,* **2016**, *51,* 8645-8654.
[http://dx.doi.org/10.1007/s10853-016-0123-1]

[82]   Devadoss, A.; Kuragano, A.; Terashima, C.; Sudhagar, P.; Nakata, K.; Kondo, T.; Yuasa, M.; Fujishima, A. Single-step electrospun TiO₂–Au hybrid electrodes for high selectivity photoelectrocatalytic glutathione bioanalysis. *J. Mater. Chem. B Mater. Biol. Med.,* **2016**, *4,* 220-228.
[http://dx.doi.org/10.1039/C5TB01740H]

[83]   Li, S.; Zhou, S.; Xu, H.; Xiao, L.; Wang, Y.; Shen, H.; Wang, H.; Yuan, Q. Luminescent properties and sensing performance of a carbon quantum dot encapsulated mesoporous silica/polyacrylonitrile electrospun nanofibrous membrane. *J. Mater. Sci.,* **2016**, *51,* 6801-6811.
[http://dx.doi.org/10.1007/s10853-016-9967-7]

[84]   Raj, S.; Shankaran, D.R. Curcumin based biocompatible nanofibers for lead ion detection. *Sens. Actuators B Chem.,* **2016**, *226,* 318-325.
[http://dx.doi.org/10.1016/j.snb.2015.12.006]

[85]   Fayemi, O. E.; Adekunle, A. S.; Ebenso, E. E. *Electrochemical detection of phenanthrene using nickel oxide doped PANI nanofiber based modified electrodes.,* **2016**.
[http://dx.doi.org/10.1155/2016/9614897]

[86]   Yang, Z.; Zhang, Z.; Nie, G.; Lu, X.; Wang, C. Palladium nanoparticles modified CoFe₂O₄ nanotubes with enhanced peroxidase-like activity for colorimetric detection of hydrogen peroxide. *RSC Advances,* **2016**, *6,* 1-25.

[87]   Sensing, N. E. *Nanoporous Carbon Nanofibers Decorated with Platinum Nanoparticles for Non-Enzymatic Electrochemical Sensing of H₂O₂,* **2015**, 1891-1905.

[88]   Liu, D.; Guo, Q.; Zhang, X.; Hou, H.; You, T. PdCo alloy nanoparticle-embedded carbon nanofiber for ultrasensitive nonenzymatic detection of hydrogen peroxide and nitrite. *J. Colloid Interface Sci.,* **2015**, *450,* 168-173.
[http://dx.doi.org/10.1016/j.jcis.2015.03.014] [PMID: 25818356]

[89]   Dai, H.; Gong, L.; Xu, G.; Li, X.; Zhang, S.; Lin, Y.; Zeng, B.; Yang, C.; Chen, G. An electrochemical impedimetric sensor based on biomimetic electrospun nanofibers for formaldehyde. *Analyst (Lond.),*

2015, *140*(2), 582-589.
[http://dx.doi.org/10.1039/C4AN02021A] [PMID: 25426499]

[90]   Xu, D.; Li, L.; Ding, Y.; Cui, S. Electrochemical hydrogen peroxide sensors based on electrospun La$_{0.7}$Sr$_{0.3}$Mn$_{0.75}$Co$_{0.25}$O$_3$ nanofiber modified electrodes. *Anal. Methods,* **2015**, 6083-6088.
[http://dx.doi.org/10.1039/C5AY01131K]

[91]   Karube, I.; Nakanishi, K. Microbial biosensors for process and environmental control. *IEEE Eng. Med. Biol. Mag.,* **1994**, *13*, 364-374.
[http://dx.doi.org/10.1109/51.294008]

[92]   Balasubramanian, K.; Burghard, M. Biosensors based on carbon nanotubes. *Anal. Bioanal. Chem.,* **2006**, *385*(3), 452-468.
[http://dx.doi.org/10.1007/s00216-006-0314-8] [PMID: 16568294]

[93]   Dzyadevych, S.V.; Arkhypova, V.N.; Soldatkin, A.P.; El'skaya, A.V.; Martelet, C.; Jaffrezic-Renault, N. Amperometric enzyme biosensors: Past, present and future. *ITBM-RBM,* **2008**, *29*, 171-180.

[94]   Ramakrishna, S.; Fujihara, K.; Teo, W.E.; Yong, T.; Ma, Z.; Ramaseshan, R. Electrospun nanofibers: Solving global issues. *Mater. Today,* **2006**, *9*, 40-50.
[http://dx.doi.org/10.1016/S1369-7021(06)71389-X]

[95]   Celebioglu, A.; Uyar, T. Electrospinning of nanofibers from non-polymeric systems: electrospun nanofibers from native cyclodextrins. *J. Colloid Interface Sci.,* **2013**, *404*, 1-7.
[http://dx.doi.org/10.1016/j.jcis.2013.04.034] [PMID: 23711660]

[96]   Canbolat, M.F.; Celebioglu, A.; Uyar, T. Drug delivery system based on cyclodextrin-naproxen inclusion complex incorporated in electrospun polycaprolactone nanofibers. *Colloids Surf. B Biointerfaces,* **2014**, *115*, 15-21.
[http://dx.doi.org/10.1016/j.colsurfb.2013.11.021] [PMID: 24316584]

[97]   Lee, S.; Jin, G.; Jang, J-H. Electrospun nanofibers as versatile interfaces for efficient gene delivery. *J. Biol. Eng.,* **2014**, *8*, 30.
[http://dx.doi.org/10.1186/1754-1611-8-30] [PMID: 25926887]

[98]   Li, M.; Mondrinos, M.J.; Gandhi, M.R.; Ko, F.K.; Weiss, A.S.; Lelkes, P.I. Electrospun protein fibers as matrices for tissue engineering. *Biomaterials,* **2005**, *26*(30), 5999-6008.
[http://dx.doi.org/10.1016/j.biomaterials.2005.03.030] [PMID: 15894371]

[99]   Rim, N.G.; Shin, C.S.; Shin, H. Current approaches to electrospun nanofibers for tissue engineering. *Biomed. Mater.,* **2013**, *8*(1), 014102.
[http://dx.doi.org/10.1088/1748-6041/8/1/014102] [PMID: 23472258]

[100]   Li, D.; Pang, Z.; Chen, X.; Luo, L.; Cai, Y.; Wei, Q. A catechol biosensor based on electrospun carbon nanofibers. *Beilstein J. Nanotechnol.,* **2014**, *5*, 346-354.
[http://dx.doi.org/10.3762/bjnano.5.39] [PMID: 24778958]

[101]   Fu, J.; Qiao, H.; Li, D.; Luo, L.; Chen, K.; Wei, Q. Laccase biosensor based on electrospun copper/carbon composite nanofibers for catechol detection. *Sensors (Basel),* **2014**, *14*(2), 3543-3556.
[http://dx.doi.org/10.3390/s140203543] [PMID: 24561403]

[102]   Marx, S.; Jose, M.V.; Andersen, J.D.; Russell, A.J. Electrospun gold nanofiber electrodes for biosensors. *Biosens. Bioelectron.,* **2011**, *26*(6), 2981-2986.
[http://dx.doi.org/10.1016/j.bios.2010.11.050] [PMID: 21196109]

[103]   Ji, X.; Su, Z.; Wang, P.; Ma, G.; Zhang, S. "Ready-to-use" hollow nanofiber membrane-based glucose testing strips. *Analyst (Lond.),* **2014**, *139*(24), 6467-6473.
[http://dx.doi.org/10.1039/C4AN01354A] [PMID: 25343161]

[104]   Ren, G.; Xu, X.; Liu, Q.; Cheng, J.; Yuan, X.; Wu, L.; Wan, Y. Electrospun poly (vinyl alcohol)/glucose oxidase biocomposite membranes for biosensor applications. *React. Funct. Polym.,* **2006**, *66*, 1559-1564.
[http://dx.doi.org/10.1016/j.reactfunctpolym.2006.05.005]

[105] Numnuam, A.; Thavarungkul, P.; Kanatharana, P. An amperometric uric acid biosensor based on chitosan-carbon nanotubes electrospun nanofiber on silver nanoparticles. *Anal. Bioanal. Chem.,* **2014**, *406*(15), 3763-3772.
[http://dx.doi.org/10.1007/s00216-014-7770-3] [PMID: 24718436]

[106] Uzun, S.D.; Kayaci, F.; Uyar, T.; Timur, S.; Toppare, L. Bioactive surface design based on functional composite electrospun nanofibers for biomolecule immobilization and biosensor applications. *ACS Appl. Mater. Interfaces,* **2014**, *6*(7), 5235-5243.
[http://dx.doi.org/10.1021/am5005927] [PMID: 24660809]

[107] Huang, S.; Ding, Y.; Liu, Y.; Su, L.; Filosa, R.; Lei, Y. Glucose Biosensor Using Glucose Oxidase and Electrospun Mn_2O_3-Ag Nanofibers. *Electroanalysis,* **2011**, *23*, 1912-1920.
[http://dx.doi.org/10.1002/elan.201100221]

[108] Tang, H.; Yan, F.; Tai, Q.; Chan, H.L. The improvement of glucose bioelectrocatalytic properties of platinum electrodes modified with electrospun TiO_2 nanofibers. *Biosens. Bioelectron.,* **2010**, *25*(7), 1646-1651.
[http://dx.doi.org/10.1016/j.bios.2009.11.027] [PMID: 20045308]

[109] Wu, J.; Yin, F. Sensitive enzymatic glucose biosensor fabricated by electrospinning composite nanofibers and electrodepositing Prussian blue film. *J. Electroanal. Chem.,* **2013**, *694*, 1-5.
[http://dx.doi.org/10.1016/j.jelechem.2013.02.003]

[110] Sapountzi, E.; Braiek, M.; Farre, C.; Arab, M.; Chateaux, J-F.; Jaffrezic-Renault, N.; Lagarde, F. One-Step Fabrication of Electrospun Photo-Cross-Linkable Polymer Nanofibers Incorporating Multiwall Carbon Nanotubes and Enzyme for Biosensing. *J. Electrochem. Soc.,* **2015**, *162*, B275-B281.
[http://dx.doi.org/10.1149/2.0831510jes]

[111] Wu, C.M.; Yu, S.A.; Lin, S.L. Graphene modified electrospun poly (vinyl alcohol) nanofibrous membranes for glucose oxidase immobilization. *Express Polym. Lett.,* **2014**, *8*, 565-573.
[http://dx.doi.org/10.3144/expresspolymlett.2014.60]

[112] Su, X.; Ren, J.; Meng, X.; Ren, X.; Tang, F. A novel platform for enhanced biosensing based on the synergy effects of electrospun polymer nanofibers and graphene oxides. *Analyst (Lond.),* **2013**, *138*(5), 1459-1466.
[http://dx.doi.org/10.1039/c2an36663k] [PMID: 23325000]

[113] Mondal, K.; Ali, A.; Singh, C.; Sumana, G.; Malhotra, B.D.; Sharma, A. Highly Sensitive Porous Carbon and Metal/Carbon Conducting Nanofiber Based Enzymatic Biosensors for Triglyceride Detection. *Sens. Actuators B Chem.,* **2017**, *246*, 202-214.
[http://dx.doi.org/10.1016/j.snb.2017.02.050]

[114] Sapountzi, E.; Braiek, M.; Vocanson, F.; Chateaux, J.F.; Jaffrezic-Renault, N.; Lagarde, F. Gold nanoparticles assembly on electrospun poly (vinyl alcohol)/poly (ethyleneimine)/glucose oxidase nanofibers for ultrasensitive electrochemical glucose biosensing. *Sens. Actuators B Chem.,* **2017**, *238*, 392-401.
[http://dx.doi.org/10.1016/j.snb.2016.07.062]

[115] Ali, M.A.; Jiang, H.; Mahal, N.K.; Weber, R.J.; Kumar, R.; Castellano, M.J.; Dong, L. Microfluidic impedimetric sensor for soil nitrate detection using graphene oxide and conductive nanofibers enabled sensing interface. *Sens. Actuators B Chem.,* **2017**, *239*, 1289-1299.
[http://dx.doi.org/10.1016/j.snb.2016.09.101]

[116] Bajaj, B.; Joh, H.I.; Jo, S.M.; Kaur, G.; Sharma, A.; Tomar, M.; Gupta, V.; Lee, S. Controllable one step copper coating on carbon nanofibers for flexible cholesterol biosensor substrates. *J. Mater. Chem. B Mater. Biol. Med.,* **2016**, *4*, 229-236.
[http://dx.doi.org/10.1039/C5TB01781E]

[117] El-Moghazy, A.Y.; Soliman, E.A.; Ibrahim, H.Z.; Marty, J.L.; Istamboulie, G.; Noguer, T. Biosensor based on electrospun blended chitosan-poly (vinyl alcohol) nanofibrous enzymatically sensitized membranes for pirimiphos-methyl detection in olive oil. *Talanta,* **2016**, *155*, 258-264.

[http://dx.doi.org/10.1016/j.talanta.2016.04.018] [PMID: 27216682]

[118]   Golshaei, R.; Karazehir, T.; Ghoreishi, S.M.; Ates, M.; Sarac, A.S. Glucose oxidase immobilization onto Au/poly[anthranilic acid-co-3-carboxy-N-(2-thenylidene)aniline]/PVAc electrospun nanofibers. *Polym. Bull.,* **2017,** *74,* 1493-1517.
[http://dx.doi.org/10.1007/s00289-016-1786-0]

[119]   Yang, J.; Li, D.; Fu, J.; Huang, F.; Wei, Q. TiO$_2$-CuCNFs based laccase biosensor for enhanced electrocatalysis in hydroquinone detection. *J. Electroanal. Chem.,* **2016,** *766,* 16-23.
[http://dx.doi.org/10.1016/j.jelechem.2016.01.030]

[120]   Fazel, R.; Torabi, S.F.; Naseri-Nosar, P.; Ghasempur, S.; Ranaei-Siadat, S.O.; Khajeh, K. Electrospun polyvinyl alcohol/bovine serum albumin biocomposite membranes for horseradish peroxidase immobilization. *Enzyme Microb. Technol.,* **2016,** *93-94,* 1-10.
[http://dx.doi.org/10.1016/j.enzmictec.2016.07.002] [PMID: 27702468]

[121]   Han, S.W.; Koh, W.G. Hydrogel-Framed Nanofiber Matrix Integrated with a Microfluidic Device for Fluorescence Detection of Matrix Metalloproteinases-9. *Anal. Chem.,* **2016,** *88*(12), 6247-6253.
[http://dx.doi.org/10.1021/acs.analchem.5b04867] [PMID: 27214657]

[122]   Dagli, U.; Guler, Z.; Sarac, A.S. Covalent Immobilization of Tyrosinase on Electrospun Polyacrylonitrile/Polyurethane/Poly (*m*-anthranilic acid) Nanofibers: An Electrochemical Impedance Study. *Polym. Plast. Technol. Eng.,* **2015,** *54,* 1494-1504.
[http://dx.doi.org/10.1080/03602559.2015.1010218]

[123]   Zhang, X.; Liu, D.; Li, L.; You, T. Direct electrochemistry of glucose oxidase on novel free-standing nitrogen-doped carbon nanospheres@carbon nanofibers composite film. *Sci. Rep.,* **2015,** *5,* 9885.
[http://dx.doi.org/10.1038/srep09885] [PMID: 25943704]

[124]   Fu, J.; Li, D.; Li, G.; Huang, F.; Wei, Q. Carboxymethyl cellulose assisted immobilization of silver nanoparticles onto cellulose nanofibers for the detection of catechol. *J. Electroanal. Chem.,* **2015,** *738,* 92-99.
[http://dx.doi.org/10.1016/j.jelechem.2014.11.025]

[125]   Li, D.; Lv, P.; Zhu, J.; Lu, Y.; Chen, C.; Zhang, X.; Wei, Q. NiCu alloy nanoparticle-loaded carbon nanofibers for phenolic biosensor applications. *Sensors (Basel),* **2015,** *15*(11), 29419-29433.
[http://dx.doi.org/10.3390/s151129419] [PMID: 26610505]

[126]   Wilson, R.; Turner, A.P. Glucose oxidase: an ideal enzyme. *Biosens. Bioelectron.,* **1992,** *7,* 165-185.
[http://dx.doi.org/10.1016/0956-5663(92)87013-F]

[127]   Seleci, M.; Ag, D.; Yalcinkaya, E.E.; Demirkol, D.O.; Guler, C.; Timur, S. Amine-intercalated montmorillonite matrices for enzyme immobilization and biosensing applications. *RSC Advances,* **2012,** *2,* 2112-2118.
[http://dx.doi.org/10.1039/c2ra01225a]

[128]   Yuksel, M.; Akin, M.; Geyik, C.; Demirkol, D.O.; Ozdemir, C.; Bluma, A.; Höpfner, T.; Beutel, S.; Timur, S.; Scheper, T. Offline glucose biomonitoring in yeast culture by polyamidoamine/cysteamine-modified gold electrodes. *Biotechnol. Prog.,* **2011,** *27*(2), 530-538.
[http://dx.doi.org/10.1002/btpr.544] [PMID: 21485034]

[129]   Jarosz-Wilkołazka, A.; Ruzgas, T.; Gorton, L. Amperometric detection of mono- and diphenols at Cerrena unicolor laccase-modified graphite electrode: correlation between sensitivity and substrate structure. *Talanta,* **2005,** *66*(5), 1219-1224.
[http://dx.doi.org/10.1016/j.talanta.2005.01.026] [PMID: 18970111]

[130]   Odaci, D.; Timur, S.; Pazarlioglu, N.; Montereali, M.R.; Vastarella, W.; Pilloton, R.; Telefoncu, A. Determination of phenolic acids using Trametes versicolor laccase. *Talanta,* **2007,** *71*(1), 312-317.
[http://dx.doi.org/10.1016/j.talanta.2006.04.032] [PMID: 19071305]

[131]   Rodriguez-Delgado, M.M.; Aleman-Nava, G.S.; Rodriguez-Delgado, J.M.; Dieck-Assad, G.; Martinez-Chapa, S.O.; Barcelo, D.; Parra, R. Laccase-based biosensors for detection of phenolic

compounds. *TrAC -. Trends Analyt. Chem.,* **2015**, *74*, 21-45.
[http://dx.doi.org/10.1016/j.trac.2015.05.008]

[132] Liu, J.; Niu, J.; Yin, L.; Jiang, F. *In situ* encapsulation of laccase in nanofibers by electrospinning for development of enzyme biosensors for chlorophenol monitoring. *Analyst (Lond.),* **2011**, *136*(22), 4802-4808.
[http://dx.doi.org/10.1039/c1an15649g] [PMID: 21961111]

[133] Damar, K.; Odaci Demirkol, D. Modified gold surfaces by poly(amidoamine) dendrimers and fructose dehydrogenase for mediated fructose sensing. *Talanta,* **2011**, *87*, 67-73.
[http://dx.doi.org/10.1016/j.talanta.2011.09.042] [PMID: 22099650]

[134] Tkáč, J.; Štefuca, V.; Gemeiner, P. *Biosensors with immobilised microbial cells using amperometric and thermal detection principles*; Appl. Cell Immobil. Biotechnol, **2005**, pp. 549-566.

[135] D'Souza, S.F. Microbial biosensors. *Biosens. Bioelectron.,* **2001**, *16*(6), 337-353.
[http://dx.doi.org/10.1016/S0956-5663(01)00125-7] [PMID: 11672648]

[136] Timur, S.; Anik, U.; Odaci, D.; Gorton, L. Development of a microbial biosensor based on carbon nanotube (CNT) modified electrodes. *Electrochem. Commun.,* **2007**, *9*, 1810-1815.
[http://dx.doi.org/10.1016/j.elecom.2007.04.012]

[137] Salalha, W.; Kuhn, J.; Dror, Y.; Zussman, E. Encapsulation of bacteria and viruses in electrospun nanofibres. *Nanotechnology,* **2006**, *17*(18), 4675-4681.
[http://dx.doi.org/10.1088/0957-4484/17/18/025] [PMID: 21727596]

[138] Klein, S.; Kuhn, J.; Avrahami, R.; Tarre, S.; Beliavski, M.; Green, M.; Zussman, E. Encapsulation of bacterial cells in electrospun microtubes. *Biomacromolecules,* **2009**, *10*(7), 1751-1756.
[http://dx.doi.org/10.1021/bm900168v] [PMID: 21197961]

[139] San, N.O.; Celebioglu, A.; Tumtas, Y.; Uyar, T.; Tekinay, T. Reusable bacteria immobilized electrospun nanofibrous webs for decolorization of methylene blue dye in wastewater treatment. *RSC Advances,* **2014**, *4*, 32249-32255.
[http://dx.doi.org/10.1039/C4RA04250F]

[140] Luppa, P.B.; Sokoll, L.J.; Chan, D.W. Immunosensors--principles and applications to clinical chemistry. *Clin. Chim. Acta,* **2001**, *314*(1-2), 1-26.
[http://dx.doi.org/10.1016/S0009-8981(01)00629-5] [PMID: 11718675]

[141] González, E.; Shepherd, L.M.; Saunders, L.; Frey, M.W. Surface Functional Poly(lactic Acid) Electrospun Nanofibers for Biosensor Applications. *Materials (Basel),* **2016**, *9*(1), 47.
[http://dx.doi.org/10.3390/ma9010047] [PMID: 28787847]

[142] Wang, X.; Wang, X.; Wang, X.; Chen, F.; Zhu, K.; Xu, Q.; Tang, M. Novel electrochemical biosensor based on functional composite nanofibers for sensitive detection of p53 tumor suppressor gene. *Anal. Chim. Acta,* **2013**, *765*, 63-69.
[http://dx.doi.org/10.1016/j.aca.2012.12.037] [PMID: 23410627]

[143] Paul, K.B.; Singh, V.; Vanjari, S.R.; Singh, S.G. One step biofunctionalized electrospun multiwalled carbon nanotubes embedded zinc oxide nanowire interface for highly sensitive detection of carcinoma antigen-125. *Biosens. Bioelectron.,* **2017**, *88*, 144-152.
[http://dx.doi.org/10.1016/j.bios.2016.07.114] [PMID: 27520500]

[144] Tripathy, S.; Krishna Vanjari, S.R.; Singh, V.; Swaminathan, S.; Singh, S.G. Electrospun manganese (III) oxide nanofiber based electrochemical DNA-nanobiosensor for zeptomolar detection of dengue consensus primer. *Biosens. Bioelectron.,* **2017**, *90*, 378-387.
[http://dx.doi.org/10.1016/j.bios.2016.12.008] [PMID: 27940241]

[145] Yang, X.; Li, X.; Zhang, L.; Gong, J. Electrospun template directed molecularly imprinted nanofibers incorporated with BiOI nanoflake arrays as photoactive electrode for photoelectrochemical detection of triphenyl phosphate. *Biosens. Bioelectron.,* **2017**, *92*, 61-67.
[http://dx.doi.org/10.1016/j.bios.2017.01.056] [PMID: 28187300]

[146] Xu, G.; Zhang, S.; Zhang, Q.; Gong, L.; Dai, H.; Lin, Y. Magnetic functionalized electrospun nanofibers for magnetically controlled ultrasensitive label-free electrochemiluminescent immune detection of aflatoxin B1. *Sens. Actuators B Chem.,* **2016**, *222*, 707-713.
[http://dx.doi.org/10.1016/j.snb.2015.08.129]

[147] Mondal, K.; Ali, M.A.; Srivastava, S.; Malhotra, B.D.; Sharma, A. Electrospun functional micro/nanochannels embedded in porous carbon electrodes for microfluidic biosensing. *Sens. Actuators B Chem.,* **2016**, *229*, 82-91.
[http://dx.doi.org/10.1016/j.snb.2015.12.108]

[148] Brince Paul, K.; Kumar, S.; Tripathy, S.; Vanjari, S.R.; Singh, V.; Singh, S.G. A highly sensitive self assembled monolayer modified copper doped zinc oxide nanofiber interface for detection of Plasmodium falciparum histidine-rich protein-2: Targeted towards rapid, early diagnosis of malaria. *Biosens. Bioelectron.,* **2016**, *80*, 39-46.
[http://dx.doi.org/10.1016/j.bios.2016.01.036] [PMID: 26803412]

[149] Prakash, M.D.; Singh, S.G.; Sharma, C.S.; Krishna, V.S. Electrochemical Detection of Cardiac Biomarkers Utilizing Electrospun Multiwalled Carbon Nanotubes Embedded SU-8 Nanofibers. *Electroanalysis,* **2016**, 380-386.

[150] Paul, B.; Reddy, R. R.; Rama, S.; Vanjari, K.; Singh, S. G. *Zinc oxide nanowire modified flexible plastic platform for immunosensing.,* **2016**.
[http://dx.doi.org/10.1109/ICSENS.2016.7808861]

[151] Hui, N.; Sun, X.; Niu, S.; Luo, X. PEGylated Polyaniline Nanofibers: Antifouling and Conducting Biomaterial for Electrochemical DNA Sensing. *ACS Appl. Mater. Interfaces,* **2017**, *9*(3), 2914-2923.
[http://dx.doi.org/10.1021/acsami.6b11682] [PMID: 28026927]

[152] Guler, Z.; Erkoc, P.; Sarac, A.S. Electrochemical impedance spectroscopic study of single-stranded DNA-immobilized electroactive polypyrrole-coated electrospun poly(ε-caprolactone) nanofibers. *Mater. Express,* **2015**, *5*, 269-279.
[http://dx.doi.org/10.1166/mex.2015.1249]

[153] Lu, Y.; Luo, L.; Ding, Y.; Wang, Y ; Zhou, M.; Zhou, T.; Zhu, D.; Li, X. Electrospun nickel loaded porous carbon nanofibers for simultaneous determination of adenine and guanine. *Electrochim. Acta,* **2015**, *174*, 191-198.
[http://dx.doi.org/10.1016/j.electacta.2015.05.165]

[154] Constantin Diculescu, V.; Chiorcea Paquim, A-M.; Maria Oliveira Brett, A. Electrochemical DNA Sensors for Detection of DNA Damage. *Sensors (Basel),* **2005**, *5*, 377-393.
[http://dx.doi.org/10.3390/s5060377]

[155] Drummond, T.G.; Hill, M.G.; Barton, J.K.; Hill, M.G.; Barton, J.K.; Barton, J.K. Electrochemical DNA sensors. *Nat. Biotechnol.,* **2003**, *21*(10), 1192-1199.
[http://dx.doi.org/10.1038/nbt873] [PMID: 14520405]

[156] Wang, H.; Wang, D.; Peng, Z.; Tang, W.; Li, N.; Liu, F. Assembly of DNA-functionalized gold nanoparticles on electrospun nanofibers as a fluorescent sensor for nucleic acids. *Chem. Commun. (Camb.),* **2013**, *49*(49), 5568-5570.
[http://dx.doi.org/10.1039/c3cc41753k] [PMID: 23673392]

[157] Tripathy, S.; Krishna Vanjari, S.R.; Singh, V.; Swaminathan, S.; Singh, S.G. Electrospun manganese (III) oxide nanofiber based electrochemical DNA-nanobiosensor for zeptomolar detection of dengue consensus primer. *Biosens. Bioelectron.,* **2017**, *90*, 378-387.
[http://dx.doi.org/10.1016/j.bios.2016.12.008] [PMID: 27940241]

[158] Graham, K.; Ouyang, M.; Raether, T.; Grafe, T.; Mcdonald, B.; Knauf, P. Polymeric Nanofibers in Air Filtration Applications. *Fifteenth Annual Technical Conference & Expo of the American Filtration & Separations Society,* **2002**, pp. 9-12.

[159] Grantz, D.A.; Garner, J.H.; Johnson, D.W. Ecological effects of particulate matter. *Environ. Int.,* **2003**,

29(2-3), 213-239.
[http://dx.doi.org/10.1016/S0160-4120(02)00181-2] [PMID: 12676209]

[160] Harrison, R.M.; Yin, J. Particulate matter in the atmosphere: which particle properties are important for its effects on health? *Sci. Total Environ.,* **2000**, *249*(1-3), 85-101.
[http://dx.doi.org/10.1016/S0048-9697(99)00513-6] [PMID: 10813449]

[161] Sundarrajan, S.; Tan, K.L.; Lim, S.H.; Ramakrishna, S. Electrospun nanofibers for air filtration applications. *Procedia Eng.,* **2014**, *75*, 159-163.
[http://dx.doi.org/10.1016/j.proeng.2013.11.034]

[162] Zhu, M.; Han, J.; Wang, F.; Shao, W.; Xiong, R.; Zhang, Q.; Pan, H.; Yang, Y.; Samal, S.K.; Zhang, F.; Huang, C. Electrospun Nanofibers Membranes for Effective Air Filtration. *Macromol. Mater. Eng.,* **2016**, •••, 1600353.

[163] Liu, C.; Hsu, P.C.; Lee, H-W.; Ye, M.; Zheng, G.; Liu, N.; Li, W.; Cui, Y. Transparent air filter for high-efficiency PM2.5 capture. *Nat. Commun.,* **2015**, *6*, 6205.
[http://dx.doi.org/10.1038/ncomms7205] [PMID: 25683688]

[164] Hsi, H-C.; Horng, R.S.; Pan, T-A.; Lee, S-K. Preparation of activated carbons from raw and biotreated agricultural residues for removal of volatile organic compounds. *J. Air Waste Manag. Assoc.,* **2011**, *61*(5), 543-551.
[http://dx.doi.org/10.3155/1047-3289.61.5.543] [PMID: 21608494]

[165] Scholten, E.; Bromberg, L.; Rutledge, G.C.; Hatton, T.A. Electrospun polyurethane fibers for absorption of volatile organic compounds from air. *ACS Appl. Mater. Interfaces,* **2011**, *3*(10), 3902-3909.
[http://dx.doi.org/10.1021/am200748y] [PMID: 21888418]

[166] Uyar, T.; Havelund, R.; Nur, Y.; Balan, A.; Hacaloglu, J.; Toppare, L.; Besenbacher, F.; Kingshott, P. Cyclodextrin functionalized poly (methyl methacrylate) (PMMA) electrospun nanofibers for organic vapors waste treatment. *J. Membr. Sci.,* **2010**, *365*, 409-417.
[http://dx.doi.org/10.1016/j.memsci.2010.09.037]

[167] Lala, N.L.; Ramaseshan, R.; Bojun, L.; Sundarrajan, S.; Barhate, R.S.; Ying-Jun, L.; Ramakrishna, S. Fabrication of nanofibers with antimicrobial functionality used as filters: protection against bacterial contaminants. *Biotechnol. Bioeng.,* **2007**, *97*(6), 1357-1365.
[http://dx.doi.org/10.1002/bit.21351] [PMID: 17274060]

[168] Jeong, E.H.; Yang, J.; Youk, J.H. Preparation of polyurethane cationomer nanofiber mats for use in antimicrobial nanofilter applications. *Mater. Lett.,* **2007**, *61*, 3991-3994.
[http://dx.doi.org/10.1016/j.matlet.2007.01.003]

[169] Xu, J.; Liu, C.; Hsu, P.C.; Liu, K.; Zhang, R.; Liu, Y.; Cui, Y. Roll-to-Roll Transfer of Electrospun Nanofiber Film for High-Efficiency Transparent Air Filter. *Nano Lett.,* **2016**, *16*(2), 1270-1275.
[http://dx.doi.org/10.1021/acs.nanolett.5b04596] [PMID: 26789781]

[170] Zhang, R.; Liu, C.; Hsu, P.C.; Zhang, C.; Liu, N.; Zhang, J.; Lee, H.R.; Lu, Y.; Qiu, Y.; Chu, S.; Cui, Y. Nanofiber air filters with high-temperature stability for efficient PM2.5 removal from the pollution sources. *Nano Lett.,* **2016**, *16*(6), 3642-3649.
[http://dx.doi.org/10.1021/acs.nanolett.6b00771] [PMID: 27167892]

[171] Wang, H.; Zheng, G.F.; Wang, X.; Sun, D.H. Study on the air filtration performance of nanofibrous membranes compared with conventional fibrous filters. *IEEE 5th Int. Conf. Nano/Micro Eng. Mol. Syst. NEMS 2010,* **2010**, pp. 387-390.

[172] Wang, Z.; Zhao, C.; Pan, Z. Porous bead-on-string poly(lactic acid) fibrous membranes for air filtration. *J. Colloid Interface Sci.,* **2015**, *441*, 121-129.
[http://dx.doi.org/10.1016/j.jcis.2014.11.041] [PMID: 25499733]

[173] Montazer, M.; Malekzadeh, S.B. Electrospun antibacterial nylon nanofibers through *in situ* synthesis of nanosilver: Preparation and characteristics. *J. Polym. Res.,* **2012**, *19*, 9980.

[http://dx.doi.org/10.1007/s10965-012-9980-8]

[174] Sundarrajan, S.; Ramakrishna, S. Fabrication of nanocomposite membranes from nanofibers and nanoparticles for protection against chemical warfare stimulants. *J. Mater. Sci.*, **2007**, *42*, 8400-8407.
[http://dx.doi.org/10.1007/s10853-007-1786-4]

[175] Choi, J.; Yang, B.J.; Bae, G.N.; Jung, J.H. Herbal Extract Incorporated Nanofiber Fabricated by an Electrospinning Technique and its Application to Antimicrobial Air Filtration. *ACS Appl. Mater. Interfaces*, **2015**, *7*(45), 25313-25320.
[http://dx.doi.org/10.1021/acsami.5b07441] [PMID: 26505783]

[176] Zhang, S.; Tang, N.; Cao, L.; Yin, X.; Yu, J.; Ding, B. Highly Integrated Polysulfone/Polyacrylonitrile/Polyamide-6 Air Filter for Multilevel Physical Sieving Airborne Particles. *ACS Appl. Mater. Interfaces*, **2016**, *8*, 29062-29072.
[http://dx.doi.org/10.1021/acsami.6b10094]

[177] Chuang, Y-H.; Cheng, F-Y.; Chang, C-T.; Chih-Ming, M.; Hong, C-B.; Shiue, A. Study on particulates and VOCs removal with TiO_2 non-woven filter prepared by electrospinning. *Proceedings of the Air and Waste Management Association's Annual Conference and Exhibition, AWMA*, **2013**, pp. 2960-2970.

[178] Nicosia, A.; Gieparda, W.; Foksowicz-Flaczyk, J.; Walentowska, J.; Wesołek, D.; Vazquez, B.; Prodi, F.; Belosi, F. Air filtration and antimicrobial capabilities of electrospun PLA/PHB containing ionic liquid. *Separ. Purif. Tech.*, **2015**, *154*, 154-160.
[http://dx.doi.org/10.1016/j.seppur.2015.09.037]

[179] Wang, L.; Zhang, C.; Gao, F.; Pan, G. Needleless electrospinning for scaled-up production of ultrafine chitosan hybrid nanofibers used for air filtration. *RSC Advances*, **2016**, *6*, 105988-105995.
[http://dx.doi.org/10.1039/C6RA24557A]

[180] Wang, S.; Zhao, X.; Yin, X.; Yu, J.; Ding, B. Electret Polyvinylidene Fluoride Nanofibers Hybridized by Polytetrafluoroethylene Nanoparticles for High-Efficiency Air Filtration. *ACS Appl. Mater. Interfaces*, **2016**, *8*(36), 23985-23994.
[http://dx.doi.org/10.1021/acsami.6b08262] [PMID: 27552028]

[181] Kayaci, F.; Sen, H.S.; Durgun, E.; Uyar, T. Electrospun nylon 6,6 nanofibers functionalized with cyclodextrins for removal of toluene vapor. *J. Appl. Polym. Sci.*, **2015**, *132*, 34-37.
[http://dx.doi.org/10.1002/app.41941]

[182] Kim, H.J.; Pant, H.R.; Choi, N.J.; Kim, C.S. Composite electrospun fly ash/polyurethane fibers for absorption of volatile organic compounds from air. *Chem. Eng. J.*, **2013**, *230*, 244-250.
[http://dx.doi.org/10.1016/j.cej.2013.06.090]

[183] Khayet, M. Membranes and theoretical modeling of membrane distillation: a review. *Adv. Colloid Interface Sci.*, **2011**, *164*(1-2), 56-88.
[http://dx.doi.org/10.1016/j.cis.2010.09.005] [PMID: 21067710]

[184] Smolders, K.; Franken, A.C. Terminology for Membrane Distillation. *Desalination*, **1989**, *72*, 249-262.
[http://dx.doi.org/10.1016/0011-9164(89)80010-4]

[185] García-Payo, M.C.; Izquierdo-Gil, M.A.; Fernández-Pineda, C. Wetting Study of Hydrophobic Membranes *via* Liquid Entry Pressure Measurements with Aqueous Alcohol Solutions. *J. Colloid Interface Sci.*, **2000**, *230*(2), 420-431.
[http://dx.doi.org/10.1006/jcis.2000.7106] [PMID: 11017750]

[186] Li, X.; Wang, C.; Yang, Y.; Wang, X.; Zhu, M.; Hsiao, B.S. Dual-biomimetic superhydrophobic electrospun polystyrene nanofibrous membranes for membrane distillation. *ACS Appl. Mater. Interfaces*, **2014**, *6*(4), 2423-2430.
[http://dx.doi.org/10.1021/am4048128] [PMID: 24467347]

[187] Chung, S.; Seo, C.D.; Choi, J-H.; Chung, J. Evaluation method of membrane performance in

membrane distillation process for seawater desalination. *Environ. Technol.,* **2014**, *35*(17-20), 2147-2152.
[http://dx.doi.org/10.1080/09593330.2014.895050] [PMID: 25145166]

[188]   Lee, E.J.; An, A.K.; He, T.; Woo, Y.C.; Shon, H.K. Electrospun nanofiber membranes incorporating fluorosilane-coated TiO$_2$ nanocomposite for direct contact membrane distillation. *J. Membr. Sci.,* **2016**, *520*, 145-154.
[http://dx.doi.org/10.1016/j.memsci.2016.07.019]

[189]   Woo, Y.C.; Tijing, L.D.; Park, M.J.; Yao, M.; Choi, J.S.; Lee, S.; Kim, S.H.; An, K.J.; Shon, H.K. Electrospun dual-layer nonwoven membrane for desalination by air gap membrane distillation. *Desalination,* **2017**, *403*, 187-198.
[http://dx.doi.org/10.1016/j.desal.2015.09.009]

[190]   Feng, C.; Khulbe, K.C.; Tabe, S. Volatile organic compound removal by membrane gas stripping using electro-spun nanofiber membrane. *Desalination,* **2012**, *287*, 98-102.
[http://dx.doi.org/10.1016/j.desal.2011.04.074]

[191]   Van Der Bruggen, B.; Vandecasteele, C.; Van Gestel, T.; Doyen, W.; Leysen, R. A review of pressure-driven membrane processes in wastewater treatment and drinking water production. *Environ. Prog.,* **2003**, *22*, 46-56.
[http://dx.doi.org/10.1002/ep.670220116]

[192]   Doyen, W.; Baée, B.; Beeusaert, L. *UF as an alternative pretreatment step for producing drinking water.,* **2000**.
[http://dx.doi.org/10.1016/S0958-2118(00)80235-9]

[193]   Le, N.L.; Nunes, S.P. Materials and membrane technologies for water and energy sustainability. *Sustain. Mater. Technol.,* **2016**, *7*, 1-28.

[194]   Yang, Y.; Li, X.; Shen, L.; Wang, X.; Hsiao, B.S. Ionic Cross-Linked Poly (acrylonitrile- *co* -acrylic acid)/Polyacrylonitrile Thin Film Nanofibrous Composite Membrane with High Ultrafiltration Performance. *Ind. Eng. Chem. Res.,* **2017**, *56*, 3077-3090.
[http://dx.doi.org/10.1021/acs.iecr.7b00244]

[195]   Gopakumar, D.A.; Pasquini, D.; Henrique, M.A.; De Morais, L.C.; Grohens, Y.; Thomas, S. Meldrum's acid modified cellulose nanofiber-based polyvinylidene fluoride microfiltration membrane for dye water treatment and nanoparticle removal. *ACS Sustain. Chem.& Eng.,* **2017**, *5*, 2026-2033.
[http://dx.doi.org/10.1021/acssuschemeng.6b02952]

[196]   Yang, Y.; Li, X.; Shen, L.; Wang, X.; Hsiao, B.S. A durable thin-film nanofibrous composite nanofiltration membrane prepared by interfacial polymerization on a double-layer nanofibrous scaffold. *RSC Advances,* **2017**, *7*, 18001-18013.
[http://dx.doi.org/10.1039/C7RA00621G]

[197]   Shen, L.; Cheng, C.; Yu, X.; Yang, Y.; Wang, X.; Zhu, M.; Hsiao, B.S. Low pressure UV-cured CS–PEO–PTEGDMA/PAN thin film nanofibrous composite nanofiltration membranes for anionic dye separation. *J. Mater. Chem. A Mater. Energy Sustain.,* **2016**, *4*, 15575-15588.
[http://dx.doi.org/10.1039/C6TA04360G]

[198]   Ramakrishna, S.; Shirazi, M. M. A. *Electrospun membranes : Next generation membranes for desalination and water/wastewater treatment,* **2014**, *1*, 46-47.

[199]   Kaur, S.; Ma, Z.; Gopal, R.; Singh, G.; Ramakrishna, S.; Matsuura, T. Plasma-induced graft copolymerization of poly(methacrylic acid) on electrospun poly(vinylidene fluoride) nanofiber membrane. *Langmuir,* **2007**, *23*(26), 13085-13092.
[http://dx.doi.org/10.1021/la701329r] [PMID: 18004889]

[200]   Wang, Y.H.; Tian, T.F.; Liu, X.Q.; Meng, G.Y. Titania membrane preparation with chemical stability for very hash environments applications. *J. Membr. Sci.,* **2006**, *280*, 261-269.
[http://dx.doi.org/10.1016/j.memsci.2006.01.027]

[201] Uyar, T.; Havelund, R.; Nur, Y.; Hacaloglu, J.; Besenbacher, F.; Kingshott, P. Molecular filters based on cyclodextrin functionalized electrospun fibers. *J. Membr. Sci.,* **2009**, *332*, 129-137.
[http://dx.doi.org/10.1016/j.memsci.2009.01.047]

[202] Bae, J.; Baek, I.; Choi, H. Mechanically enhanced PES electrospun nanofiber membranes (ENMs) for microfiltration: The effects of ENM properties on membrane performance. *Water Res.,* **2016**, *105*, 406-412.
[http://dx.doi.org/10.1016/j.watres.2016.09.020] [PMID: 27664541]

[203] Wang, Z.; Crandall, C.; Prautzsch, V.L.; Sahadevan, R.; Menkhaus, T.J.; Fong, H. Electrospun Regenerated Cellulose Nanofiber Membranes Surface-Grafted with Water-Insoluble Poly(HEMA) or Water-Soluble Poly(AAS) Chains *via* the ATRP Method for Ultrafiltration of Water. *ACS Appl. Mater. Interfaces,* **2017**, *9*(4), 4272-4278.
[http://dx.doi.org/10.1021/acsami.6b16116] [PMID: 28078887]

[204] Anka, F.H.; Balkus, K.J. Novel nanofiltration hollow fiber membrane produced *via* electrospinning. *Ind. Eng. Chem. Res.,* **2013**, *52*, 3473-3480.
[http://dx.doi.org/10.1021/ie303173w]

[205] Lee, J.; Jung, J.; Cho, Y. H.; Yadav, S. K.; Baek, K. Y.; Park, H. B.; Hong, S. M.; Koo, C. M. *Fouling-Tolerant Nano fi brous Polymer Membranes for Water Treatment.,* **2014**.

[206] Wang, J.; Zhang, P.; Liang, B.; Liu, Y.; Xu, T.; Wang, L.; Cao, B.; Pan, K. Graphene Oxide as an Effective Barrier on a Porous Nanofibrous Membrane for Water Treatment. *ACS Appl. Mater. Interfaces,* **2016**, *8*(9), 6211-6218.
[http://dx.doi.org/10.1021/acsami.5b12723] [PMID: 26849085]

[207] Ma, H.; Yoon, K.; Rong, L.; Shokralla, M.; Kopot, A.; Wang, X.; Fang, D.; Hsiao, B.S.; Chu, B. Thin-film nanofibrous composite ultrafiltration membranes based on polyvinyl alcohol barrier layer containing directional water channels. *Ind. Eng. Chem. Res.,* **2010**, *49*, 11978-11984.
[http://dx.doi.org/10.1021/ie100545k]

[208] Yoon, K.; Kim, K.; Wang, X.; Fang, D.; Hsiao, B.S.; Chu, B. High flux ultrafiltration membranes based on electrospun nanofibrous PAN scaffolds and chitosan coating. *Polymer (Guildf.),* **2006**, *47*, 2434-2441.
[http://dx.doi.org/10.1016/j.polymer.2006.01.042]

[209] Seyed Shahabadi, S.M.; Mousavi, S.A.; Bastani, D. High flux electrospun nanofiberous membrane: Preparation by statistical approach, characterization, and microfiltration assessment. *Rev. Mex. Urol.,* **2016**, *76*, 474-483.

[210] Liu, Y.; Wang, R.; Ma, H.; Hsiao, B.S.; Chu, B. High-flux microfiltration filters based on electrospun poly (vinyl alcohol) nanofibrous membranes. *Polym. (United Kingdom),* **2013**, *54*, 548-556.

[211] Karimi, E.; Raisi, A.; Aroujalian, A. TiO$_2$-induced photo-cross-linked electrospun polyvinyl alcohol nanofibers microfiltration membranes. *Polym. (United Kingdom),* **2016**, *99*, 642-653.

[212] Zhang, J.; Xue, Q.; Pan, X.; Jin, Y.; Lu, W.; Ding, D.; Guo, Q. Graphene oxide/polyacrylonitrile fiber hierarchical-structured membrane for ultra-fast microfiltration of oil-water emulsion. *Chem. Eng. J.,* **2017**, *307*, 643-649.
[http://dx.doi.org/10.1016/j.cej.2016.08.124]

[213] Soyekwo, F.; Zhang, Q.; Gao, R.; Qu, Y.; Lin, C.; Huang, X.; Zhu, A.; Liu, Q. Cellulose nanofiber intermediary to fabricate highly-permeable ultrathin nanofiltration membranes for fast water purification. *J. Membr. Sci.,* **2017**, *524*, 174-185.
[http://dx.doi.org/10.1016/j.memsci.2016.11.019]

[214] Wang, X.; Zhang, K.; Yang, Y.; Wang, L.; Zhou, Z.; Zhu, M.; Hsiao, B.S.; Chu, B. Development of hydrophilic barrier layer on nanofibrous substrate as composite membrane *via* a facile route. *J. Membr. Sci.,* **2010**, *356*, 110-116.
[http://dx.doi.org/10.1016/j.memsci.2010.03.039]

[215] He, T.; Zhou, W.; Bahi, A.; Yang, H.; Ko, F. High permeability of ultrafiltration membranes based on electrospun PVDF modified by nanosized zeolite hybrid membrane scaffolds under low pressure. *Chem. Eng. J.,* **2014**, *252*, 327-336.
[http://dx.doi.org/10.1016/j.cej.2014.05.022]

[216] Kim, H.C.; Choi, B.G.; Noh, J.; Song, K.G.; Lee, S. hyup; Maeng, S. K. Electrospun nanofibrous PVDF-PMMA MF membrane in laboratory and pilot-scale study treating wastewater from Seoul Zoo. *Desalination,* **2014**, *346*, 107-114.
[http://dx.doi.org/10.1016/j.desal.2014.05.005]

[217] Zhou, W.; Bahi, A.; Li, Y.; Yang, H.; Ko, F. Ultra-filtration membranes based on electrospun poly(vinylidene fluoride) (PVDF) fibrous composite membrane scaffolds. *RSC Adv,* **2013**, *3*, 11614-11620.

[218] Ma, H.; Burger, C.; Hsiao, B.S.; Chu, B.; Ko, F. Ultrafine polysaccharide nanofibrous membranes for water purification. *Biomacromolecules,* **2011**, *12*(4), 970-976.
[http://dx.doi.org/10.1021/bm1013316] [PMID: 21341679]

[219] Ma, Z.; Lan, Z.; Matsuura, T.; Ramakrishna, S. Electrospun polyethersulfone affinity membrane: membrane preparation and performance evaluation. *J. Chromatogr. B Analyt. Technol. Biomed. Life Sci.,* **2009**, *877*(29), 3686-3694.
[http://dx.doi.org/10.1016/j.jchromb.2009.09.019] [PMID: 19775944]

[220] Thavasi, V.; Singh, G.; Ramakrishna, S. Electrospun nanofibers in energy and environmental applications. *Energy Environ. Sci.,* **2008**, *1*, 205-221.
[http://dx.doi.org/10.1039/b809074m]

[221] Ma, Z.; Kotaki, M.; Ramakrishna, S. Surface modified nonwoven polysulphone (PSU) fiber mesh by electrospinning: A novel affinity membrane. *J. Membr. Sci.,* **2006**, *272*, 179-187.
[http://dx.doi.org/10.1016/j.memsci.2005.07.038]

[222] Ma, Z.; Masaya, K.; Ramakrishna, S. Immobilization of Cibacron blue F3GA on electrospun polysulphone ultra-fine fiber surfaces towards developing an affinity membrane for albumin adsorption. *J. Membr. Sci.,* **2006**, *282*, 237-244.
[http://dx.doi.org/10.1016/j.memsci.2006.05.027]

[223] Ma, Z.; Ramakrishna, S. Electrospun regenerated cellulose nanofiber affinity membrane functionalized with protein A/G for IgG purification. *J. Membr. Sci.,* **2008**, *319*, 23-28.
[http://dx.doi.org/10.1016/j.memsci.2008.03.045]

[224] Moheman, A.; Alam, M.S.; Mohammad, A. Recent trends in electrospinning of polymer nanofibers and their applications in ultra thin layer chromatography. *Adv. Colloid Interface Sci.,* **2016**, *229*, 1-24.
[http://dx.doi.org/10.1016/j.cis.2015.12.003] [PMID: 26792019]

[225] Clark, J.E.; Olesik, S.V. Technique for ultrathin layer chromatography using an electrospun, nanofibrous stationary phase. *Anal. Chem.,* **2009**, *81*(10), 4121-4129.
[http://dx.doi.org/10.1021/ac9004293] [PMID: 19385624]

[226] Clark, J.E.; Olesik, S.V. Electrospun glassy carbon ultra-thin layer chromatography devices. *J. Chromatogr. A,* **2010**, *1217*(27), 4655-4662.
[http://dx.doi.org/10.1016/j.chroma.2010.04.078] [PMID: 20553686]

[227] Moheman, A.; Alam, M.S.; Gupta, A.; Dhakate, S.R.; Kumar, A.; Mohammad, A. Fabrication of nanofiber stationary phases from chopped polyacrylonitrile co-polymer microfibers for use in ultrathin layer chromatography of amino acids. *RSC Advances,* **2016**, *6*, 90100-90110.
[http://dx.doi.org/10.1039/C6RA15465D]

Recent Advances in Analytical Techniques, 2019, Vol. 3, 179-227 179

CHAPTER 5

Neutron Activation Analysis: An Overview

Casimiro S. Munita[1,*], **Michael D. Glascock**[2] and **Roberto Hazenfratz**[1]

[1] Nuclear and Energy Research Institute, IPEN-CNEN/SP, SãoPaulo, SP, Brazil

[2] Research Reactor Facility, University of Missouri, Columbia, MO, USA

Abstract: An overview of neutron activation analysis (NAA) and some applications for this technique are provided. The fundamentals of the various methods of NAA (INAA, relative, k_0, large sample, prompt gamma charge particles, cyclic, molecular and radiochemical NAA, gamma-gamma coincidence NAA) are discussed in order to describe the most important scientific and technical aspects. Several problems associated with the technique are pointed out and briefly discussed. Emphasis is laid on the advantages of this technique for the determination of trace elements in geological, biological and environmental samples as an alternative analytical technique where other methods would not be the best choice. The role of NAA in quality assurance and quality control is also described.

Keywords: Biological samples, Environmental samples, Geological samples, Instrumental neutron activation analysis, INAA, Neutron activation analysis, NAA, Trace elements.

HISTORY

The discovery of neutrons by Chadwick in 1932 marked the beginning of the nuclear age [1]. There is, however, some evidence that Chadwick was not the first to come across neutron radiation [2]. Several scientists before him observed a highly penetrating type of radiation in experiments with beryllium targets and α-radiation [3, 4]. On 21 April 1910, Carl Auer von Welsbach [2] published a short paper in which "jonium" (^{230}Th) was found to induce radioactivity in samples stored in contact with the "jonium" sample. Thus, von Welsbach′s experiments would be the precursor of what today is called activation analysis, Fig. (**1**).

*Corresponding author Casimiro S. Munita: Nuclear and Energy Research Institute-IPEN-CNEN/SP, São Paulo, SP, Brazil; Tel/Fax: +55 11 31339960; E-mail: camunita@ipen.br

Atta-ur-Rahman & Sibel A. Ozkan (Eds.)

"Kurz erwähnen will ich ferner,daß viele Beobachtungen dafür sprechen, daß das Ionium andere ihm chemisch nahestehende Körper bei längerem Kontaktzu radioaktiven Emissionen anzuregen vermag. Es istwahrscheinlich, daß hiedurch eine Erschütterung des elementarenBestandes der erregten Körper und damit auch eine Veränderung ihrer chemischen Eigenschaften eintritt. Im Laufe dieser Untersuchungen habe ich auch Erscheinungen radioaktiver Art beobachtet, die mir mit den heute herrschenden Theorien nicht recht im Einklang zu stehen scheinen. Ich habe sie in der folgenden Schilderung einfach registriert. Vielleicht bilden manche von ihnen wichtige Fingerzeige für die weitere Erforschung des so geheimnisvollen Gebietes der Radioaktivität."	I would like to note that many observations indicate that, after long-lasting contact, jonium can induce radioactive emissions from other bodies, which are chemically related to the jonium. In this process, probably a concussion of the elementary inventory of the irradiated samples takes place as well as changes in their chemical properties. In the course of these investigations, I have observed phenomena of radioactive kind that are not quite in agreement with current theories. I have simply registered these phenomena. Perhaps some of them will be of importance for the further investigation of the mysterious field of radioactivity.

Fig. (1). The Welsbach´s text (left) and English translation (right) describing the activation process [2].

The most complete publication about activation analysis, however, was written by George Hevesy and his student Hilde Levi in two papers: one in 1936 [5] and another in 1938 [6]. The technique was applied to measure dysprosium and europium in rare earth matrices using thermal neutrons from a radium-beryllium source, followed by β-counting with a Geiger-Müller counter. They made their own source of neutrons by pulverizing beryllium in an agate mortar by hand, and mixing it in a glass ampoule containing ^{222}Rn (half-life 3.8 days) from a solution of radium salt. The α particles (^{4}He), emitted by ^{222}Rn, produce a nuclear reaction in beryllium that emits neutrons. At almost the same time, Seaborg and Livingood in 1938 [7] used deuteron activation at the University of California, Berkeley cyclotron, to bombard a sample of iron. When the sample was bombarded with deuterons, and after a radiochemical separation using ether extraction, the results showed a small amount of cobalt and manganese. Using this method, Seaborg and Livingood discovered and characterized a number of radioisotopes including ^{131}I, ^{55}Fe, ^{59}Fe, ^{65}Zn and ^{60}Co. The advantages of using radioactivity for qualitative and quantitative identification of the elements in a sample were recognized immediately. It was the discovery of a neutron activation analysis (NAA) method. However, it was Boyd in 1949 [8] who suggested naming the procedure the "radioactivation analysis method or more succinctly activation analysis".

Initially, until the end of the Second World War, the development of the method was slow due to the availability of only low-flux sources of neutrons and primitive radiation detectors and associated electronics. At that time, only neutron sources of (α, n) type or charged-particle accelerators with thermal neutron fluxes comprising only about 5×10^5 cm^{-2} s^{-1} were available [9]. After irradiation, tiresome radiochemical separations, followed by beta-particle detection with a Geiger-Müller or gas-proportional counter, were necessary. The next major advance occurred in the 1950s with the advent of nuclear reactors providing thermal neutron fluxes in the range of 10^{11} to 10^{12} cm^{-2} s^{-1}, scintillation detectors of NaI(Tl) and multichannel analyzers with 100-512 channels providing vastly greater efficiencies for the detection of gamma radiation. The scintillation detector produced output electrical pulses that were directly proportional to the amount of gamma-ray photon energy absorbed by the crystal in each interaction. Rapidly, the analysts saw the possibility of performing multi-element, purely instrumental analyses of a wide variety of materials of interest. However, a major limitation was the poor resolution of the NaI(Tl) detector, resulting in photopeak full-width at half maximum (FWHM) values of about 50 keV [10, 11].

In the 1960s and 1970s, the problems with detector resolution were solved with the advent of the Ge(Li) drifted detector. These new gamma radiation detectors provided energy resolutions about 20 times better than possible with NaI(Tl) scintillation detectors, producing gamma-ray pulse height spectra with FWHM values for the total absorption peaks of 1-3 keV minimizing the problem of overlapping peaks in the spectra. Although their resolution was excellent, their detection efficiencies and photofractions were poorer than those of Na(Tl) detector. However, the advantage of high resolution detectors allowed the application of INAA (instrumental neutron activation analysis) to several fields of science and opened up the entire scientific field to the study of the role of trace elements, particularly in the life and environmental sciences [12].

From the 1980s until today major advances in NAA were made due to new kinds of equipment, but nothing of the magnitude of the advances arising from the advent of the nuclear reactor, the NaI(Tl) scintillation detector or the germanium detector. The introduction of intrinsic Ge detectors, which only need to be chilled to liquid nitrogen temperature when in use, provided an advantage over Ge(Li) detectors which must be kept cold at all times. Si(Li) drifted detectors enabled the measurement of low energy gamma rays with much better efficiency than was possible previously. Significant improvements occurred in multi-channel analyzers, which changed from hard-wired to computer-based instruments, able to perform many functions which were previously only possible with an interface to an external computer. It may be considered as the period of the most extensive applications of the method, making it one of the most important tools for determining trace elements or less in many areas of science and technology [13].

Gradually, NAA acquired maturity assuming a leading place alongside other analytical methods [14]. Major recognition arrived on March 31, 2008 when it was recognized and approved by the "Comité Consultatif pour la Quantité de

Matière – Métrologie em Chimie", CCQM, that NAA had similar performance as that of other primary methods. The CCQM defines as "a primary method of measurement that measures the value of an unknown without reference to a standard of the same quantity" [15]. The primary method is a technique which is applicable to the certification of reference materials.

Principles of NAA

Neutron activation involves the irradiation of a nucleus with neutrons during which the neutron is absorbed by the target nucleus. The resulting compound nucleus is in a highly energetic state and the excess energy is immediately lost by emission of one or more gamma rays, a proton or an alpha particle, by means of radiative capture (n, γ), elastic scattering (n, n), inelastic scattering (n, n´), particle producing reactions (n, α), (n, p), and (n, 2n), and fission (n, f) [16]. The minimum amount of energy required for a specific reaction to occur is known as the threshold energy. For thermal neutron capture reactions, the incident neutron has a kinetic energy of nearly 0. For incident particles with less than 30 MeV of energy, most nuclear reactions are classified as scattering reactions or compound nucleus formation [9]. The sequence of events taking place during a typical (n, γ) reaction is illustrated in Fig. (2).

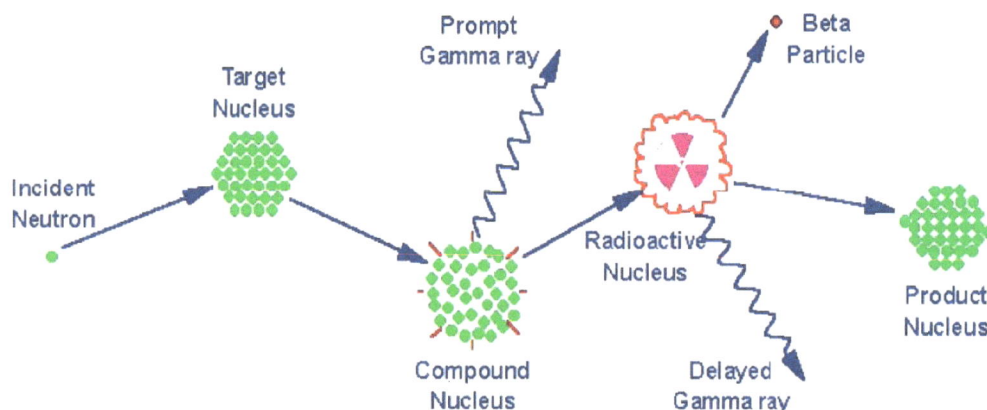

Fig. (2). Schematic diagram illustrating the sequence of events for a typical (n, γ) neutron capture reaction.

In a neutron capture reaction, the number of radioactive atoms produced depends on the number of target nuclei, number of neutrons and the cross section which defines the probability of activation. An isotope that has a high cross section is easily activated, while another isotope of the same element with a smaller cross section will be activated to a much lesser extent. In general, low energy (thermal) neutrons have larger cross sections for radiative capture (n, γ) reactions, while higher energy (fast and epithermal) neutrons, charged particles, and gamma rays have smaller cross sections and favor reactions, such as (n, p), (n, α), (p, n), and (γ, n). Cross sections are usually expressed in units of barns (b), where $1b = 10^{-24}$ cm^2.

Radioactive Decay

Radioactive decay is the spontaneous and statistical process of transformation of atoms through the emission of particles or gamma rays, after capture of neutrons by the nucleus. The number of radioactive atoms that decay per unit of time is called the activity, A, and is defined by the fundamental law of radioactivity [9]

$$A = \frac{dN}{dt} = -\lambda N \tag{1}$$

where N is the number of radioactive atoms and λ is the decay constant ($\lambda = \ln2/t_{1/2}$, where $t_{1/2}$ is the half-life). The value of λ is different for each radionuclide.

The activity for a particular radionuclide at any time t during the irradiation can be calculated by

$$A_t = \sigma_{act}\, \phi\, N\left(1 - e^{-\lambda t_i}\right) \tag{2}$$

where, A_t is the activity in number of decay per unit of time, σ_{act} is the activation cross-section, ϕ is the neutron flux, N is the number of parent atoms, λ is decay constant and t_i is the irradiation time.

The number of radioactive atoms present at any time depends on the total number of radioactive atoms produced during irradiation minus the number that decayed during the irradiation and during the interval between the end of irradiation and the start of the measurement. Therefore, for a particular radioactive nuclide, the total activity is determined by the rate of production minus the rate of decay. If the irradiation time is much longer (approximately 10 times) than the half-life of the nuclide, saturation is reached. Saturation occurs when the rate of production and rate of decay are in equilibrium and further irradiation time will not lead to an increase in activity. The ideal irradiation time depends on the type of sample and the elements to be determined. If the neutron flux in a reactor is not uniform, the total flux received by the sample must be determined using an internal or external monitor. Each radioactive nuclide is also decaying during the counting interval and corrections must be made for this decay by means of:

$$A = A_0\, e^{-\lambda t_d} \qquad\qquad (3)$$

where A is the activity at any time t, A_0 is the activity at t = 0, λ is the decay constant and t_d is the time decay.

When the term $e^{-\lambda t_d}$ is added to account for the decay of the radionuclides after the irradiation period, the equation becomes:

$$A_t = \sigma_{act}\, \phi\, N(1 - e^{-\lambda t_i})e^{-\lambda t_d} \qquad\qquad (4)$$

where the meanings of the terms were given earlier.

Neutron Energy Spectrum

There are several types of neutron sources (reactors, accelerators, radioisotope-neutron emitters), but the most suitable source of neutrons for neutron activation analysis is the nuclear research reactor due to higher possible fluxes, owing to their greater range of penetration into target materials and larger reaction cross sections [17]. The neutron fluxes in research reactors usually range from 10^{11} to 10^{15} cm^{-2} s^{-1}.

Nuclear reactors produce neutrons as a byproduct of the thermal neutron-induced fission of uranium and provide the capability of neutron activation analysis. In a reactor, the fissionable uranium isotope, ^{235}U, absorbs thermal neutrons, and the fission process results in the outputs of additional neutrons with kinetic energies ranging from 0 to 15 MeV. The average energy of neutrons from fission is about 2.5 MeV [17]. The reactor core is surrounded by materials (*e.g.*, light water, heavy water, graphite, and beryllium), which moderate or reduce the energies of the fission-produced neutrons. A typical spectrum illustrating the distribution of energies of neutrons in a reactor consists of three main components (fast, epithermal and thermal neutrons) as illustrated in Fig. (**3**) [17].

In the high-energy region, fission-produced neutrons still have most of their original energy. These neutrons are known as primary fission neutrons or fast neutrons; however, through elastic scattering with moderator nuclei, the primary fission neutrons rapidly lose their kinetic energy (KE). Activation with fast neutrons is called fast NAA (FNAA) which induce (n, p), (n, α), (n, n·), (n, 2n) reactions. Medium KE neutrons may also be used for activation. These neutrons are partially moderated and have KE of 0.5 eV to 1 MeV, commonly referred to as epithermal neutrons and are the neutrons that have been only partially moderated. In this region, the energy distribution can be approximately described by a $1/E$ slope beginning at the cadmium threshold energy and ranging up to about 10 keV. This dependence provides the most favorable conditions for activation of elements with resonances. The occurrence of resonances is associated with quasi stationary states in intermediate compound systems (nuclide nucleus + neutron) with lifetimes that are much longer than the time of flight of the neutron through the nucleus [16].

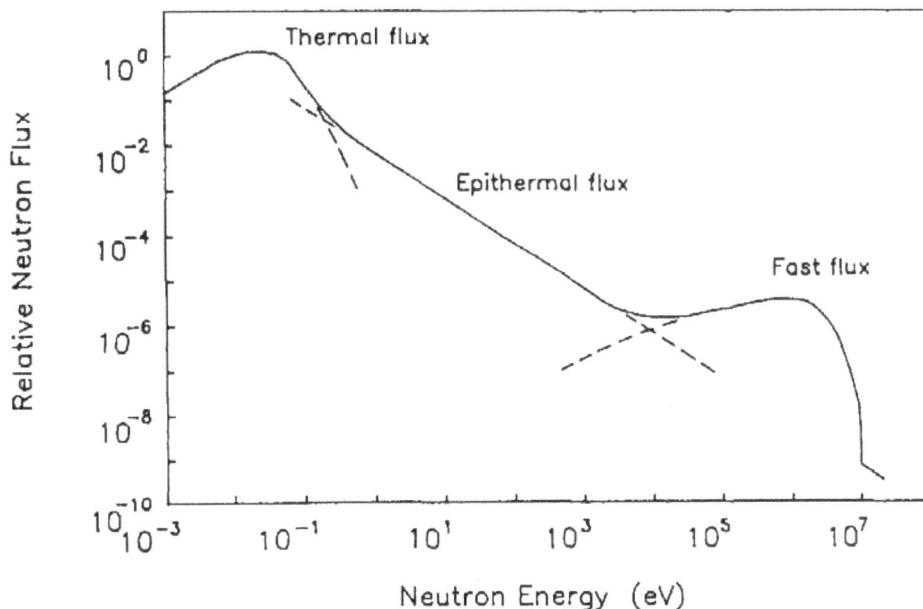

Fig. (3). Neutron energy spectrum from a nuclear fission reactor [17].

The neutrons with energies below 0.5 eV are usually referred to as thermal neutrons since they have achieved thermal equilibrium with the atoms in the reactor moderator. Thermal neutrons have approximately the same velocity *(v)* distribution as the molecules and atoms of their surroundings. At room temperature, the energy spectrum of thermal neutrons takes the form of the Maxwell–Boltzmann distribution with a mean energy of 0.0253 eV [17]:

$$\frac{dn}{dv} = \frac{4n}{v_0^3 \sqrt{\pi}} \, v^2 e^{-(v/v_0)^2} \tag{5}$$

In a typical light-water moderated reactor, the fluxes of epithermal and fast neutrons are on the order of 2–3% and 7–10% respectively of the thermal neutron flux [17]. Tabulations of cross-sections for thermal neutron reactions assume the most probable velocity of 2200 m s^{-1} at 20°C, which corresponds to a neutron energy of 0.0253 eV. The flux of thermal neutrons, ϕ_{th}, is defined as the product of the most probable velocity υ and the number density η of neutrons per unit volume.

For most nuclides, the (n, γ) reaction for low energy neutrons has a cross section inversely proportional to the neutron velocity [17]. The rate is proportional to η υ/υ, with the proportionality constant assumed to be approximately the same as the cross section at 2200 m s^{-1} [17].

Epithermal neutrons for ENAA purposes are obtained using an absorber (sometimes called filter) made of elements that have large cross sections for absorption of thermal neutrons. These properties are found in Cd, B, Gd, and others. Cadmium and boron are most often used as absorber materials. The criterion to choose the filter and its appropriate thickness have been discussed by several authors [18 - 28]. A sheet of 1 mm thick cadmium will filter out all the neutrons with energies below 0.5 eV, called the cadmium cutoff energy, in a reactor spectrum. The activity induced in the total reactor neutron spectrum divided by the activity induced under a 1 mm thick cadmium filter is called the cadmium ratio. A relation of cadmium ratios for radionuclides commonly used in activation analysis was determined experimentally [9]. The cadmium ratio of the element of interest and that of the interfering element can be compared to decide whether it would be useful to use epithermal neutron activation to enhance the element of interest. The advantage factor is a measure of the enhancement of the element of interest compared to the interfering activity. A large advantage factor indicates that the element will be greatly enhanced by epithermal activation. The background activity in a spectrum due to a matrix component may be significantly reduced using cadmium filter.

Boron also can be used to remove thermal neutrons. Boron cuts out neutrons at higher energies than cadmium, and boron or a combination of boron and cadmium can be used to optimize the effect for a particular set of interferences [29].

Activation with epithermal neutrons induces (n, γ) reactions. This type of activation analysis achieves the same reliability as conventional thermal neutron activation analysis and extends the number of elements that can be measured instrumentally in biological [25] and geological materials [30, 31].

The most commonly used description of the total reaction rate per target atom for (n, γ) reactions induced by both thermal and epithermal neutrons is given by [32]:

$$R = n(\phi_{th}\ \sigma_{th} + \phi_{epi}\ I) \tag{6}$$

where ϕ_{th} is the thermal neutron flux (cm^{-2} s^{-1}), ϕ_{epi} is the epithermal neutron flux (cm^{-2} s^{-1}), ϕ_{th} is the average thermal neutron cross section (b), and I is the effective resonance integral or epithermal cross section (b).

As the cross section curves for many nuclides are characterized by numerous resonance peaks in the epithermal neutron region, the effective resonance integral is defined by the expression:

$$I = \int_{0,5eV}^{\infty} \frac{\sigma(E)}{E}\ dE \tag{7}$$

where $\sigma(E)$ is the activation cross section as a function of energy, including the contribution from the region with the dependence $1/v$. For nuclides with the dependence $1/v$ in the epithermal region, the resonance integral is zero and the contribution from the epithermal activation is a few percent or even smaller [33], while for nuclides with high resonance integrals, the activation is high.

In general, the thermal and epithermal neutrons in a reactor are the most widely used for NAA because their fluxes are the highest and the cross sections for (n, γ) reactions are the largest; however, neutrons from thermal energy regions are much greater in number. Therefore, thermal neutron activation accounts for a majority of the induced activity and offers the greatest analytical sensitivity for most elements [34].

The neutron spectrum from 1 to 15 MeV consists of the primary fission (fast) neutrons which still have much of their original energy after the fission reaction. Fast neutrons do not induce many (n, γ) reactions compared to thermal neutrons, but are instead responsible for (n, p), (n, α), (n, n·) and (n, 2n) reactions which then become important. The cross sections for (n, γ) reactions are very small, and the nuclear reaction causes the ejection of one or more particles [17]. Fast NAA with neutron energies from 0.5 MeV and up can be helpful for determining some elements where thermal and epithermal neutrons are not useful. All these reactions occur only when the neutron energy is above the minimum threshold energy required for the specific reaction to occur.

The amount of energy released or absorbed in a reaction is defined as the Q value for the reaction. If the Q value is positive, the reaction releases energy and is exoergic. If Q is negative, the reaction requires energy needed to conserve momentum and overcome the Coulomb barrier and the reaction is endoergic [35].

At high energies, the neutron energy distribution is complex, and an average cross section for a fast neutron reaction is defined by the expression:

$$\bar{\sigma}_f = \frac{\int_{E_T}^{\infty} \sigma(E)\phi(E)dE}{\int_{E_T}^{\infty} \phi(E)dE} \tag{8}$$

Using the average cross section, the total reaction rate for a fast neutron induced reaction is given by:

$$R = n\phi_f\bar{\sigma}_f \tag{9}$$

where ϕ_f represents the average fission neutron flux. There are a few instances in which fast neutrons are important in NAA because the cross sections are generally several orders of magnitude smaller than those for thermal neutron reactions, and the detection limits are usually much higher. Materials containing high concentrations of elements which produce short-lived radionuclides, such as Al, the detection of F can be enhanced using the fast neutron reaction $^{19}F(n, p)^{19}O$.

Relative (Comparator) Method of NAA

In the relative method, a standard containing a known amount of the element to be determined is irradiated along with the unknown samples. It is assumed that the neutron flux, cross section, irradiation times, and all other variables associated with the count are identical for both the standard and the sample. It is then possible to write an equation for NAA calculations as follows [9]:

$$\frac{R_{std}}{R_{sam}} = \frac{W_{std}(e^{-\lambda t_d})_{std}}{W_{sam}(e^{-\lambda t_d})_{sam}} \tag{10}$$

where, R is the counting rates of standard (*std*) and sample (*sam*), W is the mass of the element, $\lambda = \ln2 / t_{1/2}$, and t_d is the decay time.

The best type of standard material to use is the one prepared from very pure, stoichiometrically well defined compounds, either in solid or in solution. When many elements are going to be determined in each sample, the preparation of individual standards becomes impractical because the standards occupy an excessive amount of space in the irradiation vessel. Therefore, either multi-elemental standards should be used or a single comparator for all elements becomes necessary.

Several international agencies produce a wide variety of reference materials like the U. S. National Institute of Standards and Technology (NIST), the Bureau of Reference (BCR) of the European Community, the International Standards Organization, the International Atomic Energy Agency (IAEA), the Geological Survey of Japan, the U. S. Geological Survey, the South African Bureau of Standards, the Institute of Geological and Geochemical Exploration from China and other national or international agencies in their specific fields. Sometimes two terms produce some confusion between the analysts: reference material (RM) and certified reference material (CRM). An RM is a widely distributed material intended to be used for calibrating the equipment or testing an analytical procedure [36]. A CRM is issued by a national or international organization which certifies that its composition is known within stated limits [36]. A Standard Reference Material (SRM) is a CRM issued by the U. S. National Institute of Standards and Technology. In some cases national and international organizations lend their authority to certifying the composition of the RMs, and sometimes merely prepare and distribute them and allow a consensus on the best composition to develop among the analytical community.

Flux Monitors

In some cases it may be difficult or impossible to irradiate the sample and the standard to the same neutrons flux, and in such a situation simultaneous irradiation of the sample and a monitor is performed. The purpose is to produce a flux correction factor which is the ratio of the induced activity between two positions. The relationship is measured with a flux monitor consisting of wire or foil of copper, iron, cobalt or zirconium. The masses of the monitor are determined before irradiation. The specific activity induced in the monitor beside the sample is measured and compared to the activity induced in a similar foil close to the standard irradiated at the same time, to give a flux correction factor. Any differences in decay periods, irradiation and detector efficiency must be corrected.

Geometry

The standard used for quantitative analysis of unknown sample must have a physical and chemical form that reproduces the sample geometry of the unknown as closely as possible unless major corrections are to be made. Differences in geometry may cause errors due to neutron flux, differences in self-shielding, gamma ray absorption and counting geometry. Absorption effects can be evaluated by spiking the sample with the chemical standard and comparing the result with the clean standard. The sample geometry can be evaluated using ultrapure matrices free the trace element composition of the standard. Frequently used are supports of cellulose powder, graphite and filter paper. The geological samples can be simulated using silica powder but care must be taken to avoid trace element contamination from the silica.

If the activity of an irradiated sample is compared to database values for a standard it is necessary that corrections be included.

k_0-NAA

For analysis of a large number of samples with comparable accuracy as the relative method, the k_0-standardization was developed at the Institute for Nuclear Sciences, Gent [37 - 40]. Nowadays, it is a valuable alternative for relative NAA and it is applied in several activation analysis laboratories world-wide [41]. k_0-NAA methodology requires an accurate knowledge of the neutron flux parameters, detector efficiency, sample characteristics and other nuclear parameters. The mass fractions are directly determined applying k_0 conversion factors. These k_0 factors have been experimentally determined using ^{197}Au(n, γ)^{198}Au ($E_\gamma = 411.8$ keV), but they can be converted into any comparator which is found suited to be co-irradiated with the sample [42]. Jacimovic *et al.*, in 2012 [43] give experimental details, and an extended tabulation for gamma rays of radionuclides formed by (n, γ) or (n, *f*) reaction together with evaluated values for their associated parameters such as resonance integral to thermal cross section ratios, effective resonance energies and activation decay types. The fundamental equations of the k_0-standardization method to calculate the mass fractions are as follows [42, 44]:

$$\rho_a, ppm = \frac{\left[\frac{N_p/t_m}{SDCW}\right]_a}{A_{sp,m}} \cdot \frac{k_{0,c}(m)}{k_{0,c}(a)} \cdot \frac{G_{th,m}f + G_{e,m}Q_{0,m}(\alpha)}{G_{th,a}f + G_{e,a}Q_{0,a}(\alpha)} \cdot \frac{\epsilon_{p,m}}{\epsilon_{p,a}} \times 10^6 \qquad (11)$$

or, when applying ENAA

$$\rho_a, ppm = \frac{\left[\left(\frac{N_p/t_m}{SDCW}\right)_{Cd}\right]_a}{\left[(A_{sp})_{Cd}\right]_m} \cdot \frac{k_{0,c}(m)}{k_{0,c}(a)} \cdot \frac{F_{Cd,m}G_{e,m}Q_{0,m}(\alpha)}{F_{Cd,a}G_{e,a}Q_{0,a}(\alpha)} \cdot \frac{\epsilon_{p,m}}{\epsilon_{p,a}} \times 10^6 \qquad (12)$$

where a – analyte in sample, with $k_{0,c}(a) = k_{0,c}(s)$,

m – coirradiated monitor, with experimentally determined $k_{0,c}(m)$ [if $m = c$, then $k_{0,c}(m) \equiv 1$].

$$A_{sp} = \frac{N_p/t_m}{SDCw} \tag{13}$$

with N_p – peak area corrected for pulse losses (true and random coincidence; dead time),

t_m – measuring time,

S, D, C – saturation, decay and counting factor, to be modified in case of complex activation and decay,

w – mass of irradiated element (g),

W – sample mass (g),

$f = \Phi_s/\Phi_e$, thermal (subcadmium) to epithermal neutron flux-ratio,

G_{th} – thermal neutron self-shielding correction factor.

$$Q_0(\alpha) = (Q_0 - 0.429)\bar{E}_r^{-\alpha} + 0.429/[(0.55)^{\alpha}(2\alpha + 1)] \tag{14}$$

with α – representing the non-ideal $1/E^{1+\alpha}$ epithermal neutron flux distribution,

$\bar{E}_r$ – effective resonance energy,

$Q_0 = I_0/\sigma_0$,

I_0 – (n, γ) resonance integral,

σ_0 – 2200 $m \cdot s^{-1}$ (n, γ) cross-section, to be modified in case of complex activation-decay,

G_e – epithermal neutron self-shielding correction factor,

F_{Cd} – correction factor for Cd transmission of epithermal neutrons,

ϵ_p – full energy peak detection efficiency, including gamma-ray attenuation.

The k_0 factor is defined as a compound nuclear constant:

$$k_{0,c}(s) = \frac{M_c \Theta_s \sigma_{0,s} \gamma_s}{M_s \Theta_c \sigma_{0,c} \gamma_c} \tag{15}$$

with M – molar mass;

Θ – isotopic abundance;

γ – absolute gamma-ray intensity.

Sample Preparation

The quality of the results is strongly dependent on the rigor and caution taken during the preparation of the sample. Impurities that may be introduced into the sample during preparation need to be avoided. The container used to collect the samples can be a source of contamination and much use is made polyethylene and high-purity quartz *vials*. Specific problems of collection, representative of the material as a whole and preparation of several types of samples were discussed in several publications to avoid losses and contamination [45]. In general, the procedures for preparing geological samples for INAA involve crushing and grounding the material to a fine powder, and they are simpler than the procedures for preparation of biological materials, as the latter vary according to the nature and physical form of the sample [46, 47]. A number of steps are required to prepare solid biological materials, such as washing, drying, ashing and homogenization. Hair, nail and autopsy samples need to be washed prior to analysis to remove external contamination. Freeze-drying (lyophilization) is a process that involves freezing the sample to about -20 °C, and it is preferable because when the temperature exceeds 100 °C during oven-drying, a loss of Sn and Hg may happen [48]. The homogenization of biological materials can be achieved for both wet and dry specimens by means of the cryogenic homogenization technique [49, 50].

INAA Procedure

In general, the analytical procedure for INAA involves a series of steps to optimize the sensitivity and accuracy of measurements. Each step will influence the overall quality of the results and can be summarized in the following steps:

Irradiation Procedures and Decay Time

The irradiation time should be adjusted to provide the required sensitivity for the analyte of interest and to minimize matrix activities that are longer lived than the analyte activity [51, 52]. The benefits of high activity are that, the measurement can be made at a large distance to the detector, minimizing the error caused by counting geometry variations, and that short measurement times can be used. In general, short-lived radionuclides require short irradiations and the long-lived ones require long irradiations. This can be varied, for example, using a large sample with shorter irradiation times, or small samples with longer irradiation times. Almost all applications of INAA are based upon analysis of samples with masses from several tens to several hundred of milligrams. Thus said, it also needs to be noted that even when gram amounts of geological, biological and environmental samples are being analyzed, these are mostly transparent to the irradiating neutrons. Similarly the attenuation of gamma radiation during the measurement is negligible. Special attention is needed to avoid radiation damage to the sample and their containers. Generally, polyethylene containers are used for solid samples and quartz glass ampoules for liquid samples for thermal neutron irradiation due to their high purity grade [52].

The decay time should be adjusted to minimize the matrix activity at the time of counting, while retaining sufficient activity for the required sensitivity. The relative statistical error is inversely proportional to the square root of the number of counts in the photopeak. The optimum decay time depends on the half-life of the nuclide and on the half-lives of the major spectral interferences. In the case of longer lived interferences, waiting times should be as short as possible. If the half-life of the nuclide of interest is much longer than that of the interferences, waiting times can be long, so that measurements can be performed almost free of interferences [53]. The best decay time can be determined experimentally when the elemental composition is approximately known.

Choice of Measurement Time

The counting time must be chosen in each case so as to be consistent with the delay time (t_d), $t_{1/2}$, statistical errors, *etc*. The optimum decay period before counting will depend on the elements to be determined in a sample and on the major components causing background activity and interferences. In some cases, one radionuclide of interest may also interfere with another radionuclide of interest. Often samples are counted several times in a program analysis. In the absence of interferences, the limit of detection may be set by comparing the count rate with the noise from the matrix. In the case of interfering elements, the previous study concerning t_d holds for the counting time since we can choose it according to whether we want to privilege one isotope. The decision of how long to count will depend on the activity in the peak of interest and the precision required.

Data Processing and Data Reduction

In modern gamma ray spectroscopy system, all the electronic components associated to the detector are combined into a single module. A computer is used to visually show the spectrum and to do calculations. Various algorithms are used to determine the shape and energy of each gamma ray peak present in the spectrum and to determine the area encompassed by the peak. Subsequent corrections for decay times, geometry, flux variation, interferences, fission product corrections, and comparison with a standard lead to a quantitative analysis.

Interference in NAA

In practice the spectra are not so simple, contain many peaks and there are interferences. Therefore, the resolution of the detector is important. For multielement analysis, a typical detector resolution is in the range 1.7-2.2 keV full width half maximum at the 1.33 MeV gamma ray line of ^{60}Co. The resolution at the low energy of the spectrum is generally quoted for the 122 keV gamma ray line of ^{57}Co.

When uranium and thorium are irradiated they will not only be activated to produce (n, γ) products, but also will undergo induced fission. Uranium and thorium also undergo spontaneous fission; however, the contribution for practical purposes, can be considered insignificant [54]. Slow neutrons cause the fission of ^{235}U and not of ^{234}U and ^{238}U. Fast neutrons with energies greater than 1 MeV cause fission of ^{234}U, ^{235}U, ^{238}U and ^{232}Th [54]. The fission of ^{235}U produces fission fragments that for some elements are the same nuclides used to determine the element *via* INAA.

A typical example lies in geological samples where isotopes ^{131}Ba, ^{140}La, ^{141}Ce, ^{147}Nd, ^{153}Sm, ^{177}Lu, ^{51}Cr and ^{181}Hf are commonly used in activation analysis of the corresponding elements; however, they are produced directly or indirectly as fission products of ^{235}U (~ 0.72% natural isotopic abundance) resulting from spectral and nuclear interference [54]. In the analysis of geological samples, the principal fission products are Ba and La.

Barium is frequently determined *via* the ^{130}Ba(n, γ)^{131}Ba reaction using 496.3 gamma ray. In addition, ^{131}Ba is not produced by fission, but can be interfered by the gamma ray peak of 497.0 keV of ^{103}Ru which is a product of fission. The interference increases over time because the half-life of ^{103}Ru exceeds that of ^{131}Ba [55, 56]. The contribution of 497.0 keV gamma ray from ^{103}Ru cannot be readily calculated by using another peak from ^{103}Ru, because it is the only peak with sufficient intensity to be measured. ^{103}Ru has a half-life different from ^{131}Ba and the M_x /M_U factor for ^{131}Ba, which is 3.37 at the end of irradiation, needs to be adjusted for ^{103}Ru decay by the factor $e^{0.0402 \times t_d}$ where 0.0402 is $\lambda_{131Ba} \rightarrow \lambda_{103Ru}$ and the t_d is the time after irradiation, both in units of days [56]. The correction becomes greater with time because the half-life of ^{103}Ru is larger than that of ^{131}Ba.

Another issue is the determination of lanthanum *via* the [139]La(n, γ)[140]La reaction with a half-life of 40.27 h. [140]La is not produced directly by uranium fission but rather as a daughter of the [140]Ba ($t_{1/2}$ = 11.7 d) fission product [57]. This interference is rarely significant because geological samples have much higher mass fractions of lanthanum compared to uranium, but the interference can be very significant after a long decay time [55]. Since [140]La is daughter of the fission produced by [140]Ba, the mass fraction of [140]La needs to be corrected for the time of decay following irradiation. When t_d = 0, the ratio M_x/M_U is 0.0028. This value increases with time by a factor of $e^{0.35893 \times t_d}$ -1 [56]. Large corrections can be avoided by counting for [140]La as soon as possible after irradiation when the adjustment is usually less than 1%. The correction for [140]La produced during the irradiation is independent of the duration of irradiation at least when the [140]Ba nears saturation. In this case, corrections for [140]La produced during irradiation need to be applied [55, 56].

To obtain the correct mass fraction in the presence of these fission products, the amount of uranium in the geological sample must be determined first and then multiplied by the M_x/M_U factor. This value is then subtracted from the measured mass fraction for the element (Table **1**) [56].

Table 1. Correction factors determined experimentally for contribution of [235]U fission [56].

Isotope	Gamma ray energy, keV	Correction factor, M_x/M_U
[95]Zr	756.7	11.3 ± 0.2
[97]Zr	743.4	26.6 ± 0.9
[99]Mo	140.5	1.41 ± 0.05
	739.5	1.39 ± 0.05
[103]Ru	497.1	0.126 ± 0.003
[131]Ba	496.2	$3.37 \pm 0.07 \times e^{0.0402 \times t_d}$
[140]La	1596.5	$0.0028 \pm 0.0002 \times e^{0.35893 \times t_d}$ -1
[141]Ce	145.4	0.287±0.008
[143]Ce	293.3	1.35 ±0.04
[147]Nd	91.1	0.21±0.01
	103.2	0.004±0.001*

* Korotev and Lindstrom [55].

For analysis of [95]Zr and [97]Zr, the contributions from uranium fission are extremely great, and the correction factors are respectively 11.3 and 26.6. The corrections for [99]Mo and the rare-earth elements, REE, isotopes are much smaller, but may be significant if the true amount in the geological samples is comparable to uranium.

These isotopes are direct products of ^{235}U fission and corrections factors are necessary for any decay time [55, 56].

Uncertainty Evaluation

The accuracy of an analysis by NAA depends first on the nuclear constants associated with the element under analysis: the reaction cross section, gamma radiation yield, decay constant, *etc.* The uncertainty source of INAA can be evaluated in three main groups [58]:

1. preparation of the sample and comparator. This type of uncertainties include uncertainties from standard, k_0-factors, neutron fluence rate monitor;
2. irradiation characteristics. These comprise neutron spatial and spectral variations, absorption effects and interfering nuclear reactions; and
3. gamma spectrometric measurement. Including the counting geometry and statistics, the decay time and the live-time corrections, the radiation background, the spectrum fitting and spectral interferences.

For radiochemical neutron activation analysis (RNAA) the uncertainty source are associated with the radiochemical separation procedure and yield determination using stable or radioactive tracers.

A good overview and details of these uncertainties have been discussed by Kucera *et al.* in 2004 [58].

Zahn *et al.* in 2015 [59] using a comparative NAA in biological and geological matrices found that the largest part of the uncertainties come from the concentration of the elements and from the counting statistics in both the sample and the standard.

The following section describes detectors and electronics most commonly used for gamma-ray spectrometry.

Detectors and Associated Electronics

There are many papers, books and manuals in the literature which describe gamma-ray detection in detail [9, 60 - 63]. For most NAA applications, detection is performed using a high purity intrinsic germanium detector, which consists of a crystal mounted in a vacuum cooled cryostat. Cooling is achieved by insertion of the cryostat in a Dewar vessel filled with liquid nitrogen, or by electrically powered cryogenic refrigerators. The cryostat Cryo-Pulse 5 Plus is an electrically powered cryostat for use with HPGe radiation detector. Cryo-Circle II is described as "hybrid" cryostat because it combines the advantages of electric cooling with the reliability of liquid nitrogen, has a 25 liter LN_2 reservoir for continued operation in case of power failure, and it can typically operate for over a year before adding more gaseous or liquid nitrogen. Following is a brief description of the operating of three types of detectors: semiconductor, coaxial and planar, commonly utilized in the NAA laboratories.

The performance of many types of semiconductor detectors regarding detection efficiency and energetic resolution capacity depends on their size and the energy of the incident radiation to be measured. The output of a radiation detector is a current pulse. The amount of energy deposited is reflected by the integral of this pulse with respect to time, which is directly proportional to the amplitude or height of the pulse. It is assumed that the shape of the output pulse does not change from one event to another. The detector response represents the relation between the energy of the incident radiation and the total amount of electric charges deposited in the crystal. For semiconductor detectors, the response is linear for different energies.

The basic element of a semiconductor detector is a crystal of semiconductor material with a *p-i-n* diode structure. In such a configuration, *n* and *p* refers to the nature of the impurities in the crystal. The *n*-impurities are pentavalent atoms, acting as electron donors, *p*-impurities are trivalent atoms, acting as electron acceptors, and *i* is the intrinsic layer. There are two forms of semiconductor detectors: the *p-n* diode detectors and blocked impurity band detectors.

The main advantages of semiconductor detectors are good resolution, low dead time and linearity between the energy and pulse height over a large range of energy.

The majority of coaxial Ge-detectors have *p* and *n* contacts on the outer surface of the cylindrical crystal and on the axial hole drilled in at the bottom of the cylinder. The *n*-contacts are relatively fine (0.5 to 1 mm) diffused layer, whereas the *p*-contacts are ion-implanted, and therefore very fine. There are two different detector configurations: 1) *n*-contact outside of the crystal and *p*-contact inside of the drilled core, and 2) *p*-contact outside of the crystal and *n*-contact inside of the drilled core. The *n*-contact outside of the crystal is the detector geometry for intrinsic Ge detectors, and the majority of the HPGe detectors. This detector type can be used in INAA for measurement of photons with energies higher than 60 keV [64].

The detectors with the *p*-contact outside the crystal can be used to measure gamma rays with energies as low as 10 or even 3 keV, depending on the type of window (Al or Be).

In the planar detector, the crystal is placed in a cryostat with the thin *p*-contact facing the cryostat's beryllium entrance window. The absorption of low energy gamma rays is minimized and the detector type has performance optimized for X-ray and gamma-ray energies from 5 keV to 200 keV. Planar detectors may be useful to minimize the effect of interferences with high-energy radiation when measuring radionuclides which emit X-rays and low-energy gamma rays.

Detector efficiency relates to how well the detector absorbs gamma rays and is defined as intrinsic efficiency and geometric efficiency, both types are energy dependent. The intrinsic efficiency is the ratio of signal pulses divided by the number of photons arriving at the detector.

Geometric efficiency is the ratio of the signal pulses recorded divided by the number emitted by the source and depends on the detector type, shape, sensitive volume of detector, the sample size and shape, attenuation of the radiation in the source itself and the distance between the sample and the detector.

The product of the interaction efficiency and the geometrical factor is denoted as total efficiency, while the product of total efficiency and peak-to-total ratio results in the full energy photopeak efficiency.

In general, for INAA it is desirable to have the efficiency as high as possible; however, detectors with relative efficiency ranging from 1 to 20% are frequently adequate.

$LaBr_3$ and $CeBr_3$ are the new generation of high resolution scintillation detectors with maximum sizes up to 3" x 3" which are used with a new algorithm to post-process scintillator detector spectra to render photopeaks with high accuracy and efficiencies up to about 75% [60]. These detectors are an excellent choice for applications resulting in relatively simple gamma-ray spectra.

Voltage is delivered to the detector by means of a high voltage power supply which has typical output characteristics of 3 to 5000 volts, at low currents and with a low noise level.

The pulses produced in the detector are very small and so the charge sensitive preamplifier, usually a field-effect transistor - FET, is mounted as close to the detector as possible. Normally this preamplifier forms part of the cryostat system to avoid electronic noise. To prevent saturation of the preamplifier, when the output level reaches a certain voltage a short pulse is sent to a light-emitting diode producing a light pulse which discharges the FET. Some preamplifiers have a visual warning such as a red light to indicate that the count rate is too high for the preamplifier circuitry.

The signal is sent to the shaping amplifier where it is shaped and amplified, usually up to a maximum of 10 V, depending on the gain setting. The resulting pulse, which still has amplitude corresponding to the original energy of the gamma ray, is sorted by a pulse height analyzer. The basic schematic diagram of the electronics of a gamma-ray spectrometer for NAA is given in Fig. (**4**).

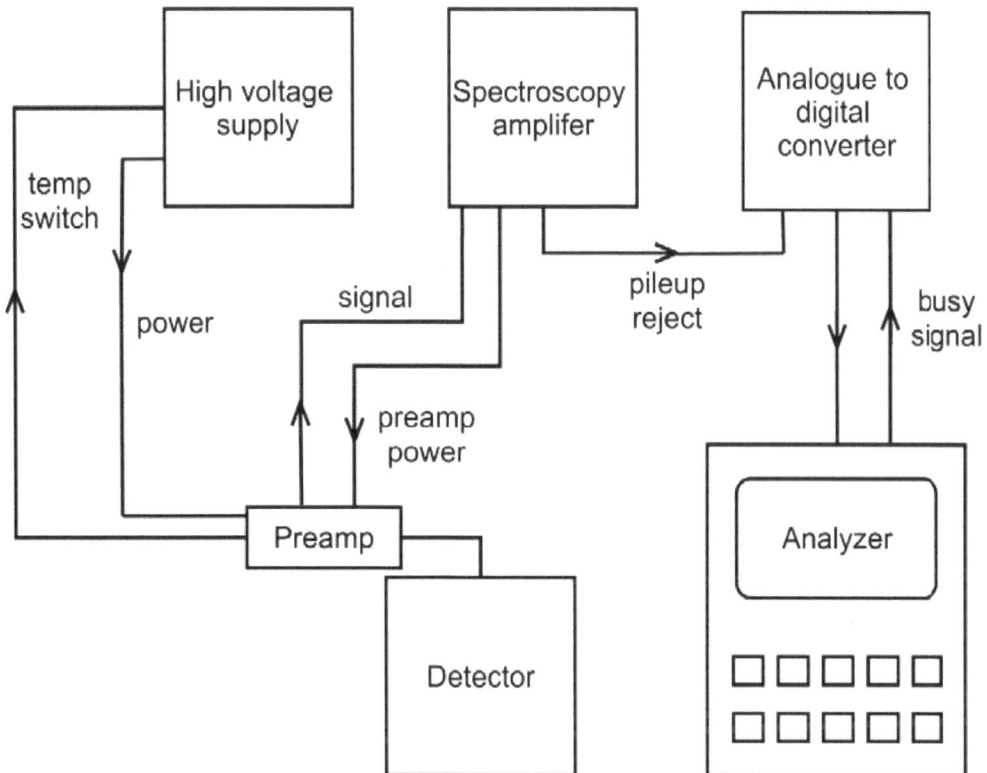

Fig. (4). Schematic of a gamma-ray spectrometry system showing the electronics required for NAA [62].

The amplifier has two purposes: 1) to increase the size of the small signal from the output of the preamplifier to a pulse of amplitude up to about 10 V. An output pulse from the amplifier will have amplitude in the range of 0-10 V, dependent on the gain setting. Normally there will be a coarse gain and a fine control for a very precise setting of the amplifier gain. 2) To improve the signal to noise ratio of the signal pulses. Semi-Gaussian shaping is generally used, with differentiation and integration of the signal.

Amplifiers can work with quite high counting rates before pulse pile-up begins to occur. Pile-up is consequence of two or more gamma rays being detected almost simultaneously. Without pile-up rejection or a loss free counting system, some pile-up occurs even at low count rates. In the absence of hard-wired pile-up rejection, pile-up must be calibrated as a function of count rate. In the case of high counting rates, it is possible for pulses to overlap and to be seen as one peak whose energy is the sum of the energies of the individual pulses. Amplifiers can be obtained with pile-up rejection circuits built into them. The circuit will reject any pulses which are observed as being added together and a signal from the amplifier will tell the analyzer to ignore that pulse.

The output from the amplifier is a peak of nearly Gaussian shape with amplitude proportional to the gamma-ray energy which enters the detector. The analog-to-digital converter (ADC) changes this pulse into a digital signal proportional to the pulse height which is deposited as a count in the appropriate channel number of the analyzer. The time for the ADC to process a pulse increases linearly with the channel number of the pulse that is being processed. A typical clock time in an ADC processing time is 10 ns per channel. The processing time which is lost during counting is called the dead time. The dead time is the time at which the ADC is busy processing one pulse and cannot accept another. This loss of pulses can be corrected by measuring the time for which the ADC is busy. The analyzer interrogates the ADC busy signal at a very high frequency while incrementing a live time clock. The live time clock only operates when the ADC is operating. The difference between the time incremented on the live-time clock and the real-time clock in the analyzer is the dead time and it is indicated in percentage.

Gamma-Ray Spectrum

The gamma-ray spectrum is characteristic for each radionuclide and its shape is due to the interactions that take place in the detector. If the electronics are operating normally, the relationship between the analyzer channel number and the corresponding gamma-ray energy should be linear. Therefore, the plot of gamma-ray energy against channel number should be a straight line with a slope dependent on the amplifier gain, which may or may not go through zero depending on the way that the ADC is set up.

Data Processing

Today the analyzer not only carries out pulse height analysis and controls the timing of the analysis, but is also responsible for the data processing system. The data processing consists of a number of stages, including storage of spectra, peak search, radionuclide identification, peak area determination, decay and efficiency corrections.

The main advantages of such systems are that the programs are good and the analyzers are designed to allow several users to operate a variety of tasks simultaneously. In addition, the system can be interfaced with computers to be more flexible to access other softwares, such as statistical packages, graphics and chemical data management programs.

Automation System for NAA

One of the difficulties involved with the NAA method is the time that the operator spends in all steps of the analysis. This may become a hindrance for a large series of samples or for the simultaneous analysis of several different batches of samples. In addition, when different elements are determined from isotopes with diverse half-lives, each sample must be measured at different decay times. An automatic system for each step of the analysis minimizes the problem because it shortens the time analysis by several hours and reduces tedious manual operation [65 - 67]. There are a few laboratories that have automatic systems in operation. Some systems are sold commercially; however, most laboratories cannot acquire them because they are costly. To implement a system, two aspects must be considered: creation of devices and systems for the automation analysis (hardware), and the development of computer technology and software, both for hardware control and treatment of analytical data [65]. All of the operations are controlled by means of the spectrum analysis software.

The software for the automatic quantitative identification of elements in a sample was developed at the FLNP JINR [65]. The program uses values of isotope activities recorded in output files of the Genie 2000 analysis of spectra of samples, standards, and monitors to calculate the value of element concentration in the sample using the comparative NAA.

Compton Suppressor System

Compton scattering of gamma ray photons in the active volume detector leads to distortion of the spectrum resulting in an increased background – the so-called Compton effect. The purpose of the Compton suppression system is to decrease the background of Compton-scattered events and cascades emitting radionuclides by the use of anti-coincidence electronics. This method has been successfully used to increase the sensitivity in decay scheme analysis for the spectra [68]. The principal advantages of using Compton suppression systems are the reduction in the detection limit and the improvement in the accuracy of the analysis. Landsberger and Peshev in 1996 [69] made a detailed study to reduce the Compton effect by using Compton suppression system in INAA.

The suppression factor is defined as the ratio of the peak to Compton ratios for suppressed and unsuppressed spectra. The suppression of coinciding gamma rays depends on the configuration of the spectrometer and on the distance between the sample and the detector. Changing the sample to detector distance appreciably changes the probability for coinciding hits of gamma rays in the semiconductor detector, affecting the Compton suppression efficiency [70].

The principal interfering short-lived nuclides in INAA of environmental, geological and biological samples are ^{28}Al ($t_{1/2}$= 2.3 min), ^{56}Mn ($t_{1/2}$= 2.58 h) and ^{24}Na ($t_{1/2}$= 15 h), which can be considerably suppressed [71]. Long-lived nuclides in neutron irradiated soils, silicate rocks, and other environmental samples are ^{140}La ($t_{1/2}$= 40.2 h), ^{58}Fe ($t_{1/2}$= 44.6 d), ^{46}Sc ($t_{1/2}$= 83.9 d), and ^{60}Co ($t_{1/2}$= 5.26 y). These nuclides emit coinciding gamma rays, which can be substantially suppressed by the Compton suppression system, except for ^{59}Fe, in the course of measurements.

Wu and Landsberger in 1994 [72] analyzed this effect in three SRMs 2709, 2710 and 2711 from NIST, which were analyzed for 43 major and trace elements by INAA using thermal and epithermal neutrons. The analysis was carried out with and without the Compton suppression system for ten elements. The authors found that there were no distinct advantages in soil samples using and/or the Compton suppression system for the determination of As, Sb or Sm. Gallium was best determined using a 5-minute epithermal irradiation time and a 1-day decay period. For Cd, Au, Mo, K, and W, the use of the Compton suppression systems offered significant advantages in sensitivity and precision. Lanthanum was best determined using routine NAA. In other paper [73], the authors determined 48 major, minor and trace elements in the SRM 2781 using Compton suppression system using different neutrons energy, irradiations, and counting systems. The results found by the authors are in agreement with the certified or consensus values. The Compton suppressor system is not suitable for all elements, and in some cases the detection limits are even worse. However, improvements in the detection limits were found for I, Hg, Ni, Sr, and Zn.

Large Sample neutron Activation Analysis, LSNAA

All the multi-elemental analysis methods described above involve studying a small portion of material, a few milligrams of solids or a few milliliters of liquids. The limitation on the size of the sample is one of the biggest concerns that the analyst faces when dealing with a large sample. For example, soil, rocks, plants, *etc.* can be more easily and representatively sampled at quantities of around hundreds of grams to kilograms than at quantities of less than 1 gram, because a sample is considered to be representative only if it can represent the average composition of the material [74].

The sensitivity of NAA elemental determinations depends mainly on the induced activities. This activity is proportional to neutron flux and sample weight. The usual mass of samples for activation analysis is in the range of 100 mg using reactors with high neutron fluxes. In addition, the lower activity generated by irradiation inside nuclear reactors with lower fluxes can be compensated by irradiation of large samples, up to 250 times the weight of the samples normally used in INAA. INAA of large samples can be advantageous for materials that are not homogeneous such as contaminated soil, industrial raw materials or solid wastes [75]. The sensitivity of the analysis may increase by using scintillation detectors in combination with innovative spectrum deconvolution technique [76].

The main problem in large sample NAA is determining the neutron self-shielding and gamma-ray self-attenuation corrections [77, 78].

In large samples, neutron absorption and scattering result in substantial self-shielding, causing depression of the neutron flux at the center of the sample compared to the periphery [76]. Neutron self-thermalization may cause substantial changes in the neutron spectrum throughout the sample especially for samples of high density and high Z [78].

After these corrections are determined, the analysis proceeds as in normal NAA, as if the sample was small.

The basic equation of NAA to determine the mass of the unknown element is given as [77]:

$$A_0 = \phi_{th}\sigma_{eff}\frac{N_{AV}\theta\,m}{M}\left(1 - e^{-\lambda t_{ir}}\right)e^{-\lambda t_d}\frac{(1-e^{-\lambda t_m})}{\lambda}\gamma\varepsilon \qquad (16)$$

where, A_0 is the area of the peak, ϕ_{th} is the thermal neutron flux (cm^{-2} s^{-1}), σ_{eff} is the effective absorption cross section, N_{AV} is Avogadro's number (mol^{-1}), θ is the isotopic abundance, m is the mass of the irradiated element (g), M is the atomic mass number (g mol^{-1}), λ is the decay constant of the radioisotope formed (s^{-1}), and t_{ir} is the duration of irradiation time (s).

Eq. 16 can be used in LSNAA as in eq. 17 after calculating the ratio of neutron self-shielding and the gamma ray self-attenuation inside the sample.

$$A_0 = \phi_{th}\sigma_{eff}\frac{N_{AV}\theta m}{M}\left(1 - e^{-\lambda t_{ir}}\right)e^{-\lambda t_d}\frac{(1-e^{-\lambda t_m})}{\lambda}\gamma\varepsilon f_{n,\gamma} \qquad (17)$$

where, $f_{n,\gamma}$ is the neutron and gamma attenuation factor.

There are many approaches for these calculations such as pure theoretical modeling, and Monte Carlo modeling using *a priori* available information about the test sample composition [76].

14-MeV Fast Neutron NAA

Neutrons sources not involving the nuclear reactor, namely accelerators and isotopic sources are used in only a small proportion of applications in neutron activation analysis. These sources are often used in commercial applications [79]. Commercially available D-T neutron generators, based on the reaction of a deuteron beam on tritium ^{3}H(d, n)^{4}H, are generally used for fast neutron activation analysis. FNAA with neutron generator produced neutrons, however, is capable of determining many light elements that cannot be done with either thermal or epithermal neutrons from a reactor. Oxygen analyses are often difficult to perform with conventional analytical methods, but quite simple with 14-MeV FNAA in a wide variety of matrices including metals, geologic materials, coal, liquid fuels, petroleum derivatives and fractions [80]. The reaction used is ^{16}O(n, p)^{16}N. The same reaction is also used for *in vivo* NAA determination for oxygen.

Another element that is easily determined using 14-MeV neutrons is nitrogen. The reaction is ^{14}N(n, 2n)^{13}N. ^{13}N is a positron emitter, and the 511 keV annihilation photons are detected. The same reaction is used in *in vivo* activation analysis. Other elements that are determined by FNAA are F, Mg, Si, Cu, Fe, P, and Zn [81]. Table **2** shows a summary of the most commonly used 14-MeV energy determinations [9].

Table 2. Some 14-MeV FNAA determinations.

Reaction	Product, half-life	Gamma rays, MeV
^{16}O (n, p) ^{16}N	7.13 s	6.13; 7.12
^{14}N(n, 2n) ^{13}N	9.97 min	0.511*
^{28}Si(n, p) ^{28}Al	2.25 min	1.78
^{31}P(n, α) ^{28}Al	2.25 min	1.78
^{63}Cu(n, 2n) ^{62}Cu	9.74 min	0.511*
^{90}Zr(n, 2n) ^{89m}Zr	4.18 min	0.588

*β+ annihilation radiation.

Prompt Gamma Neutron Activation Analysis, PGNAA or PGAA

When a nucleus captures a neutron, it gains energy, which results in excited nuclear states, and it de-excites very quickly, in less than 10^{-13} s, by emission of one or more prompt gamma rays. PGNAA is based on the measurement of gamma rays emitted during nuclear reactions rather than gamma rays emitted from radioactive decay. The half-lives of these excited states are too short for the samples to be transferred as is usually done with delayed-gamma NAA. Therefore, these gamma rays are detected during sample irradiation in order to be used analytically [82].

Using PGNAA, the irradiation and the counting of the sample need to be performed simultaneously. If the sample is irradiated in the reactor, the detector needs to be very close to the reactor, which is typically an environment with a very high background of gamma rays and neutrons. In many PGNAA procedures, a beam of neutrons is extracted from the reactor and a great deal of shielding is required to protect the detector from both neutrons and scattered gamma radiation [9]. Since the extracted beams have lower intensities, long irradiation times are often required.

The actual practice of PGNAA involves a few experimental difficulties. To avoid some of the difficulties associated with PGNAA and to improve the sensitivity of the method, some reactors produce cold neutrons by placing a cold source, usually a refrigerated tank of liquefied deuterium, inside the nuclear reactor in an area of high neutron flux. The neutrons are cooled in this tank and guide tubes are provided for the neutrons to leave the cold source. The energy of the cold neutron is very low, about 0.005 eV [9].

PGNAA is especially useful for the determination of low Z elements like B, H, N, P, S, and is also useful for measuring high cross-section elements such as Cd, Sm and Gd in various types of samples (geological, biological, and alloys) [9, 83].

Charged Particle Activation Analysis, CPAA

In CPAA, the particles used for activation are charged particles and the most frequently used are protons, deuterons, tritons, and ^{3}He and α particles generated by accelerators [84]. The type of nuclear reaction induced depends on the identity and energy of the charged particle. There are many applications that use protons with 15 MeV in energy because they are easily accelerated and have low Coulomb barriers resulting in a (p, n) reaction. The high energy of the protons can also induce (p, α), (p, d), or (p, γ) reactions [9, 84]. With deuterons, the principal reactions used are (d, n) and (d, p). Alpha particles and ^{3}He ions have higher Coulomb barriers than the protons, so higher initial energy to induce nuclear reactions is needed [9].

The elements that can be determined by CPAA are different from those commonly determined by NAA. Table **3** summarizes a few of the most useful CPAA determinations [9]. The light elements that have low Coulomb barriers and low capture cross sections for neutrons can be easily determined using CPAA.

Table 3. Common elements determined by CPAA.

Element	Reaction	Product, half-life	Gamma rays, MeV
B	$^{11}B(p, n)^{11}C$	20.3 min	0.511*
	$^{10}B(d, n)^{11}C$	20.3 min	0.511*
	$^{10}B(p, \alpha)^{7}Be$	53.28 d	0.478
C	$^{12}C(d, n)^{13}N$	9.97 min	0.511*
	$^{12}C(He, \alpha)^{11}C$	20.3 min	0.511*
N	$^{14}N(p, \alpha)^{11}C$	20.3 min	0.511*
	$^{14}N(p, n)^{14}O$	70.6 s	2.31; 1.18
	$^{14}N(d, n)^{15}O$	122 s	0.511*
O	$^{16}O(p, \alpha)^{13}N$	9.97 min	0.511*
	$^{16}O(^{3}He, p)^{18}F$	1.83 h	0.511*
P	$^{31}P(\alpha, n)^{34m}Cl$	32.2 min	2.13; 1.18; others
Ca	$^{40}Ca(\alpha, p)^{43}Sc$	3.89 h	0.373; others
Ti	$^{48}Ti(p, n)^{48}V$	15.98 d	0.984; 1.31; others
Fe	$^{56}Fe(p, n)^{56}Co$	77.3 d	0.847;1.24; others
Zr	$^{90}Zr(p, n)^{90}Nb$	14.6 h	1.13;2.32; others
Ag	$^{107}Ag(^{3}He,2n)^{108m}In$	57 min	0.633; 0.786; others
Cd	$^{111}Cd(p, n)^{111}In$	2.80 d	0.245; 0.171
W	$^{182}W(p, n)^{182m}Re$	12.7 h	1.12; 1.22; others
Pb	$^{206}Pb(p, n)^{206}Bi$	6.24 d	0.803; 0.881; others

$^{a}\beta^{+}$ annihilation radiation.

The equation used for calculations applicable for NAA is not applicable for CPAA because the charged particles are not able to penetrate the entire sample and the particles lose energy, so the cross section for a given reaction is not constant [9]. The activation equation for a given element, modified for use in CPAA is given as:

$$A = \frac{CN_a I}{M} \int_0^R \sigma(x)dx \qquad (18)$$

where, A is the absolute activity, C is the concentration, N is the Avogadro's number, a is the isotopic abundance, I is the beam intensity (particles/s), M is the atomic weight, R is the range in sample (cm), x is the depth in sample (cm), and σ

is the cross section (cm^2).

As with NAA, the fundamental equation is not used directly in practice. The equation using the comparator method is used and is given by the equation [5]

$$C_x = \frac{C_s A_x I_s R_s}{A_s I_x R_x}$$ (19)

where, C is the concentrations of an unknown (x) and standard (s), A is the measured activity, I is the beam intensity, and R is the particle range dependent on matrix (from tabled values).

Cyclic Neutron Activation Analysis, CNAA

CNAA is a modification of NAA and it can be useful when the element of interest has a short-lived nuclide and it is interfering with other activation products in INAA [85 - 87]. In this case, the sample is irradiated repeatedly, and the spectra are summed together to give one final total spectrum. The advantages are the reduction of detection limit, better precision and accuracy and short total determination time [88]. For example, Se in biological samples can be determined by INAA using 264.6 keV gamma ray from long-lived ^{75}Se ($t_{1/2}$ = 121 d) produced *via* the ^{74}Se(n, γ)^{75}Se reaction. The time of the analysis is quite long and there is the possibility of interference from ^{203}Hg ($t_{1/2}$ = 46.9 d). The alternate approach is CNAA using ^{76}Se(n, γ)^{77m}Se reaction. The half-life of ^{77m}Se is 17.5 s and gamma ray energy is 161.9 keV. The potential interference is ^{116m}In ($t_{1/2}$ = 54 min), but indium is rarely present in biological matrices [88].

The use of the cyclic procedure results only in enhanced counting statistics for the short half-life radionuclide while keeping signal to noise ratios low [89]. This technique has been widely applied in biological, environmental, geological, and industrial studies, and the most often measured elements include Se, F, Pb, Hf, Sc, O, Ag, and Rh. The detector response in cyclic activation analysis is given by the equation:

$$DR = \frac{N\phi\sigma k}{\lambda}\left(1 - e^{-\lambda t}\right)\left(1 - e^{-\lambda t'}\right)\left(e^{-\lambda t''}\right)\left[\frac{n}{1 - e^{-\lambda T}} - \frac{\left(e^{-\lambda T}\right)\left(1 - e^{-n\lambda T}\right)}{\left(1 - e^{-\lambda T}\right)^2}\right]$$ (20)

where DR is the detector response in terms of accumulative counts recorded in n counting period, N is the number of stable target nuclei of interest in the sample, ϕ is the flux density of incident particles (cm^{-2} s^{-1}), k is the factor containing the detector efficiency and branching ratio for the decay path, λ is the decay constant for the indicator radionuclide (s^{-1}), t is the irradiation time for each cycle (s), t' is the counting time for each irradiation (s), t'' is the delay time prior to counting for each irradiation (s), T is the delay period from the end of a previous delay period to the end of the next delay period (cycle time, s), and n is the number of cycles.

Radiochemical Neutron Activation Analysis, RNAA

Activation analysis can also be performed in conjunction with chemical separation of the activated elements to obtain radiochemically pure states. This form of activation analysis is called radiochemical neutron activation analysis. The usual procedure is to add known amounts of inactive isotopic carriers to the neutron irradiated and dissolved sample and after assuring that the complete isotope exchange has occurred, proceed with the chemical manipulations to separate the given element from all or most of the others. Chemical separation is necessary in cases in which the main components of the sample develop high activity due to irradiation, thereby interfering with the analysis of minor or trace elements. In INAA on biological samples, the very strong activities due to ^{24}Na, ^{80}Br, ^{36}Cl, and the bremsstrahlung of ^{32}P induced from the abundant elements in biological materials often restrict the determination of As, Cd, Cu, Mo, *etc.* with short and medium half-lives, which cannot usually be performed *via* the INAA determination. Subsequently, radiochemical separations are required of these elements. The main radiochemical separation is not the removal of matrix effects occurring in other analytical methods, but the elimination of interfering radionuclides with strong activities or overlapping gamma rays, and the analysis is more sensitive and accurate than the instrumental method.

Radiochemical separations may be classified into two groups: individual radiochemical separation; and group radiochemical separation.

Individual separations are more time-consuming and often require the determination of the chemical yield. For example, the hydrated antimony pentoxide (HAP) showed excellent selectivity for Na with a decontamination factor exceeding 10^{10}. Detailed procedures on the individual isolation of the elements may be found in the Nuclear Sciences Series [90], and in many papers [91].

Group separations imply the separation of the radionuclides into one or a few groups. Care should be taken to choose a group-separation scheme appropriate to the particular matrix involved, for a radionuclide with an intense activity may completely mask the others in the group. Several group separation procedures were developed in conjunction with distillation, ion exchangers and solvent extraction [92 - 94].

Molecular Activation Analysis (or Speciation NAA), MAA

The MAA is a combination of traditional nuclear analysis methods with physical, chemical or biological separations procedures. The MAA procedure refers to an activation analysis method that is able to give information about the chemical species of elements in samples of interest [95 - 97]. Essential and toxic trace elements play an important role in life and environmental sciences which are highly related to their chemical species. In many cases, MAA plays a unique role because it has unique features, such as high sensitivity, good accuracy and precision, non-destructiveness, none or reduced matrix effect, and freedom from contamination from acids or reagents used in the analysis [98 - 100]. Chai *et al.*, in 2004 [101] described the advantages and limitations of MAA technique for chemical speciation study in biological and environmental samples. Hou *et al.* in 1999 [102] developed a method for the determination of iodine, iodate, organic iodine and total iodine in seawater using a strongly basic anion-exchange column. Chen *et al.* in 2000 [95] determined the mass fraction of 24 trace elements in liver organelles by means of MAA based on biological separation techniques combined with INAA. The procedure requires the compounds of interest to have high neutron cross sections and to form radionuclides of suitable half-lives. While no acid digestion procedure can be employed since this may change the chemical form of the species, it is important for the chemical species to be in a form that is amenable to chemical separation.

Two terms are frequently found in the literature: species and speciation and both are used sometimes interchangeably. Strictly speaking, the term "species" refer to a chemical form, state or valence of an element in a medium, for example Cr^{3+}, CrO_4^{2-}, $Cr_2O_7^{2-}$, whereas "speciation" means an action resulting in transformation, alteration or variation from one species to another.

Gamma-Gamma Coincidence Activation Analysis, γγ-cAA

Gamma-gamma coincidence (γγ-c) spectrometry is the technique for identifying and/or quantifying nuclear decay events based on the observation of unique multiple γ-ray signatures [103]. Coincidence counting, as it is commonly named, uses at least two γ-ray detectors to measure the coincidence γ-rays emitted in cascade by a nuclide, γ_a and γ_b, in which case the equation to describe the number of coincidence events measured by the system is [17]:

$$N_{ab} = \varepsilon_a \ \varepsilon_b \ A \ T_c \qquad \textbf{(21)}$$

where N_{ab} is the total number of coincidence events which were counted, ε_a and ε_b are the efficiencies of detectors 1 and 2, respectively, A is the activity of the radionuclides (desintegrations/s), and T_c is the total counting time. The technique is commonly used in nuclear structure analysis to build up radioactive decay schemes.

In principle, experiments confirm that γγ-coincidence counting has advantages to offer over single counting [103, 104]. The application of coincidence counting in NAA was reported several years ago [105]. Since then, some papers have been published describing the use of this technique. However, it continues being one of the most rare techniques employed in NAA. One possible reason of this infrequent use to analytical measurements is that nuclear structure studies require the determination of coincidence relationships between gamma rays to construct nuclear level schemes. On the other hand, nuclear spectrometry as used in NAA, depends on the knowledge of the elemental origins and, to a lesser extent, the intensities of gamma rays, but does not require a deep knowledge of gamma-ray coincidence relationships. Drescer *et al.* in 2017 [104] used two cerium-doped lanthanum bromide (LaBr$_3$:Ce) scintillation detectors in γγ-coincidence configuration to demonstrate the advantages that coincident detection provides relative to a single detector.

An element to be determined by cNAA must be an isotope formed by neutron capture that emits a gamma-ray cascade of reasonable intensity. Many nuclides have gamma-ray cascade, but often the β-branch feeding is weak or the gamma-ray absolute intensities are small. Another major difficult may be the technically tedious task of acquiring, processing and analyzing coincidence data. Only in recent years, improvements in electronic and computing have turned such a predominantly digital data acquisition system attractive [106 - 109]. Recently, NIST has developed software in C++, the *qpx-gamma* software, available for download, to facilitate routine acquisition, data processing and spectrum construction compatible with Linux, OS X and Windows [106 - 110].

The number of elements that can be determined by cNAA is smaller than the number that can be determined by conventional NAA. Cooper in 1971 [105] provided a list of nuclides that may benefit from cNAA.

The major advantage of cNAA technique is to minimize the spectral interferences, improving the the detection limit due to the virtual elimination of all the background peaks.

Alvarez *et al.* in 1980 [111] used γγ-cNAA technique to measure iridium and other platinum groups in sediment samples from Creataceous-Tertiary boundary layers around the world. The authors found high levels of iridium, which would be similar to that of the Creataceous-Paleogene (approximately 66 million years ago) boundary layer, but significantly different from the ratio in the Earth´s crust but abundant in asteroids. The results would indicate that a large asteroid impact with the Earth took place, which was probably responsible for the mass extinction of some three-quarters of plant and animal species, including the dinosaurs. The theory was confirmed by Schulte *et al.* in 2010 [112]

Zeisler *et al.* in 2017 [113], showed a comparison of INAA with γγ-cNAA to determine Ag, Co, Cs, Se and U in three certified reference materials to test the performance of the methods The results obtained with the coincidence system showed improvements in the counting uncertainties.

Quality Assurance and Quality Control

In any analytical laboratory of gamma-ray spectrometry, two concepts should be considered: quality assurance (QA) and quality control (QC), which are directly related with the quality of the results. QA and QC are components of the laboratory´s quality management system [114]. Quality assurance is the part of quality management focused on providing confidence that quality requirements will be fulfilled. Is the operation to minimize the probability of errors in the analyses. It also includes quality control, which is the part of the quality management system focused on fulfilling quality requirements. Quality control is the operation by which the possible occurrence of errors is inspected once the analysis has been completed [114]. It is a planned activity designed to verify the quality of the measurement, *e.g.* analyzing blanks or samples of known concentration [115]. There are two types of QC: internal and external. Internal QC is the operation carried out by the staff as part of the measurements process providing evidence that the system operates satisfactorily and the results can be accepted. Certified Reference Materials are used to evaluate any bias in the results produced. In external QC, it is require that the performance of the laboratory is comparable with other laboratories carrying out similar measurements. These activities are achieved by means of *proficiency testing*, PT, also called *inter-laboratory comparison*. There are two types of PT: 1) there are those set up to assess the competence of a group of laboratories to undertake a very specific analysis, *e.g.* iron in hair samples, and 2) those used to evaluate the performance of laboratories across a certain sector for a particular type of analysis, *e.g.* determination of trace levels of metals in soil samples.

There are many PT schemes available operated by different organizations worldwide, including the International Atomic Energy Agency, IAEA. Further information on the PT can be found on EPTIS, a web-based PT information system [116].

The reliability, precision and accuracy of the results must be checked by analytical quality control.

The majors aspects of QA are related with:

1. Calibration of gamma-ray spectrometer. The principal types of calibration are: a) energy-channel, b) energy-resolution (FWHM), and c) energy-efficiency calibration;
2. Determination of spectral background, with the purpose to apply background subtraction during quantification;
3. Selection of suitable standards and flux monitors to make a batch;
4. Acquisition of gamma-ray spectra of each sample, standards and monitor;
5. Calculation of the activity of unknown sample;
6. Statistical evaluation of the data and use of control charts. Different parameters, such as the stability of the analytical procedure may be monitored by control charts [117]. For this a standard material is measured regularly and the analytical responses are plotted in time order on a chart;
7. Technical competence of the analyst.

A good laboratory practice requires the maintenance of records at each step, such as: a) sample related information, b) intermediate process, including name of the standard material, flux monitor, irradiation and counting conditions, c) results with uncertainties.

Reference materials, standards and control samples are used to validate, to establish the measurement traceability of the method and to enhance the reliability of the result. If the results of the CRM or RM materials agrees with the certified values within a reasonable uncertainty (generally within 95% confidence intervals) the procedure used is assumed accepted. Details of QA and QC using NAA is given by Bode in 2010 [115] and in Barwick 2008 [114].

APPLICATIONS OF NAA

It is generally accepted that the NAA method is a well-established analytical technique. Today there is a trend to use NAA to verify and to validate other methods of elemental analysis. The principles of the method are well understood. In the recent years, it has remained strong as a powerful analytical technique with applications in a broad range of disciplines such as: geology, human sample analysis, environmental monitoring, archaeology, materials science and many others [118].

Geological Samples

Geological samples such as soils, minerals, sediments and rocks are the most frequent type of sample analyzed by NAA as part of geological mapping. In the study of geochemistry of igneous rocks, the rare earth elements (REE) that can be determined by NAA are of special interest. In the case of silicate rocks, the determination of aluminum and magnesium is difficult because the fast neutron induces $^{28}Si(n, p)^{28}Al$ and $^{27}Al(n, p)^{27}Mg$ reactions. The effect of the interference can be significant, depending on the Al/Si ratio and the thermal to fast neutron flux ratio in the reactor. Several authors described and discussed all aspects of the analytical procedures for NAA in geological samples [119 - 124].

In general, analyses of geological samples fall into three categories [125, 126]: 1) multi-elemental analysis where the total elemental composition of a sample is required; 2) evaluation of REE as a group to elucidate the processes during fractionation, and 3) analysis of specific elements of interest such as halides. In geological processes, NAA is widely used for the determination of REE. The similar chemical behavior of the REE stems from the similarity of their ionic radius and charge (+3). The mass fractions of the REE are normalized to values found in chondritic (unfractionated) meteorites obtaining a smooth pattern. However, as europium is also able to exist as Eu^{2+}, it may not behave in the same way as the rest of the REE. The normalized plot may show an anomaly, where the mass fraction may be higher or lower than expected. In addition, cerium also may show anomalous results since it can also exist as Ce^{4+}. Obtaining information from the europium anomaly is necessary to have the precise data on both samarium and gadolinium values as well as europium. In the case of ^{153}Gd whose interference is from ^{153}Sm, the samarium must be allowed to decay before gadolinium is measured. An additional interference for the REE analysis is the interference due to other radionuclides produced during the fission of uranium in the sample during irradiation. In the determination of La, Ce and Nd, the analyst must consider that such radionuclides can be produced during the fission process. Corrections can be made by irradiating uranium and analyzing the resulting spectrum for these interferences.

Human Sample Analysis

Neutron activation analysis plays an important role in the study of trace elements in areas related to health. The samples can be divided into hard tissue (hair, nails, bones), soft tissue (breast tissue, neural tissue, brain, liver and thyroid) and body fluids (blood, urine, serum, feces) [16]. Some trace elements such as As, Ca, Cl, Co, Cu, Cr, F, Fe, I, K, Mg, Mn, Mo, Ni, Se, Si, Sn, V, and Zn are considered to be essential because they have an important role in the biochemical processes of the body. Cadmium, mercury, and lead may also have some physiological functions besides being toxic [127]. The main problem in the analysis of human samples is the very low concentrations of some elements, where NAA has a valuable contribution because it is capable of determining very small quantities of elements [127, 128].

Hair and nails analyses are most frequently used for assessing human environmental and occupation exposure health studies to detect any absorption of toxic elements such as Hg, Cd, Pb, Se, Sb and As [129 - 131]. A complete description of trace-element composition, sampling problems and analytical techniques for hair and nails is given by Valkovic in 1988 [132]. The main problem with hair is the external contamination which must be removed before the analysis. Usually the hair samples are washed with water, acetone and Milli Q water [133]. Hair analysis has been used to estimate the correlation between concentration of elements in hair and the health conditions. For instance, it was found that children suffering from asthma showed increased mass fraction of Au, Cs and Sb in hair [130].

The determination of elemental mass fractions in the brain is often used to study diseases, for example Alzheimer´s [131, 132, 134 - 136]. Some studies showed an increased mass fraction of Fe and Zn in one part of brain, so that the authors surmised that the elevated Fe could possibly play a role in neuronal degeneration in Alzheimer´s disease.

Another application of NAA is the determination of trace elements in human body fluids like blood, serum, urine and feces. There are many purposes of blood and urine analysis, one of them is the occupational exposure assessment, like chromium and manganese determination in workers of stainless steel industry [129]. NAA is also used to detect abnormalities that may be due to clinical problems. The mass fraction of elements in human plasma or serum was described fully by Versieck and Cornelis in 1989 [137].

Forensic NAA

The use of NAA in criminal investigations is well established. The basic idea is that, a criminal often leaves something at the site of crime such as bullets, hair, pieces of glass, trace elements of Sb and Ba deposited on the hand by the gun and so on. Sample wipes taken from a suspect´s hand can reveal not only if he or she has fired a gun recently, but also the type of gun and ammunition used. Such identification is rapid and non-destructive [138 - 142].

Although air samples show similarity for the major constituents, there can be a great variation in trace elements from one person to another because of differences in diet, body metabolism, occupation and environment.

Environmental Samples

Environmental chemistry covers a wide area of study concerned with air, water, plants, soil and animals [143]. NAA is a routine technique in the study of environmental problems.

Air samples play an important role in the investigation of air pollution by means of the analysis of aerosol samples and ash particulate matter. There are several comprehensive papers on the subject of analysis of atmospheric pollutants using NAA [144]. They describe the problems of sampling air pollution particulates and provide details of the impurities found in filter materials. Air samples are usually collected over a period of time on a filter medium such as cellulose paper, polystyrene, glass fiber or an organic membrane. The filter often contains trace elements such as Al, Ba, Fe, Mg and Zn that may affect the analysis. Cellulose filter has the lowest mass fraction of these trace elements and it is the preferred filter material for activation analysis [145, 146]. For the collection of air particulates, cascade impactors made of polyethylene or polycarbonate are frequently used. A typical scheme using irradiations with a thermal neutron flux of about 10^{13} n cm^{-2} s^{-1} to determine 32 elements for short-, medium- and long-lived radionuclides is presented in Table **4** [147].

Table 4. Neutron activation analysis of aerosol samples [147].

Irradiation time	Decay time	Counting	Determined element
20 s	5 min	6 min	Al, Ca, Cl, Mg, Na, Ti, V
	60min	10 min	Ba, Cl, Mn, Na
20 min	30 h	2h	As, Au, Br, Ga, K, La, Na, Sb, W, Zn
80 min	10 d	7 h	Au, Ba, Br, Ce, Co, Cr, Cs, Eu, Fe, Hf, Hg, La, Th, Rb, Sb, Sc, Se, Sm, Ta, Zn

An alternative to the characterization of air particulate matter is the use of plants as environmental pollution biomonitors. Biological materials used for air pollution monitoring include mosses, lichens, tree leaves/needles and bark [148]. The sample preparation in the analysis of plants is rather simple and involves washing, drying and grinding.

Water samples normally are more difficult to handle for activation analysis because this usually requires a pre-concentration step to ensure adequate sensitivity. If the analysis is focused on dissolved material, the water can be irradiated as a liquid or freeze-dried to remove the liquid and the residue can be irradiated as a solid [149]. Preconcentration on an ion exchange resin or activated charcoal are commonly used to collect the analytes of interest. Seawater is a matrix particularly difficult for NAA, since the sodium and chloride activity induced in the matrix makes analysis of short-lived radionuclides difficult.

Archaeological Samples

The contribution of the physical and chemical methods in archaeological samples is found in the field of archaeometry. NAA is among the methods that provide information about the chemical composition of artifacts (pottery, glass, ceramics, *etc.*), but it may also be used for dating using the K-Ar method [150, 151]. The chemical composition reflects the geochemical features of the raw materials from which an artifact was made. Furthermore, it is possible to identify the source of raw materials or to find relationships between the samples leading, in some cases, to the identification of the origin of the excavated material [152]. The chemical analysis may support the understanding of ancient technology and also provide information for conservation and restoration of the findings.

Many analytical methods are used for the analysis of archaeological materials. However, the applicability of an analytical method is determined by several criteria such as: 1) the method should ensure the reliable determination of 10 to 20 elements; 2) it should provide high sensitivity for trace elements; 3) it should have high precision and accuracy; and finally 4) the method should not be laborious, since the solution of an archaeological problem demands analysis of a large number of samples [153 - 155]. In light of such requirements, INAA is among the most widely used methods in archaeometric studies.

For INAA applications in archaeometry, the samples are usually irradiated one [156], two [157] or three times [158] in a nuclear reactor, followed by one to six gamma-ray measurements after different cooling times. Without serious difficulty, the content of Al, As, Au, Ba, Br, Ca, Ce, Co, Cr, Cs, Dy, Eu, Fe, Hf, K, La, Lu, Mg, Mn, Na, Nd, Ni, Rb, Sb, Sm, Sc, Ta, Tb,Ti, Th, U, V, Yb, and Zn can be determined by INAA.

CONCLUSION

This overview shows that neutron activation analysis is currently a mature analytical technique. The principles of the NAA method are well understood and the major advances occurred at the beginning of the technique development, in the last century. In the recent years, there have been no fundamental aspects in the method which imply further development, although improvements in the method will continue in the future, as it is a situation of maturity but not stagnation. Furthermore, it has remained steady as a powerful analytical technique with applications in practically all fields of science. In comparison with other techniques, INAA in general does not require any special sample pretreatment, it is a non-destructive technique for both qualitative and quantitative analysis, and it typically involves no reagent blank. It can detect and measure many elements simultaneously. In some fields, it remains irreplaceable, like in the analysis of solid materials which are difficult to dissolve, and in the analysis of samples with low trace element concentrations, where the risk of contamination is high and an analysis with very high accuracy is necessary.

CONSENT FOR PUBLICATION

Not applicable.

CONFLICT OF INTEREST

The author confirms that he has no conflict of interest to declare for this publication.

ACKNOWLEDGEMENTS

Declared none.

REFERENCES

[1]   Chadwick, J. Possible existence of a neutron. *Nature,* **1932**, *129*, 312.
       [http://dx.doi.org/10.1038/129312a0]

[2]   Steinhauser, G.; Loffler, G.; Adunka, R. The possible discovery of neutron activation in 1910. *J. Radioanal. Nucl. Chem.,* **2013**, *296*, 157-163.
       [http://dx.doi.org/10.1007/s10967-012-2065-7]

[3]   Rutherford, E. Nuclear constitution of atoms. *Proc. R. Soc. Lond., A Contain. Pap. Math. Phys. Character,* **1920**, *97*, 374-400.
       [http://dx.doi.org/10.1098/rspa.1920.0040]

[4]   Webster, H.C. The artificial production of nuclear γ-radiation. *Proc. R. Soc. Lond., A Contain. Pap. Math. Phys. Character,* **1932**, *136*, 428-453.
       [http://dx.doi.org/10.1098/rspa.1932.0093]

[5]   Hevesy, G.; Levi, H. Kjl. The action of neutrons on the rare earth elements. *Danske Videnskab. Selskab. Math-Fys. Medd.,* **1936**, *14*(5), 3-34.

[6]   Hevesy, G.; Levi, H. Artificial activity of hafnium and some other elements. *Danske Videnskab. Selskab Math.-. Fys. Medd.,* **1938**, *15*(11), 1-18.

[7]   Seaborg, G.T.; Livingood, S. Artificial radioactivity as a test for minute traces of elements. *J. Am. Chem. Soc.,* **1938**, *60*, 1784-1786.
       [http://dx.doi.org/10.1021/ja01275a018]

[8]   Boyd, G.E. Method of activation analysis. *Anal. Chem.,* **1949**, *21*(3), 335-347.
       [http://dx.doi.org/10.1021/ac60027a005]

[9]   Ehmann, W.D.; Vance, D.E. *Radiochemistry and Nuclear Methods of Analysis*; Wineforder, J.D., Ed.; John Wiley & Sons, Inc., **1991**.

[10]   Filby, R.H. The determination of bromine in rocks by neutron activation analysis. *Anal. Chim. Acta,* **1964**, *31*, 434-440.
       [http://dx.doi.org/10.1016/S0003-2670(00)88852-2]

[11]   Chauvin, R.; Levegue, P. Utilisation d'un discriminatuer d'impulsions associé à um detecteur gamma a scintillation pour l'analyse par activation. *Int. J. Appl. Radiat. Isot.,* **1956**, *1*, 115-122.
       [http://dx.doi.org/10.1016/0020-708X(56)90024-2]

[12] Gordon, G.E.; Randel, K.; Goles, G.G.; Corliss, J.B.; Benson, M.H.; Oxley, S. Instrumental activation analysis of standards rocks with high resolution gamma-ray detectors. *Geochim. Cosmochim. Acta,* **1968**, *32*(4), 369-396.
[http://dx.doi.org/10.1016/0016-7037(68)90074-4]

[13] Guinn, V.P. Rapid growth, maturity, current problems, future prospects of NAA. *J. Radioanal. Nucl. Chem.,* **2000**, *244*(1), 23-25.
[http://dx.doi.org/10.1023/A:1006746200149]

[14] De Bruin, M. Activation analysis: future development in a historical perspective. *J. Radioanal. Nucl. Chem.,* **1998**, *234*(1-2), 5-7.
[http://dx.doi.org/10.1007/BF02389738]

[15] Greenberg, R.R.; Bode, P.; Fernandes, E.A. Neutron activation analysis: a primary method of measurement. *Spectrochim. Acta B At. Spectrosc.,* **2011**, *66*, 193-241.
[http://dx.doi.org/10.1016/j.sab.2010.12.011]

[16] Frontasyeva, M.V. Neutron activation analysis in the life sciences. *Phys. Part. Nucl.,* **2011**, *42*(2), 332-378.
[http://dx.doi.org/10.1134/S1063779611020043]

[17] Glascock, M.D. Nuclear spectroscopy.*Treatise on Geochemistry,* 2nd ed; Holland, H.D.; Turckian, K.K., Eds.; Elsevier: Oxford, **2014**, Vol. 15-16, pp. 273-290.
[http://dx.doi.org/10.1016/B978-0-08-095975-7.01419-4]

[18] Sato, T.; Kato, T. Estimates of iodine in biological materials by epithermal neutron activation analysis. *J. Radioanal. Nucl. Chem.,* **1982**, *68*(1-2), 175-180.
[http://dx.doi.org/10.1007/BF02517620]

[19] Perry, S.J. Evaluation of boron for the epithermal neutron activation analysis of short-lived radionuclides in geological and biological matrices. *J. Radioanal. Nucl. Chem. Articles,* **1984**, *81*(1), 143-151.
[http://dx.doi.org/10.1007/BF02132929]

[20] Davies, J.A.; Hart, P.A.; Jefferies, A.C. Trace multielement analysis in boron materials by epithermal neutron. *J. Radional. Nucl.,Chem. Articles,* **1986**, *98*(2), 275-287.

[21] Rowe, J.J.; Steinnes, E. Determination of 30 elements in coal and fly ash by thermal and epithermal neutron-activation analysis. *Talanta,* **1977**, *24*(7), 433-439.
[http://dx.doi.org/10.1016/0039-9140(77)80125-2] [PMID: 18962115]

[22] Rossito, F.; Terrani, M.; Terrani, S. Choice of neutron filters in activation analysis. *Nucl. Instrum. Methods,* **1972**, *103*, 77-83.
[http://dx.doi.org/10.1016/0029-554X(72)90462-4]

[23] Gladney, E.S.; Perrin, D.R. Determination of bromine in biological, soil, and geological standard reference materials by instrumental epithermal neutron activation. *Anal. Chem.,* **1979**, *51*(12), 2015-2018.
[http://dx.doi.org/10.1021/ac50048a028] [PMID: 539656]

[24] Elnimr, T.; Ela-Assaly, F.M. Determination of the attenuation of epicadmium neutrons using the method of varying Cd-thickness. *J. Radional. Nucl. Chem. Articles,* **1987**, *109*(1), 3-9.
[http://dx.doi.org/10.1007/BF02117518]

[25] Landsberger, S.; Simsons, A. Chromium, nickel, and arsenic determinations in human samples by thermal and epithermal neutron activation analyses. *Biol. Trace Elem. Res.,* **1987**, *13*(1), 357-362.
[http://dx.doi.org/10.1007/BF02796646] [PMID: 24254690]

[26] Glascock, M.D.; Tian, W.Z.; Ehmann, W.D. Utilization of a boron irradiation vessel for neutron activation analysis of short-lived radionuclides in biological and geological materials. *J. Radioanal. Nucl. Chem.,* **1985**, *92*(2), 379-390.
[http://dx.doi.org/10.1007/BF02219768]

[27]   Bem, H.; Ryan, D.E. Choice of boron shield in epithermal neutron. determinations. *Anal. Chim. Acta,* **1981**, *124*, 373-380.
[http://dx.doi.org/10.1016/S0003-2670(01)93585-8]

[28]   Chisela, F.; Gawlik, D.; Bratter, P. Advantages of boron filters in instrumental epithermal neutron activation analysis of biological materials. *J. Radioanal. Nucl. Chem.,* **1987**, *112*(2), 293-308.
[http://dx.doi.org/10.1007/BF02132362]

[29]   Holzbecher, J.; Chatt, A.; Ryan, D.E. Slowpoke epi-cadmium neutron flux in activation analysis of trace elements. *Can. J. Spectros.,* **1985**, *30*, 67-72.

[30]   Steinnes, E. Determination of nickel in rocks by instrumental neutron activation analysis. *Anal. Chim. Acta,* **1969**, *68*, 25-30.
[http://dx.doi.org/10.1016/S0003-2670(01)85142-4]

[31]   Steinnes, E.; Brune, D. Determination of uranium in rocks by instrumental activation-analysis using epithermal neutrons. *Talanta,* **1969**, *16*(9), 1326-1329.
[http://dx.doi.org/10.1016/0039-9140(69)80010-X] [PMID: 18960639]

[32]   Glascock, M.D.; Neff, H. Neutron activation analysis and provenance research in archaeology. *Meas. Sci. Technol.,* **2003**, *14*(9), 1516-1526.
[http://dx.doi.org/10.1088/0957-0233/14/9/304]

[33]   Brune, D. Epithermal neutron activation analysis for iodine in small aqueous samples. *Anal. Chim. Acta,* **1969**, *46*(1), 17-21.
[http://dx.doi.org/10.1016/0003-2670(69)80036-X]

[34]   Parry, S.S. Detection limits in epithermal neutron activation analysis of geological materials. *J. Radioanal. Nucl. Chem.,* **1980**, *59*(2), 423-427.
[http://dx.doi.org/10.1007/BF02517293]

[35]   Tsipenyuk, Y.M. Nuclear reactions.*Nuclear methods in science and technology*; Bradley, D.A., Ed.; , **1997**.

[36]   Tsipenyuk, Y.M. Nuclear reactions.*Nuclear methods in science and technology*; Bradley, D.A., Ed.; , **1997**.

[37]   De Corte, F.; Simonits, A.; Bellemans, F.; Freitas, M.C.; Jovanovic, S.; Smodis, B.; Erdtmann, G.; Petri, H.; De Wispelaere, A. Recents advances in the k_0-standardization of neutron activation analysis: extensions, applications, prospects. *J. Radioanal. Nucl. Chem.,* **1993**, *169*(1), 125-158.
[http://dx.doi.org/10.1007/BF02046790]

[38]   De Corte, F. k_0 and comparator NAA: influences of interactions. *J. Radioanal. Nucl. Chem.,* **2000**, *245*(1), 157-161.
[http://dx.doi.org/10.1023/A:1006785417003]

[39]   Brushwood, J.; Beeley, P.A.; Nielsen, K.; Bennet, L.G. Analysis of two standard reference matrials using the shape independent model for k0 based INAA. *J. Radioanal. Nucl. Chem.,* **2000**, *245*(1), 205-207.
[http://dx.doi.org/10.1023/A:1006715717428]

[40]   De Corte, F.; Simonits, A.; De Weispelaere, A.; Hoste, J. Accuracy and applicability of the k_0-standardization method. *J. Radioanal. Nucl. Chem. Articles,* **1987**, *113*(1), 145-161.
[http://dx.doi.org/10.1007/BF02036056]

[41]   De Corte, F. A journal through the pre-Budapest-2013 k(0) users workshops. *J. Radioanal. Nucl. Chem.,* **2014**, *300*(2), 453-455.
[http://dx.doi.org/10.1007/s10967-013-2838-7]

[42]   Xiao, C.; Yao, Y.; Jin, X.; Hua, L.; Wang, P.; Ni, B. k_0-NAA for determination of REE in reference materials of ores sources. *J. Radioanal. Nucl. Chem.,* **2017**, *311*(1), 1287-1289.
[http://dx.doi.org/10.1007/s10967-016-5157-y]

[43] Jacimovic, R.; De Corte, F.; Kennedy, G.; Vermaerche, V.; Revay, Z. The 2012 recommended k(0) database. *Radioanal. Nucl. Chem.,* **2014,** *300*(2), 589-592.
[http://dx.doi.org/10.1007/s10967-014-3085-2]

[44] De Corte, F.; Moens, L.; Jovanovic, S.; Simonits, A.; De Weispelaere, A. Applicability of the $1/E^{1-\alpha}$ epithermal spectrum representation and the effective resonance energy E_r in NAA. *J. Radioanal. Nucl. Chem. Articles,* **1986,** *102*(1), 37-57.
[http://dx.doi.org/10.1007/BF02037948]

[45] International Atomic Energy Agency. *Practical aspects of operating a neutron activation analysis laboratory*; , **1990**.

[46] Iyengar, G.V.; Sansoni, B. Sample preparation of biological material., **1980**.

[47] Sansoni, B.; Iyenga, G.V. Sampling and storage of biological materials., **1980**.

[48] Bock, R. *A handbook of decomposition methods in analytical chemistry*; John Wiley & Sons Inc., **1979**.

[49] Donev. I.Y. Rapid Homogenisation and drying of biological materials. *J. Radioanal. Nucl. Chem.,* **1977,** *39*(1/2), 317-322.

[50] De Broer, J.L.; Maessen, F.J. Optimum Experimental conditions of the brittle fracture technique for homogenization of biological materials. *Anal. Chim. Acta,* **1980,** *117*, 371-375.
[http://dx.doi.org/10.1016/0003-2670(80)87041-3]

[51] Watterson, J.I. Optimisation of irradiation and decay times in nuclear activation analysis. *J. Radioanal. Nucl. Chem.,* **1975,** *26*, 135-150.
[http://dx.doi.org/10.1007/BF02516520]

[52] Zikovsky, L. The optimum irradiation, decay and counting times in activation analysis for single and double counting and for different minimization functions**.** *J. Radioanal. Nucl. Chem.,* **1974,** *22*(1), 165-172.
[http://dx.doi.org/10.1007/BF02518102]

[53] Okada, M. Optimal "cooling time" to minimize interfering activity in non-destructive activation analysis. *Anal. Chim. Acta,* **1961,** *24*, 410-412.
[http://dx.doi.org/10.1016/0003-2670(61)80084-6]

[54] Ila, P.; Jagam, P.; Muercke, G.K. Multielement analysis of uraniferous rocks by INAA: Special reference to interferences due to uranium and fission of uranium. *J. Radioanal. Nucl. Chem.,* **1983,** *79*(2), 215-232.
[http://dx.doi.org/10.1007/BF02518934]

[55] Korotev, R.L.; Lindstrom, D.J. Interferences from fission of ^{235}U in INAA of rocks. *Trans. Am. Nucl. Soc.,* **1985,** *49*, 177-178.

[56] Glascock, M.D.; Nabelek, P.I.; Weinrich, D.D.; Coveney, R.M., Jr Correcting for uranium fission in instrumental neutron activation analysis of high-uranium rocks. *J. Radioanal. Nucl. Chem.,* **1986,** *99*(1), 121-131.
[http://dx.doi.org/10.1007/BF02060832]

[57] Kennedy, G.; Fowler, A. Interference from uranium in neutron activation analysis or rare-earths in silicate rocks. *J. Radioanal. Nucl. Chem.,* **1983,** *78*(1), 165-169.
[http://dx.doi.org/10.1007/BF02519761]

[58] Kucera, J.; Bode, P.; Stepanek, V. Uncertainty evaluation in instrumental and radiochemical NAA.*Quantifying uncertainty in nuclear analytical measurements, IAEA-TECDOC 1401*; , **2004**, pp. 77-102.

[59] Zahn, G.S.; Ticianelli, R.B.; Lange, C.N.; Favaro, D.I.; Figueiredo, A.M. Uncertainty analysis in comparative NAA applied to geological and biological matrices. *International Nuclear Atlantic Conference ,* **2015**. October 4-9 Sao Paulo, Brazil

[60]   Meng, L.J.; Ramsden, D. An intercomparison of three spectral-deconvolution algorithms for gamma-ray spectroscopy. *IEEE Trans. Nucl. Sci.,* **2000**, *47*(4), 1329-1336.
[http://dx.doi.org/10.1109/23.872973]

[61]   Knoll, G.F. *Radiation detection and measurement,* 4th ed; Wiley & Sons Inc: New York, **2010**.

[62]   Parry, S.J. *Handbok of neutron activation analysis*; Viridian: UK, **2003**.

[63]   www.canberra.com

[64]   Paulus, T.J.; Raudorf, T.W.; Trammell, R.; Coyne, B. Comparative timing performance of large volume HPGe germanium detectors. *IEEE Transact. Nucl. Sci.,* **1981**, *Ns-28*(1), 544-548.

[65]   Dmitriev, A.Y.; Pavlov, S.S. Automation of the quantitative determination of elemental content in samples using neutron activation analysis on the IBR-2 reactor at the Frank Laboratory for neutron physics, Joint Institute for Nuclear Research. *Phys. Part. Nucl. Lett.,* **2013**, *10*(1), 33-36.
[http://dx.doi.org/10.1134/S1547477113010056]

[66]   Yonggang, Y.; Caijin, X.; Long, H. Development of INAA automation at CARR. *J. Radioanal. Nucl. Chem.,* **2016**, *307*(3), 1651-1656.
[http://dx.doi.org/10.1007/s10967-015-4557-8]

[67]   Pavlov, S.S.; Dmitriev, A.Y. A.Y. Automation system for NAA at the reactor IBR-2 Frank Laboratoty of neutron physics, Joint Intitute for Nuclear Research, Dubna, Russia. *J. Radioanal. Nucl. Chem.,* **2016**, *309*(1), 27-38.
[http://dx.doi.org/10.1007/s10967-016-4864-8] [PMID: 27375310]

[68]   Popescu, G.; Herman, S.; Glover, S.; Spitz, H.C. Compton background suppression with a multi-element scintillation detector using high speed data acquisition and digital signal processing. *J. Radioanal. Nucl. Chem.,* **2016**, *307*(3), 1949-1955.
[http://dx.doi.org/10.1007/s10967-015-4355-3]

[69]   Landsberger, S.S.; Peshev, S. Comptom suppression neutron activation analysis: past, present and future. *J. Radioanal. Nucl. Chem.,* **1996**, *202*(1-2), 201-224.
[http://dx.doi.org/10.1007/BF02037943]

[70]   Bacchi, M.A.; Santos, L.G.; Fernandes, E.A.; Bode, P.; Tagliaferro, F.S.; França, E.J. INAA with Compton supression: how much can the analysis of plant materials be improved? *J. Radioanal. Nucl. Chem.,* **2007**, *271*(2), 345-351.
[http://dx.doi.org/10.1007/s10967-007-0215-0]

[71]   Landsberger, S. Compton suppression neutron activation methods in environmental analysis. *J. Radioanal. Nucl. Chem.,* **1994**, *179*(1), 67-79.
[http://dx.doi.org/10.1007/BF02037926]

[72]   Wu, D.; Landsberger, S. Comparison of NAA methods to determine medium-lived radionuclides in NIST soil standard reference materials. *J. Radioanal. Nucl. Chem.,* **1994**, *179*(1), 155-164.
[http://dx.doi.org/10.1007/BF02037936]

[73]   Landsberger, S.; Wu, D. D. A comprehensive study for the determination of forty eight elements in the certification of a hazardous standard reference material. *J. Radioanal. Nucl. Chem.,* **1995**, *193*(1), 49-59.
[http://dx.doi.org/10.1007/BF02041916]

[74]   Tzika, F.; Stamatelatos, I.E.; Kalef-Ezra, J. Neutron activation analysis of large volume samples: the influence of inhomogeneity. *J. Radioanal. Nucl. Chem.,* **2007**, *271*(1), 233-240.
[http://dx.doi.org/10.1007/s10967-007-0135-z]

[75]   Shakir, N.S.; Jervis, R.E. Corrections factors required for quantitative large volume instrumental neutron activation analysis. *J. Radioanal. Nucl. Chem.,* **2001**, *248*(1), 61-68.
[http://dx.doi.org/10.1023/A:1010669806644]

[76]   Bode, P. Opportunities for innovation in neutron activation analysis. *J. Radioanal. Nucl. Chem.,* **2012**,

291(2), 275-280.
[http://dx.doi.org/10.1007/s10967-011-1193-9] [PMID: 26224911]

[77] Mandour, M.A.; Badawi, A.; Mohamed, N.M.; Emanm, A. Towards a methodology for bulk sample neutron activation analysis, *J. Taibah Univ.Sci.,* **2016**, *10*, 235-241.
[http://dx.doi.org/10.1016/j.jtusci.2015.04.009]

[78] Gwozdz, R.; Grass, F. Activation analysis of large samples. *J. Radioanal. Nucl. Chem.,* **2000**, *233*(3), 523-529.
[http://dx.doi.org/10.1023/A:1006778814030]

[79] Schneider, E.W.; Yusuf, S.O. Applications of nuclear methods in the automotive industry. *J. Radioanal. Nucl. Chem. Articles,* **1996**, *203*(2), 489-503.
[http://dx.doi.org/10.1007/BF02041525]

[80] James, W.D. 14 MeV fast neutron activation analysis in the year 2000. *J. Radioanal. Nucl. Chem.,* **2000**, *243*(1), 119-123.
[http://dx.doi.org/10.1023/A:1006775432158]

[81] Jonah, S.A.; Okunade, I.O.; Jimba, B.W.; Umar, I.M. Application of low-yield neutron generator for rapid evaluation of alumino-silicate ores from Nigeria by FNAA. *Nucl. Instrum. Methods Phys. Res. A,* **2001**, *463*(1-2), 321-323.
[http://dx.doi.org/10.1016/S0168-9002(01)00287-X]

[82] Szentmiklósi, L.; Kasztovzky, Zs.; Belgya, T. Fifteen years of success: user access program at the Budapest prompt-gamma activation analysis laboratory. *J. Radioanal. Nucl. Chem.,* **2016**, *309*(1), 71-77.
[http://dx.doi.org/10.1007/s10967-016-4774-9]

[83] Borsaru, M.; Berry, M.; Biggs, M.; Rojc, A. *In situ* determination of sulphur in coal seams and overburden rock by PGNAA. *Nucl. Instrum. Methods Phys. Res. B,* **2004**, *213*, 530-534.
[http://dx.doi.org/10.1016/S0168-583X(03)01623-9]

[84] Strijckmans, K. Charged particle activation analysis.*Surface characterisation: a practical approach*; Hellborg, R.; Brune, D., Eds.; Scandinavian Scientific Press & VCH: Weinheim, **1997**.

[85] Khmis, I.; Othman, I.; Nasri, M.; Bakkour, M. Transfer time optimization of a rapid cyclic instrumental neutron activation analysis for trace element detection. *Rev. Sci. Instrum.,* **2001**, *172*, 1492-1494.
[http://dx.doi.org/10.1063/1.1340027]

[86] Rodríguez, N.; Yoho, H.; Lansdberger, S. Determination of Ag, Au, Cu and Zn in ore samples from two Mexican mines by various thermal and epithermal NAA techniques. *J. Radioanal. Nucl. Chem.,* **2016**, *307*(2), 955-961.
[http://dx.doi.org/10.1007/s10967-015-4277-0]

[87] Spyrou, N.M. Cyclic activation analysis – a review. *J. Radioanal. Nucl. Chem.,* **1981**, *261*(1), 211-242.
[http://dx.doi.org/10.1007/BF02517411]

[88] Zhang, H.; Chai, Z.F.; Qing, W.Y.; Chen, H.C. Cyclic neutron activation analysis for determination of selenium in food samples using ^{77m}Se. *J. Radioanal. Nucl. Chem.,* **2009**, *281*(1), 23-26.
[http://dx.doi.org/10.1007/s10967-009-0087-6]

[89] Parry, S.J.; Benzing, R.; Bolstad, K.L.; Steinnes, E. Epithermal/fast neutron cyclic activation analysis for the determination of fluorine in environmental and industrial materials. *J. Radioanal. Nucl. Chem.,* **2000**, *244*(1), 67-72.
[http://dx.doi.org/10.1023/A:1006770905601]

[90] Nuclear Sciences Series. *The Radiochemistry of the elements. Nat. Acad. Sci. USA,*

[91] Zinovyed, V.; Ablesimov, N.E.; Egov, A.I. Instrumental and radiochemical neutron activation analysis of the quartz adularia veins from the deposit Milogradovka, the Far east, Primorye. *J. Radioanal. Nucl. Chem.,* **2017**, *311*(1), 141-153.

[http://dx.doi.org/10.1007/s10967-016-4948-5]

[92]   Kucera, J.; Kamenik, J.; Povinec, P.P. Radiochemical separation of mostly short-lived neutron acativation products. *J. Radioanal. Nucl. Chem.,* **2017**, *311*(2), 1299-1307.

[93]   Dybzynski, R. 50 years of adventures with neutron activation analysis with the special emphasis on radiochemical separations. *J. Radioanal. Nucl. Chem.,* **2015**, *303*(2), 1067-1090.
[http://dx.doi.org/10.1007/s10967-014-3822-6]

[94]   Kucera, J.; Zeisler, R. Do we need radiochemical separation in activation analysis? *J. Radioanal. Nucl. Chem.,* **2004**, *262*(1), 255-260.
[http://dx.doi.org/10.1023/B:JRNC.0000040883.15153.db]

[95]   Chen, Ch.; Lu, X.; Zhang, P.; Hou, X.; Chai, Z. Subcellular distribution patterns of twenty four elements in the human liver samples studied by molecular activation analysis. *J. Radioanal. Nucl. Chem.,* **2000**, *244*(1), 199-205.
[http://dx.doi.org/10.1023/A:1006701206935]

[96]   Stone, S.F.; Hancock, D.; Zeisler, R. Characterization of biological macromolecules by electrophoresis and neutron activation. *J. Radioanal. Nucl. Chem.,* **1987**, *112*(1), 95-108.
[http://dx.doi.org/10.1007/BF02037280]

[97]   Das, H. *Speciation of trace elements with special reference to the use of radioanalytical methods*; ECN: Petten, **1993**.

[98]   Chifang, Ch.; Xueying, M.; Yupi, W. Molecular activation analysis for chemical species studies. *Fresenius J. Anal. Chem.,* **1999**, *363*(5), 477-480.

[99]   Opelanio, L.R.; Rack, E.P.; Blotcky, A.J.; Crow, F.W. Determination of chlorinated pesticides in urine by molecular neutron activation analysis. *Anal. Chem.,* **1983**, *55*(4), 677-681.
[http://dx.doi.org/10.1021/ac00255a022] [PMID: 6859538]

[100]   Chen, Ch.; Zhang, P.; Lu, X. Subcellular distribution of Al, Cu, Mg, Mn and other elements in the human liver. *Fresenius J. Anal. Chem.,* **1999**, *363*(5), 512-516.
[http://dx.doi.org/10.1007/s002160051235]

[101]   Chai, Z.F.; Zhang, Z.Y.; Feng, W.Y. Study of chemical speciation of trace elements by molecular activation analysis and other nuclear techniques. *J. Anal. At. Spectrom.,* **2004**, *19*(1), 26-33.
[http://dx.doi.org/10.1039/b307337h]

[102]   Hou, X.; Dahigaard, H.; Rietz, B.; Jacobsen, U. Determination of chemical species of iodine in seawater by radiochemical neutron activation analysis combined with ion-exchange preparation. *Anal. Chem.,* **1999**, *71*(14), 2745-2750.
[http://dx.doi.org/10.1021/ac9813639]

[103]   Metwally, W.A.; Gardner, R.P.; Mayo, C.W. Elemental PGNAA analysis using gamma-gamma coincidence counting with the library least-squares approach. *Nucl. Instrum. Methods Phys. Res. Sect B,* **2004**, *213*, 394-399.
[http://dx.doi.org/10.1016/S0168-583X(03)01660-4]

[104]   Drescher, A.; Yoho, M.; Landsberger, S.; Durbin, M.; Biegalski, S.; Meier, D.; Schwantes, J. Gamma-gamma coincidence performance of LaBr3:Ce scintillation detectors *vs* HPGe detectors in high count-rate scenarios. *Appl. Radiat. Isot.,* **2017**, *122*, 116-120.
[http://dx.doi.org/10.1016/j.apradiso.2017.01.012] [PMID: 28130979]

[105]   Cooper, J.A. Radioanalytical applications of gamma-gamma coincidence techniques with lithium-drifted germanium detectors. *Anal. Chem.,* **1971**, *43*(7), 838-845.
[http://dx.doi.org/10.1021/ac60302a017]

[106]   Shetty, M.; Şahin, D. Data acquisition and analysis software for gamma coincidence spectrometry. *J. Radioanal. Nucl. Chem.,* **2016**, *309*(1), 243-247.
[http://dx.doi.org/10.1007/s10967-016-4762-0] [PMID: 27325905]

[107] Yoho, M.; Landsberger, S. Determination of selenium in coal fly ash *via* gamma-gamma coincidence neutron activation analysis. *J. Radioanal. Nucl. Chem.,* **2016**, *307*(1), 733-737.
[http://dx.doi.org/10.1007/s10967-015-4209-z]

[108] Howard, C.; Ferm, M.; Cesaratto, J.; Daigle, S.; Iliadis, C. Radioisotope studies of the farmville meteorite using γγ-coincidence spectrometry. *Appl. Radiat. Isot.,* **2014**, *94*, 23-29.
[http://dx.doi.org/10.1016/j.apradiso.2014.07.001] [PMID: 25063942]

[109] Tomli, B.E.; Zeisler, R.; Lindstrom, R.M. γγ coincidence spectrometer for instrumental neutron activation analysis. *Nucl. Instr. Meth. Phys. Res. Sect A,* **2008**, *589*, 243-249.
[http://dx.doi.org/10.1016/j.nima.2008.02.094]

[110] https://github.com/usnistgov/qpx-gamma

[111] Alvarez, L.W.; Alvarez, W.; Asaro, F.; Michel, H.V. Extraterrestrial cause for the cretaceous-tertiary extinction. *Science,* **1980**, *208*(4448), 1095-1108.
[http://dx.doi.org/10.1126/science.208.4448.1095] [PMID: 17783054]

[112] Schulte, P.; Alegret, L.; Arenillas, I.; Arz, J.A.; Barton, P.J.; Bown, P.R.; Bralower, T.J.; Christeson, G.L.; Claeys, P.; Cockell, C.S.; Collins, G.S.; Deutsch, A.; Goldin, T.J.; Goto, K.; Grajales-Nishimura, J.M.; Grieve, R.A.; Gulick, S.P.; Johnson, K.R.; Kiessling, W.; Koeberl, C.; Kring, D.A.; MacLeod, K.G.; Matsui, T.; Melosh, J.; Montanari, A.; Morgan, J.V.; Neal, C.R.; Nichols, D.J.; Norris, R.D.; Pierazzo, E.; Ravizza, G.; Rebolledo-Vieyra, M.; Reimold, W.U.; Robin, E.; Salge, T.; Speijer, R.P.; Sweet, A.R.; Urrutia-Fucugauchi, J.; Vajda, V.; Whalen, M.T.; Willumsen, P.S. The Chicxulub asteroid impact and mass extinction at the Cretaceous-Paleogene boundary. *Science,* **2010**, *327*(5970), 1214-1218.
[http://dx.doi.org/10.1126/science.1177265] [PMID: 20203042]

[113] Zeisler, R.; Cho, H.; Ribeiro, I.S., Jr; Shetty, M.; Turkoglu, D. On neutron activation analysis with γγ coincidence spectrometry. *J. Radioanal. Nucl. Chem.,* **2017**.
[http://dx.doi.org/10.1007/s10967-017-5342-7]

[114] Barwick, E.P. *Quality assurance in analytical chemistry*; Ando, D.J., Ed.; John. Wiley & Sons Inc., **2008**.

[115] Bode, P. Quality control and assurance of neutron activation analysis. In: *Encyclopedia of Analytical Chemistry*; , **2010**.

[116] www.aptis.bam.de

[117] Mullins, E. Introduction to control charts in the analytical laboratory. *Analyst (Lond.),* **1994**, *119*, 369-375.
[http://dx.doi.org/10.1039/an9941900369]

[118] Witkowska, E.; Szczepaniak, K.; Biziuk, M. Some applications of neutron actination analysis: a review. *J. Radioanal. Nucl. Chem.,* **2005**, *265*(1), 141-150.
[http://dx.doi.org/10.1007/s10967-005-0799-1]

[119] Cristache, C.; Duliu, O.G.; Ricman, C.; Toma, M.; Dragolici, F.; Bragea, M.; Done, L. Determination of elemental content in geological samples. *Rom. J. Phys.,* **2008**, *53*(7-8), 941-946.

[120] El-Taher, A.; Khater, A.E. Elemental characterization of Hazm El-Jalamid phosphorite by instrumental neutron activation analysis. *Appl. Radiat. Isot.,* **2016**, *114*, 121-127.
[http://dx.doi.org/10.1016/j.apradiso.2016.05.012] [PMID: 27235886]

[121] Ermolaeva, V.N.; Mikhailova, A.V.; Kogarko, L.N.; Kolesov, G.M. Leaching rare-earth and radioactive elements from alkaline rocks of the Lovozero Massif, Kola Peninsula. *Geochem. Int.,* **2016**, *54*(7), 633-639.
[http://dx.doi.org/10.1134/S001670291607003X]

[122] Das, H.A.; Faahof, A.; Van der Sloot, H.A. *Radioanalysis in geochemistry (developments in geochemistry)*; Elsevier Science: Amsterdam, **1989**.

[123] Olise, F.O.; Oladejo, O.F. Instrumental neutron activation analysis of uranium in samples from tin mining and processing sites. *J. Geochem. Explor.,* **2014**, *142*, 36-42. [http://dx.doi.org/10.1016/j.gexplo.2014.01.004]

[124] Bounouira, H.; Choukri, A.; Cherkaoui, R.; Chakiri, S.; Bounakhla, M.; Embarch, K. Utilization of instrumental neutron activation analysis (INAA) for geochemical behavior study of major and trace elements in Bouregreg river basin (Morocco). *J. Radioanal. Nucl. Chem.,* **2012**, *292*(3), 1049-1058. [http://dx.doi.org/10.1007/s10967-011-1597-6]

[125] El-Taher, A. Rare-earth elements in Egyptian granite by INAA. *Appl. Radiat. Isot.,* **2007**, *65*(4), 458-464. [http://dx.doi.org/10.1016/j.apradiso.2006.07.014] [PMID: 17208446]

[126] Alnour, I.; Wagiran, H.; Ibrahim, N.; Hamzah, S.; Wee, B.S.; Elias, M.S. Rare earth elements determination and distribution patterns in granite rock samples by using INAA absolute method *J. Radioanal. Nucl. Chem.,* **2015**, *303*(3), 1999-2009.

[127] Cesareo, R., Ed. *Nuclear analytical techniques in medicine,* 1st ed; Elsevier Science: Amsterdam, **1988**.

[128] Heydorn, K. *Neutron activation analysis for clinical trace element research*; CRC Press: Boca Raton, Florida, **1984**.

[129] Kucera, J.; Bencko, V.; Tejral, J.; Borska, L.; Soukal, L. Biomonitoring of occupational exposure: neutron activation determination of selected metals in the body tissues and fluids of workers manufacturing stainless steel vessels. *J. Radioanal. Nucl. Chem.,* **2004**, *259*(1), 7-11. [http://dx.doi.org/10.1023/B:JRNC.0000015798.00316.4f]

[130] lekseeva, O.A.; Belov, A.G.; Frontasyeva, M.V.; Gundorina, S.F.; Gustova, M.V.; Kusmenko, L.G.; Perelygin, V.P.; Zaverioukha, O.S. Neutron, gamma and Roentgen fluorescent activation analysis of hair of children suffering from bronchial asthma. *Radiat. Meas.,* **2001**, *34*, 521-525. [http://dx.doi.org/10.1016/S1350-4487(01)00220-7]

[131] Ramakrishna, V.V.; Singh, V.; Garg, A.N. Occupational exposure amongst locomotive shed workers and welders using neutron activation analysis of scalp hair. *Sci. Total Environ.,* **1996**, *192*(3), 259-267. [http://dx.doi.org/10.1016/S0048-9697(96)05316-8] [PMID: 9025319]

[132] Valkovic, V. *Human hair*; CRC Press: Boca Raton, Florida, **1988**, Vol. I and II, .

[133] Mazzilli, B.; Munita, C.S. Mercury determination in dentist's hair and nails by instrumental neutron activation analysis. *Cienc. Cult.,* **1986**, *38*(3), 522-526.

[134] Deibel, M.A.; Ehmann, W.D.; Markesbery, W.R. Copper, iron, and zinc imbalances in severely degenerated brain regions in Alzheimer's disease: possible relation to oxidative stress. *J. Neurol. Sci.,* **1996**, *143*(1-2), 137-142. [http://dx.doi.org/10.1016/S0022-510X(96)00203-1] [PMID: 8981312]

[135] Ehmann, W.D.; Markesbery, W.R.; Kasarskis, E.J.; Vance, D.E.; Khare, S.S.; Hord, J.D.; Thompson, C.M. Applications of neutron activation analysis to the study of age-related neurological diseases. *Biol. Trace Elem. Res.,* **1987**, *13*(1), 19-33. [http://dx.doi.org/10.1007/BF02796618] [PMID: 24254662]

[136] Samudralwar, D.L.; Diprete, C.C.; Ni, B.F.; Ehmann, W.D.; Markesbery, W.R. Elemental imbalances in the olfactory pathway in Alzheimer's disease. *J. Neurol. Sci.,* **1995**, *130*(2), 139-145. [http://dx.doi.org/10.1016/0022-510X(95)00018-W] [PMID: 8586977]

[137] Versieck, J.; Cornelis, R.T. *Trace elements in human plasma for serum*; CRC Press: Boca Raton, Florida, **1989**.

[138] Caldwell, R.J. The use of neutron activation analysis in forensic science. *Aust. J. Forensic Sci.,* **2009**, *1*(2), 11-15. [http://dx.doi.org/10.1080/00450616809410284]

[139]   Moensens, A. *Analysis: neutron activation*; Willey Encyclopedia of Forensic Science, **2009**.

[140]   Suzuki, Y. Application of neutron activation analysis to forensic science. *Radioisotopes,* **1994**, *43*(7), 437-440.
[http://dx.doi.org/10.3769/radioisotopes.43.437]

[141]   Kishi, T. Forensic neutron activation analysis the Japanese scene. *J. Radioanal. Nucl. Chem.,* **1987**, *114*(2), 275-280.
[http://dx.doi.org/10.1007/BF02039801]

[142]   Moauro, A.; Carconi, P.L.; Casadio, S. Instrumental neutron activation analysis applications in materials science and in forensic surveys. *J. Radioanal. Nucl. Chem.,* **1987**, *216*(2), 171-178.
[http://dx.doi.org/10.1007/BF02033774]

[143]   Zeiner, M.; Cindric, I.J. Review-trace determination of potentially toxic elements in (medicinal) plant materials. *Anal. Methods,* **2017**, *9*(10), 1550-1574.
[http://dx.doi.org/10.1039/C7AY00016B]

[144]   Wimolwattanapun, W.; Bunprapob, S.; Ho, M.D.; Sutisna, ; Oura, Y.; Ebihara, M. Quality assessment of INAA data for small-sized environmental reference samples. *Anal. Sci.,* **2014**, *30*(8), 787-792.
[http://dx.doi.org/10.2116/analsci.30.787] [PMID: 25109639]

[145]   Garimella, S.; Deo, R.N. Neutron activation analysis of atmospheric aerosols from a small pacific Island Country: a case of Suva, Fiji Islands. *Aerosol Air Qual. Res.,* **2007**, *7*(4), 500-517.
[http://dx.doi.org/10.4209/aaqr.2007.05.0028]

[146]   Martinez, T.; Lartigue, J.; Lopez, C.; Beltran, C.; Navarrete, M.; Cabrera, L.; Riveroll, M. INAA of aerosol samples in Mexico City. *J. Radioanal. Nucl. Chem.,* **2000**, *241*(1), 127-131.
[http://dx.doi.org/10.1023/A:1006776501483]

[147]   International Atomic Energy Agency. Co-Ordinated research programme on the use of nuclear and nuclear-related techniques in the study of environmental pollution associated with solid wastes. *Report on the Third Research Co-Ordination Meeting, Vienna, Austria,* **1991**.

[148]   International Atomic Energy Agency. *Report on the Intercomparison run for the determination of trace and minor elements in Lichen material. IAEA/AL/079, NAHRES-33, Vienna, Austria*; , **1999**.

[149]   Chutke, N.L.; Ambulkar, M.N.; Garg, A.N. An environmental pollution study from multielemental analysis of pedestrian dust in Nagpur city, Central India. *Sci. Total Environ.,* **1995**, *164*(3), 185-194.
[http://dx.doi.org/10.1016/0048-9697(95)04465-D]

[150]   Guillou, H. Hémond, Ch.; Singer, B.S.; Dyment, J. Dating young MORB of the Central Indian Ridge (19°S): unspike K-Ar technique limitations *versus* $^{40}Ar/^{39}Ar$ incremental heating method. *Quat. Geochronol.,* **2017**, *37*, 42-54.
[http://dx.doi.org/10.1016/j.quageo.2016.10.002]

[151]   Pinti, D.L.; Gholeb, B.; Samper, A.; Gillot, P.Y. Non-destructive potassium measurement in minerals by gamma-ray spectrometry: toward an enhanced K-Ar dating method. *Geochem. J.,* **2015**, *49*(6), 16-19.
[http://dx.doi.org/10.2343/geochemj.2.0392]

[152]   Glascock, M.D.; Neff, H. Neutron activation analysis in archaeology. *Meas. Sci. Technol.,* **2003**, *14*, 1516-1526.
[http://dx.doi.org/10.1088/0957-0233/14/9/304]

[153]   Santos, J.O.; Reis, M.S.; Munita, C.S.; Silva, J.E. Box-Cox transformation on dataset from compositional studies of archaeological potteries. *J. Radioanal. Nucl. Chem.,* **2017**, *311*(2), 1427-1433.
[http://dx.doi.org/10.1007/s10967-016-4987-y]

[154]   Hazenfratz, R.; Munita, C.S.; Glascock, M.D. Study of exchange networks between two Amazon archaeological sites by INAA. *J. Radioanal. Nucl. Chem.,* **2016**, *309*(1), 195-205.

[http://dx.doi.org/10.1007/s10967-016-4758-9]

[155] Santos, J.O.; Munita, C.S.; Soares, E.A. Provenance studies in Amazon basin by means of chemical compocition obtained by INAA. *J. Radioanal. Nucl. Chem.,* **2015**, *306*(3), 713-719.
[http://dx.doi.org/10.1007/s10967-015-4243-x]

[156] Baria, R.; Cano, N.F.; Silva-Carrera, B.N.; Watanabe, S.; Neves, E.G.; Tatumi, S.H.; Munita, C.S. Archaeometric studies of ceramic from the São Paulo II archaeological site. *J. Radioanal. Nucl. Chem.,* **2015**, *306*(3), 721-727.
[http://dx.doi.org/10.1007/s10967-015-4183-5]

[157] Yeh, S.J.; Harbottle, G. Intercomparison of the Asara-Perlman and Brookhaven archaeological ceramic analytical standards. *J. Radioanal. Nucl. Chem.,* **1986**, *97*(2), 279-291.
[http://dx.doi.org/10.1007/BF02035673]

[158] Kuleff, I.; Djingova, R.; Penev, I. Instrumental neutron activation analysis of pottery from provenience study. *J. Radioanal. Nucl. Chem.,* **1986**, *99*(2), 345-358.

CHAPTER 6

Non-commercial Polysaccharides-based Chiral Selectors in Enantioselective Chromatography

Chexu Wang, Ali Fouad, William Maher and **Ashraf Ghanem**[*]

Chirality Program, Faculty of Science and Technology, University of Canberra, ACT, 2601, Australia

Abstract: Enantioselective chromatography is yet well documented as a powerful technique for the enantioselective resolution of racemates. Such potency has been empowered by advances in chiral stationary phases (CSPs). Many polysaccharide derivatives has been utilized as CSP in enantioselective chromatography and several reviews have been prepared focusing on the commercially available CSPs. In this review, we shed some light on the use of non-commercial polysaccharides namely chitosan, glycogen, chitin, pectin and amylopectin and their derivatives as chiral selectors in enantioselective chromatography and alternative polysaccharides derivatives exhibiting better analytical ability than para-substituted derivatives of polysaccharides.

Keywords: Chiral Selectors, HPLC, Polysaccharides.

INTRODUCTION

Many pharmaceutics and herbicides are chiral. They exist as two incongruent stereoisomers called enantiomers. As optical isomers, they rotate linearly polarized light in opposite directions although they are generally known to have similar physical properties (*e.g.*, melting point, hydrophobicity, *etc.*) and they can behave quite differently to one another in a chiral (asymmetric) environment. Since biological processes tend to involve chiral chemicals (*e.g.* enzymes), chirality constitutes an important topic in drug development [1]. The United States Food and Drug Administration (FDA) require toxicology testing for racemates only, even if a company plans to market a single isomer. If unexpected or significant toxicity is found in the racemate, FDA suggests querying the agency on whether similar studies are required of the enantiomers. In such cases, the FDA requires that only the active drug enantiomer (the eutomer) is produced by an enantioselective process (*e.g.*, *via* asymmetric synthesis, resolution *via* diastereomers, kinetic resolution, enzyme catalysis, chirality pool approach). The

---

[*] **Corresponding author Ashraf Ghanem:** Chirality Program, Faculty of Science and Technology, University of Canberra, ACT, 2601, Australia; Tel/Fax: +61 2 6201 2089; E-mail: ashraf.ghanem@canberra.edu.au

Atta-ur-Rahman & Sibel A. Ozkan (Eds.)

inactive enantiomer (the distomer) constitutes 'isomeric ballast' or it may be highly toxic. In the case of thalidomide, one enantiomer possessed the required therapeutic effect, while the other was eventually shown to be teratogenic causing birth defects in the unborn babies. While the use of enantiomerically pure drugs may appear to be a viable solution to such a problem, configurationally unstable stereoisomers like thalidomide may interconvert (known variously as enantiomerization, enantiomeric inversion or racemisation) [2]. The thalidomide tragedy was entirely avoidable, had the physiological properties of the individual thalidomide forms been identified, separated and tested prior to commercialization.

Enantioselective chromatography has been well documented as a powerful, contemporary and practical technique for the chiral separation of racemic drugs, food additives, agrichemicals, fragrances and chiral pollutants [1, 2]. This technique is several steps ahead of other previously reported methods to access pure enantiomers; including synthesis from a chirality pool, asymmetric synthesis from pro-chiral substrates and the resolution of racemic mixtures [3]. The separation of racemic mixtures has been considered as the most feasible method for industrial applications compared to the time consuming and expensive synthetic approaches [4].

Remarkable developments have been achieved in enantioselective chromatography since the first chiral separation of enantiomers using CSP in the mid-sixties [5]. Following this development, several subclasses have emerged as well established chromatographic techniques with outstanding applications in chiral separation such as electrochromatography (EC), supercritical fluid chromatography (SFC), counter current chromatography (CCC), gas chromatography (GC), and high performance liquid chromatography (HPLC) [6].

Most enantioselective separations are performed by direct resolution using chiral stationary phases (CSPs) where the chiral selector is adsorbed, attached, bound, encapsulated or immobilized to an appropriate support to make a CSP. The enantiomers are resolved by the formation of transient diastereomeric complexes between the CSP and the analyte. Thousands of CSPs have been reported, with more than one hundred commercialized [7]. Among the existing CSPs, those prepared from polysaccharides such as cellulose and amylose, attract more attention due to their separation capability. In general, the development of chemically modified polysaccharides are the mainstream trend in commercial and non-commercial CSP development. Commercially available polysaccharide-based chiral stationary phases mainly include cellulose and amylose as chiral selectors adsorbed, bonded, encapsulated or immobilized [7 - 14]. Herein, we focus on the non-commercial reported polysaccharide derivatives as chiral selectors including chitosan, chitin, glycogen, pectin, amylopectin and others. In particular, special emphasis given to 1) the separation ability of the chiral selectors and 2) factors that influence the efficiency of separation and possible ways to improve the ability of separation.

CHIRAL SELECTORS

Commercial CSPs derived from cellulose tris(3,5-dimethylphenylcarbamate) (CDMPC) and amylose tris(3,5-dimethylphenylcarbamate) (ADMPC) were the best for the enantioselective separation of racemates [15 - 22]. To match the trend in chiral separations, several non-commercial CSPs have been developed, among which, chitosan-, chitin-, glycogen-, pectin-, amylopectin-based and others showed promising results.

Chitosan

Chitosan has a similar structure to cellulose with an amino group at the C2-position in chitosan but a hydroxyl group at C2-position in cellulose. The similarity in structure implies possible similar satisfactory enantioselective separation capability. It is a linear polysaccharide consisted of β-(1→4)-linked D-glucoseamine and N-acetyl-D-glucosamine generally prepared *via* the deacetylation of chitin. In 1997, Kuravchi *et al.* prepared N-carboxymethylated chitosan (NCMC) by N-carboxymethylation followed by modification with 6-amino-6-deoxy-β-cyclodextrin (Fig. **1**). The prepared chitosan derivatives were attached to a macroporous silica gel and used as a stationary phase for chiral HPLC separations of 2,4-dinitrophenyl-α-amino acids. The low substitution degree of the stationary phase β-CD-NCMC exhibited lower enantioselectivity than the highly substituted β-CD-NCMC in spite of similar amount of the immobilized β-cyclodextrin (β-CD) moiety [23].

In a further report, Okamoto *et al.* used and compared the 3, 5-dimethylphenyl-carbamates of chitosan, xylan and the 3,5-dichlorophenylcarbamate of galactosamine as chiral stationary phases in HPLC. The results revealed relatively high chiral recognition ability when compared with nine new 3,5-dichloro- and 3,5-dimethylphenylcarbamates of polysaccharides as CSPs in HPLC. Their results suggest that the selection of the best CSPs are unpredictable. It demonstrated that the new CSPs used to separate some enantiomers were better resolved on xylan bis(3,5-dimethyl- or 3,5dichlorophenylcarbamate) (Fig. **2**) [24].

In 1998, Franco *et al.* compared the enantioselectivity of three mixed 10-undecenoyl / 3, 5-dimethylphenylaminocarbonyl derivatives of amylose and chitosan immobilized on allyl silica gel of an analogous cellulose-derived chiral stationary phase previously prepared. The chitosan derived CSP (CSP3) showed a superior ability to discriminate enantiomers when using both heptane–2-propanol and heptane chloroform mixtures as mobile phases. The authors asserted that the new polysaccharides based CSPs can be further developed, because of their ease of preparation and their high stereoselectivity (Fig. **3**) [25].

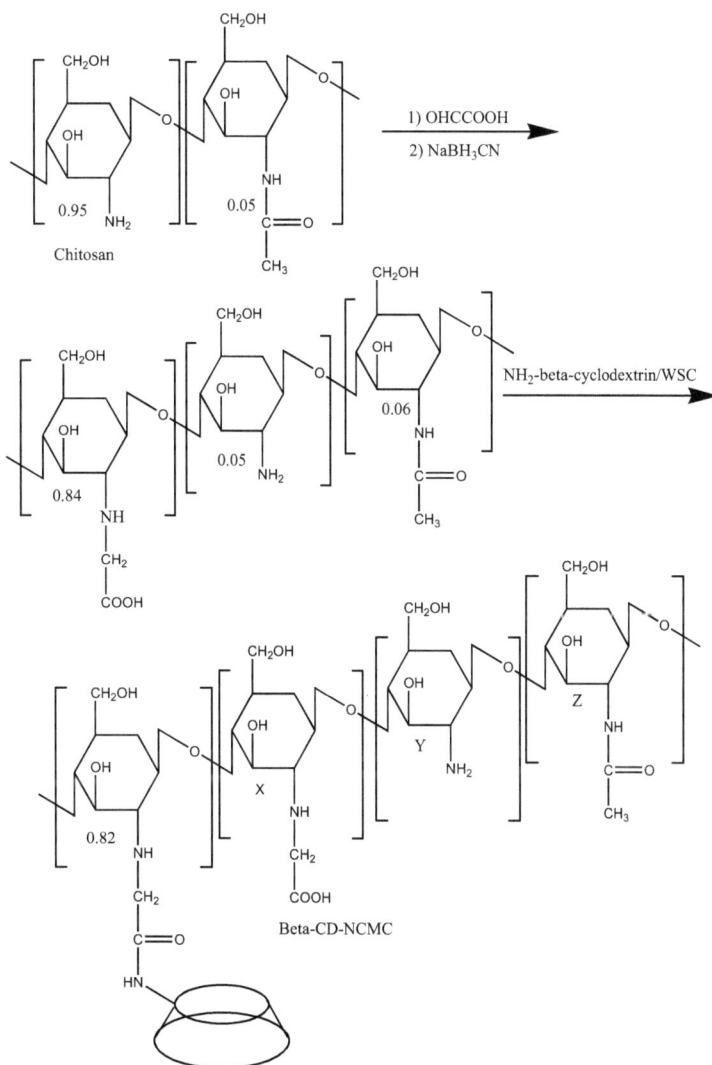

Fig. (1). Synthesis of β-CD-NCMC. NCMC is assumed to contain N,N-dicarboxymethylated units as the remaining units and thus, β-CD moieties in β-CD-NCMC may be bonded in part to the disubstituted units (see the text for the details). The contents of the unmodified units (x, y, z) in β-CD-NCMC could not be determined [23].

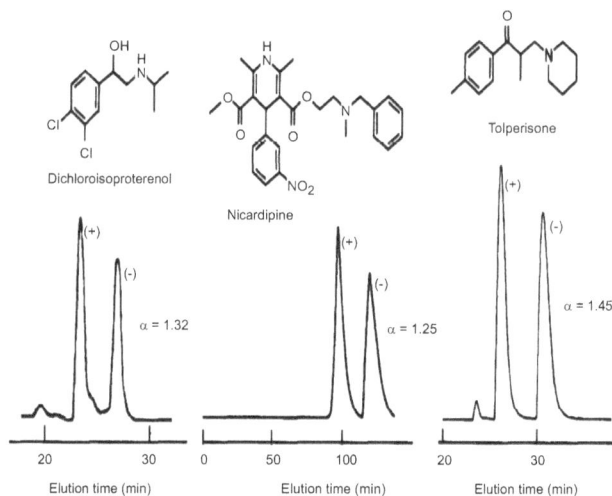

Fig. (2). Separation of enantiomers of dichloroisoproterenol, nicardipine, and tolperisone on bis(3,5-dimethylphenylcarbamate) and bis(3,5dichlorophenylcarbamate) of xylan. Eluent: hexane-2-propanol (98/2%v/v) and hexane-2-propanol (90/10%v/v). Flow rate: 0.5 ml/mm [24].

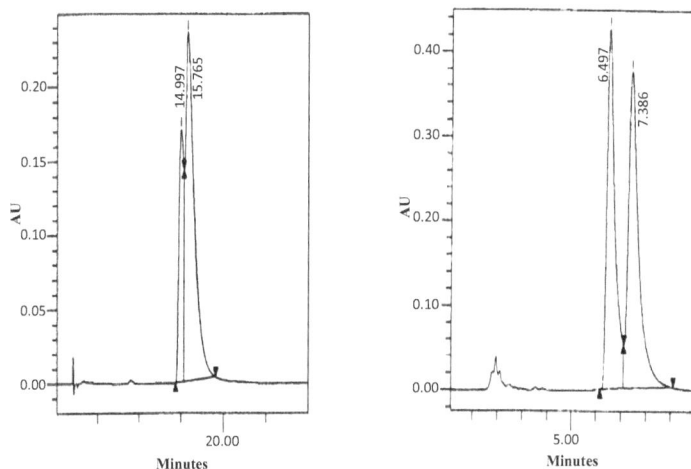

Fig. (3). Resolution of lorazepam on CSP3 using as mobile phase (a) heptane–2-propanol (90:10%, v/v), (b) 100% chloroform [25].

A series of mixed chitosan derivatives (10-undecenoyl/phenylaminocarbonyl or benzoyl) were prepared, characterized and immobilized on allyl silica gel. The influence of the starting polysaccharide material, as well as that of the solvents used as mobile phase modifiers on the preparation and performance of the resulting chiral stationary phases (CSPs) was discussed. Highly substituted derivatives showed better enantioselectivity due to improved solubility of chiral selectors while the amount of acetyl group had an insignificant effect on enantiorecognition (Fig. **4**) [26].

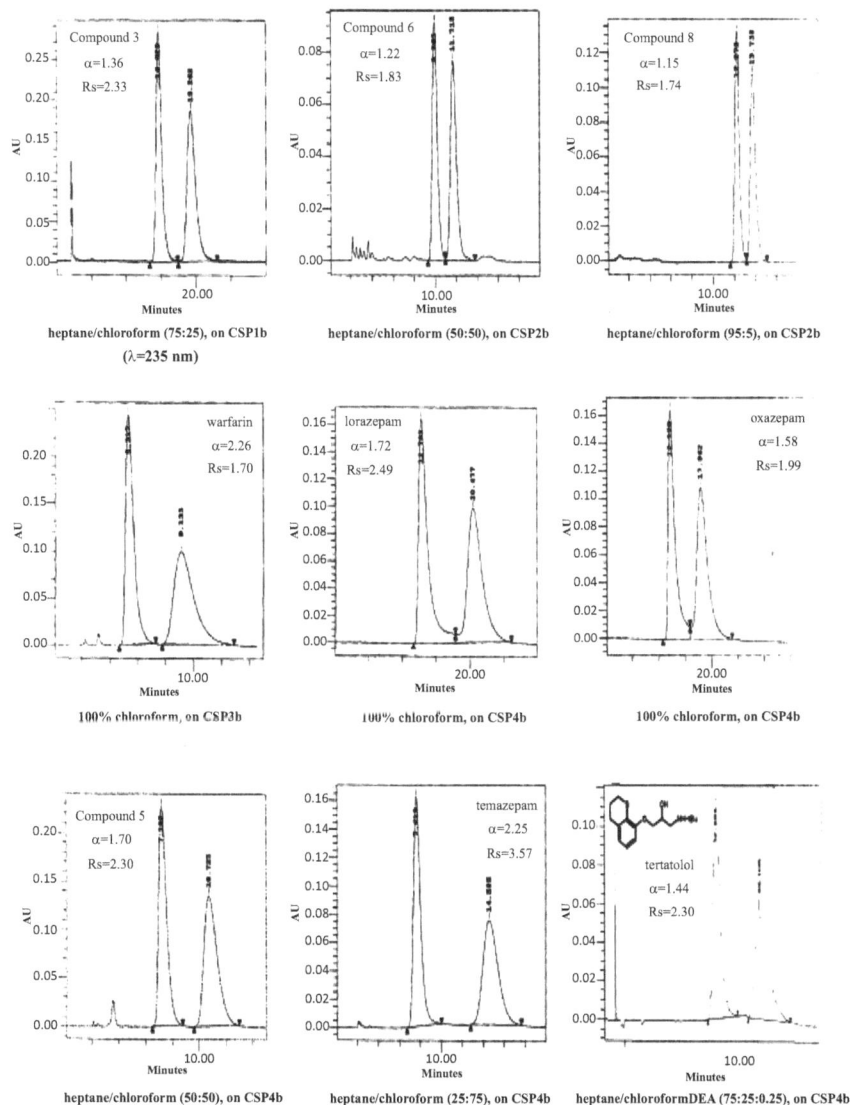

Fig. (4). Resolution of some racemic test compounds on different chitosan-based CSPs (flow rate: 1 ml /min; UV detection: 254 nm, except otherwise indicated) [26].

Li *et al*. prepared a chiral stationary phase with 1,1-carbonyldiimidazol. A film of hydrophilic chitosan was covalently coupled on to the aminopropyl silica before immobilization of bovine serum albumin which weakend the hydrophobic interaction between enantiomers and albumin and decreased the retention time of the enantiomers. The retention time was decreased from 39 min to 8.74 min. Due to the presence of the hydrophilic chitosan; the retention factor of racemic tryptophan (D-tryptophan and L-tryptophan) was significantly reduced. The time required shortened and a large amount of mobile phase was saved. The optimum chromatographic conditions was determined by examining various mobile phases and injection conditions (Fig. **5**) [27].

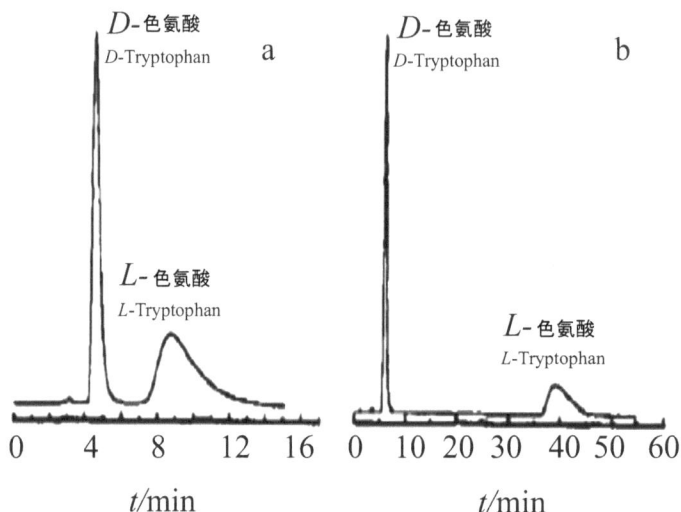

Fig. (5). Comparison of the retention of D, L-tryptophan on the column I (a) and column II (b) [27].

In 2007, Son and Jegal reported the use of chitosan to modify N-nicotinoyl-L-phenylalanine and 3, 5-dimethylphenylisocyanate to prepare a chiral stationary phases for HPLC use, especially for the separation of optical isomers of a series of α-amino acids (Fig. **6**). The modified chitosan prepared in this study showed good solubility in several organic solvents and was easy to handle when coating the silica particles to prepare chiral HPLC columns with good chiral separation capabilities over a wide range of temperature range [28].

In a related report, Potekhina *et al.* investigated the enantioselectivity for several basic compounds, including fluoxetine, chlorcyclizine, pindolol, and other medicines by a new sorbent based on silica gel modified with N-(3-sulfo-3-carboxy)-propionylchitosan (SCPC modified silica gel). This sorbent was synthesized, and its chromatographic properties studied. The selectivity and resolution in polar-organic HPLC (PO-HPLC) is in general better than in reversed-phase HPLC (RP-HPLC). The high resolution values are related to the high performance of the column.(Fig. **7**) [29]

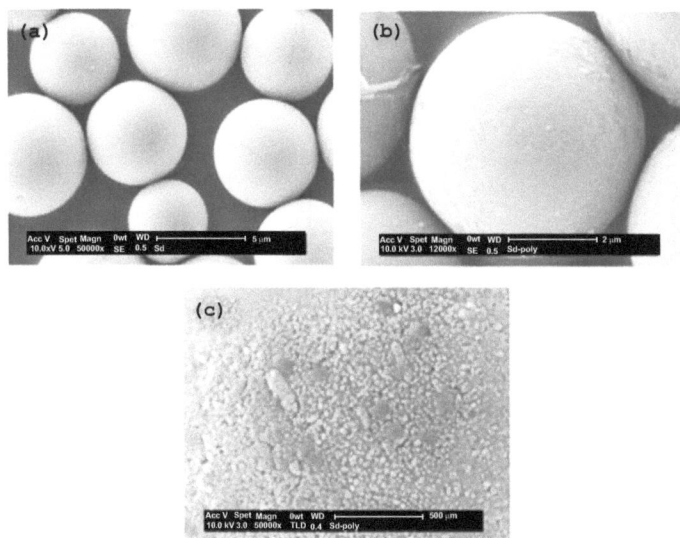

Fig. (6). SEM photographs of silica particles before and after coating with the CSP based on chitosan; (a) before coating, (b) after coating, and (c) the magnification of the coated surface [28].

Yammoto *et al.* prepared deacetylated chitosan by treatment of commercial chitosan with 50% w/v aqueous NaOH. This was then derivatized into several new chitosan phenylcarbamate derivatives having a urea and an imide moiety at the 2-position of the glucosamine ring by reaction with isocyanate and phthalic anhydride/isocyanate, respectively. The chitosan derivatives were coated on macroporous silica gel and evaluated as CSPs for HPLC demonstrating improved chiral recognition ability by using the completely deacetylated chitosan (Fig. **8**) [30].

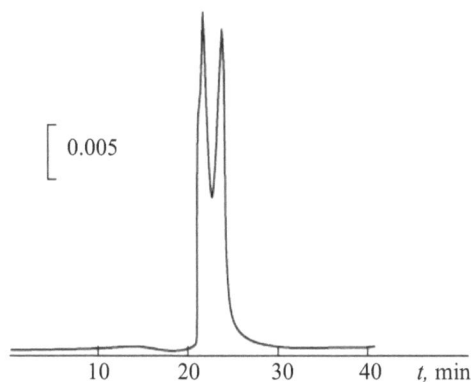

Fig. (7). Chromatogram for fluoxetine enantioseparation (mobile phase, methanol: 0.05% TEAA (50: 50%v/v); pH 4.6; flow rate, 1 ml/min; $\lambda = 230$ nm) [29].

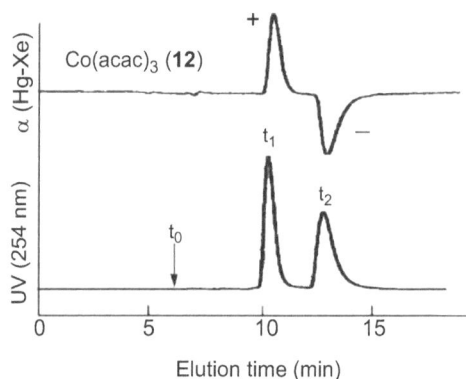

Fig. (8). Resolution of cobalt(III) tris(acetylacetonate) on chitosan carbamate-imide derivative [30].

In a related article, a novel cationic hydrophilic interaction monolithic stationary phase has been prepared by Lü *et al.* where carboxymethyl chitosan (CMCH) was attached to a monolithic silica skeleton using carbodiimide as an activation reagent then used in capillary liquid chromatography (Fig. **9**). The amino and hydroxyl moieties of CMCH acted as both ion-exchange and polar interaction sites. The performance of the column was studied by the separation of polar acidic compounds. The chitosan functionalized monolithic silica column showed good selectivity for aromatic acids and aliphatic acids. The results revealed that these compounds were separated primarily based on hydrophilic interaction [31].

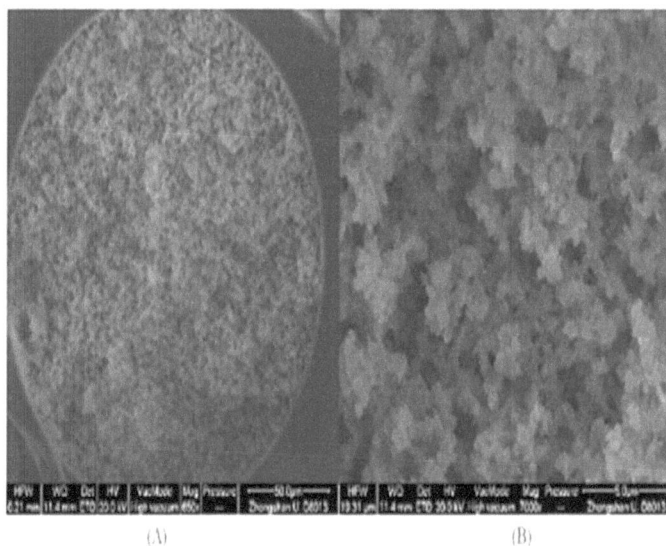

Fig. (9). SEM photographs of the cross-section of the monolithic column at magnification of (A) 650× and (B) 7000× [31].

Chen and Hsieh subsequently reported the copolymerization of glycidyl methacrylate-modified nano-chitosan with methacrylamide (MAA) and bis-acrylamide cross-linkers (forming the MAA-CS (methacrylamide chiral stationary) capillary monolith) rather than the attachment of nano-chitosan to the copolymer of glycidyl methacrylate, MAA, and bisacrylamide (forming the MAA-CS capillary monolith). For tryptophan, because of the poor loading of CS chiral selectors in the capillary during fabrication, the MAA-CS capillary showed only satisfactory resolution. The new chiral stationary phase; however, has its limitations albeit the enantiomer separation of the two groups of α-tocopherol stereoisomers (Fig. **10**) [32].

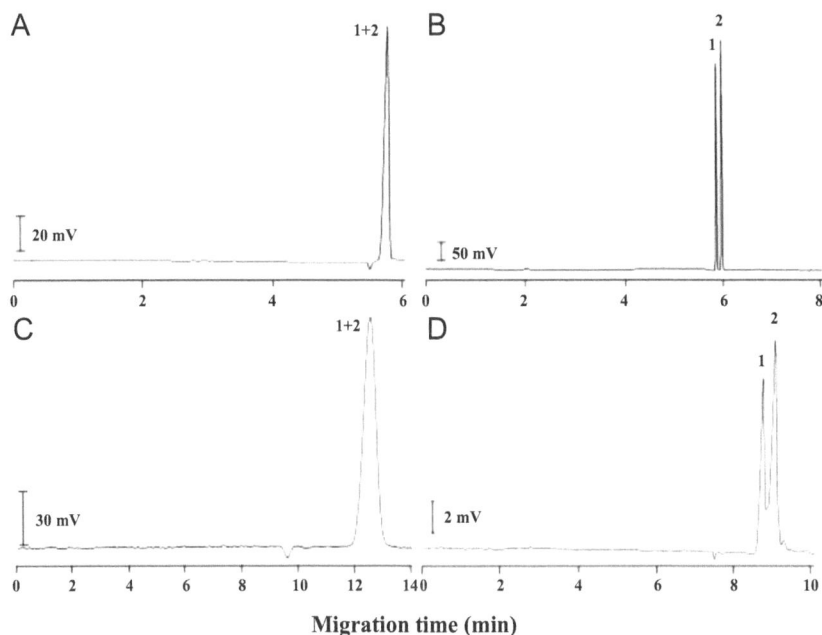

Fig. (10). Enantioseparations of the tryptophans using the MAA-CS capillary of (60 cm (55 cm) X 75 mm id). Conditions: BGE, Tris buffer, 100mM at pH equals (A) 8.5, (B) 9.5, (C) 10.5, and (D) 9.5 with the addition of 10% v/v MeOH. The applied voltage was 15 kV. Samples: hydrostatic injection of 10 cm for 5 s and detection at 214 nm. Peak corresponds to (1) D-tryptophan and (2) L-tryptophan [32].

Two new monoliths (3-glycidyloxypropyl)trimethoxysilane-chitosan (GTS–CS) and hydrosilation-chitosan (SiH–CS–s) were prepared by Chen *et al* followed by a comparative study between GTS–CS, SiH–CS–s, (3-glycidyloxypropyl) trimethoxysilanebovine serum albumin (GTS–BSA) and methacrylamide-chitosan (MAA–CS) monolith prepared by the opolymerization of chitosan (CS) with methacrylamide. The crosslinking agent (succinic acid) was crucial for CS loading and chiral resolution. Noteworthy, no resolution of the tryptophan enantiomers was observed for the GTS–CS capillary without adding succinic acid.

The GTS–CS–s monolith exhibited better resolution than the SiH–CS–s capillary. Although the column efficiency and the resolution of the tryptophans separated in both monoliths were not superior to MAA–CS monolith, the performance of the GTS–CS–s monolith was close to that of the GTS–BSA monolith, in which the fixed chiral selector, BSA, demonstrated stronger chiral recognition of tryptophan (α= 4.0) than CS (α = 1.4-1.6). For the chiral separation of ($\pm$)-catechin, the electrostatic repulsions between the anionic catechins and the succinic acid-modified phases greatly affected the performance of the GTS–CS–s and SiH–CS–s monoliths while the performance of the MAA–CS monolith was hindered by the hydrogen-bonding between the five catechin hydroxyl groups and the polyacrylamide backbone. For the acidic racemates, 2-phenylpropionic acid and its derivative ibuprofen, their chiral separations were achieved with satisfactory resolution (Fig. **11**) [33].

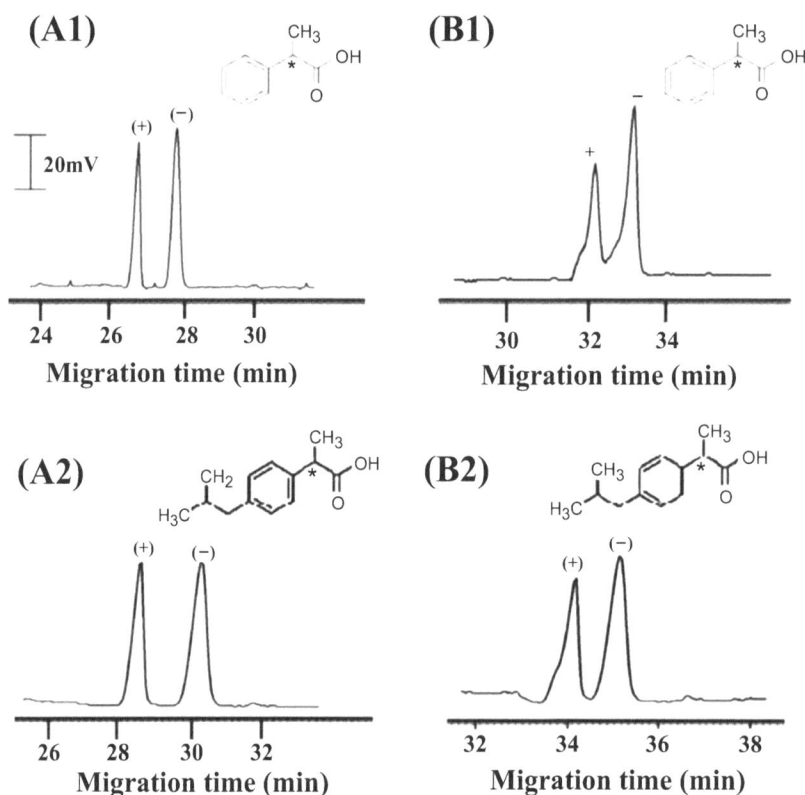

Fig. (11). Enantioseparations of L-phenylcarboxylic acids in the GTS–CS–s and SiH–CS–s capillaries. (A) BGE used in the GTS–CS–s capillary (60 cm (55 cm) ×75 m I.D.): borate buffer, pH 7.5, 30 mM. (B) BGE used in the SiH–CS–s capillary (60 cm (55 cm) ×75 m I.D.): borate buffer, pH 8.0, 10 mM. The applied voltage was 25 kV. Samples: hydrostatic injection of 10 cm for 5 s and detection at 214 nm. Samples: (1) 2-phenylpropionic acid; (2) ibuprofen [33].

Carbamoylation is a potent modification that has been widely documented to improve enantioselective recognition of naturally occurring polysaccharides for further use as chiral selectors [34]. The chitosan was first carbamylated by Han *et al.*, and a silica-based chitosan tris(3-chlorophenylcarbamate) derivative CSP was used for the enantioseparation of nine racemic compounds using normal phase mode by HPLC. The effect of the modifier type and percentage on the optimization of the resolution was studied. The separation of Epoxiconazole using this CSP was comparable to those attained by Chiralcel OD-H and Chiralpak AD-H columns. Among the tested racemates, seven racemic compounds were separated rapidly on the CSP of chitosan tris(3-chlorophenylcarbamate). Baseline or near-baseline separation was achieved for benzoin, penconazole, hexaconazole, and epoxiconazole, while the others were only partially separated (Fig. **12**) [35].

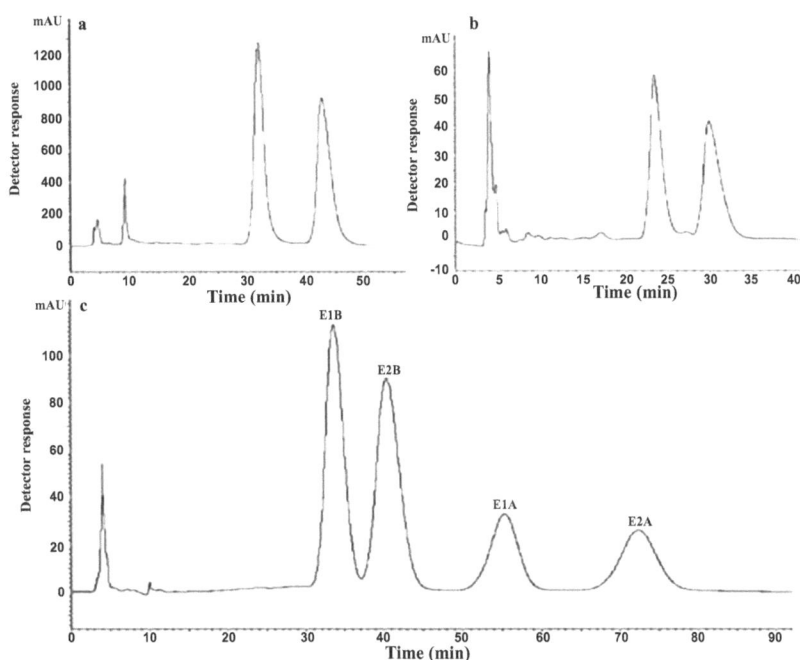

Fig. (12). Enantioseparation chromatograms for chiral compounds: a benzoin, b penconazole, and c epoxiconazole using n-hexane/2-propanol 95:5%, v/v for a & b and n-hexane/1-propanol 95:5%, v/v for c [35].

An open-tubular column coated with chitosan–silica hybrid using chitosan and silane-coupling agent (γ-glycidoxypropyltrimethoxysilane, GPTMS) was developed for CEC. The stationary phase was hydrophilic due to the amine and hydroxyl functional groups of the chitosan–silica hybrid. The seperation mechanism for was based on hydrophilic and electrostatic interactions combined with an electrophoretic mechanism. The CEC method was compared to CE method using a bare capillary where CS/GPTMS-coated OT column and proved to be as good alternative to CE [36].

A similar CSP has been introduced by Yao *et al. via* covalently immobilizing bovine serum albumin to the column (Fig. **13**). The covalent binding of proteins onto a monolithic matrix was investigated to overcome the drawback of loss and/or denaturing of the biomolecules from physical adsorption and the encapsulation method. The enantioseparation of D, L-tryptophan by capillary electro chromatography (CEC) was investigated using the prepared column. A resolution of 2.44 was achieved using 20 mmolL^{-1} phosphate buffer at pH 7.5 [37].

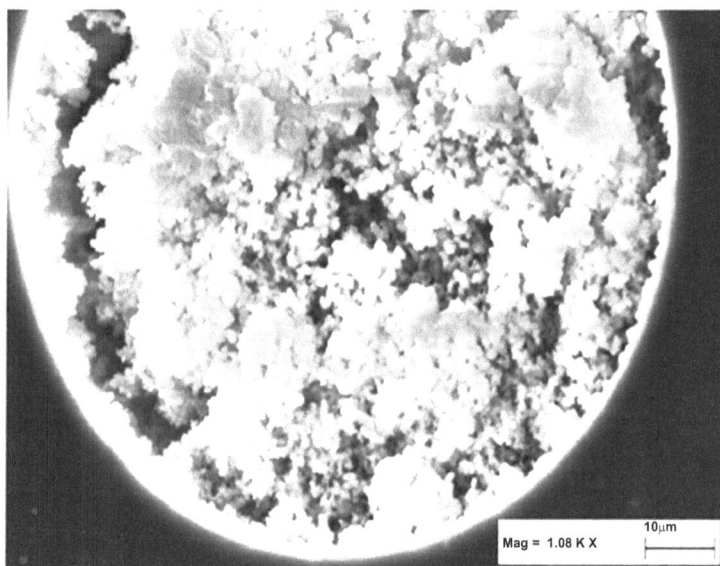

Fig. (13). SEM photograph of a oilica monolithic column with covalently linked to BSA. 20 kV, WD 4 mm, and magnification 1080X [37].

In 2012, Chen and Syu compared three different approaches for immobilizing cross-linked chitosan molecules (CS-s) in sol–gel phases to form chiral open-tubular CEC (OT-CEC) capillaries. Column I was a bare capillary was first silanized with triethoxysilane (TEOS) and then reacted with the reaction product of 3-glycidyloxypropyltrimethoxysilane (GTS) and CS-s. Column II was prepared by the silanization of a bare capillary with a mixture of TEOS and GTS silanes followed by reaction with CS-s. Column III was prepared using, all the reagents, including TEOS, GTS, and CS-s that were reacted together in a bare capillary. The hydrophobic tryptophan and catechin enantiomers were better separated by columns I and II. (Fig. **14**) [38]

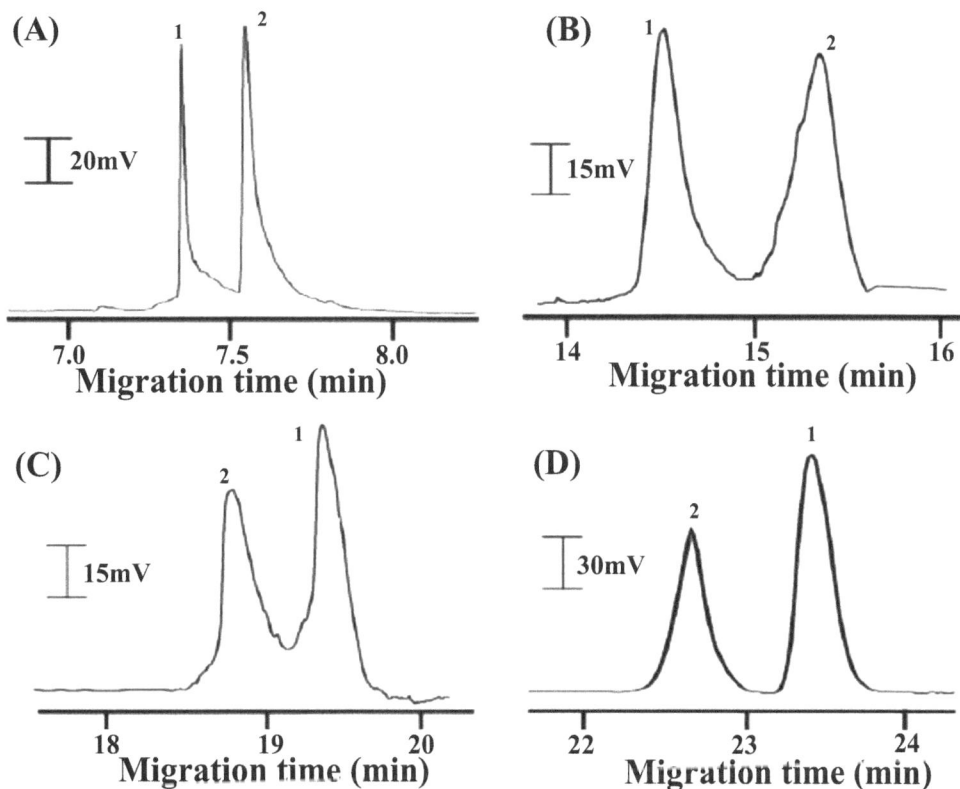

Fig. (14). Enantioseparations of (d/l)-Trp in the CS-immobilized sol–gel capillaries. (A) column I (60 cm (55 cm)×75 um I.D.); (B) column II (60 cm (55 cm)×75 um I.D.); (C) column III (44 cm (38 cm)×75 um I.D.); and (D) GTS–CS-s (60 cm (55 cm)×75 um I.D.). BGE: Tris buffer, 50 mM, (A) pH 10.0; (B) pH 9.5 with 20% MeOH; (C) pH 9.0 with 50% MeOH; and (D) pH 8.5. The applied voltage was 10 kV except for (D), which was 15 kV. Samples: hydrostatic injection of 15 cm for 10 s and detection at 214 nm. Peaks correspond to (1) l-Trp, and (2) d-Trp [38].

In 2014, ring opening metathesis mediated polymerization (ROMP) for the continuous assembly of polymers (CAP) was used to fabricate immobilized-type chiral stationary phases (CSPs) for enantiomeric separations. The chiral recognition abilities of the CSPs were explored using liquid chromatography (LC). The results revealed that higher amount of polysaccharide cross-linked films have better chiral separation capabilities. The re-initiation reactions can respectively, improve the polysaccharide content by 8 and 44 weight % for chitosan- and amylose-based macro-cross-linkers respectively (Fig. **15**) [39].

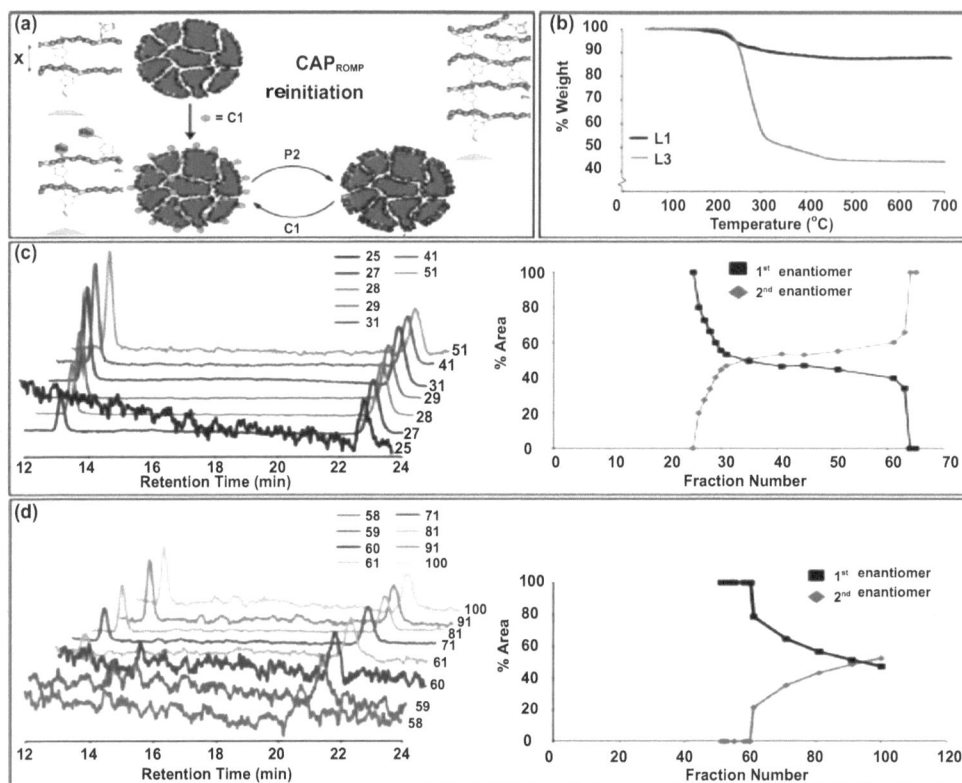

Fig. (15). a) General scheme of CAP ROMP reinitiation on mesoporous silica particles with amylose-based macrocross-linker P2. b) TGA analysis of the CSPs fabricated *via* single CAP ROMP step (L1) (red) and after two reinitiation steps (L3) (blue) with P2. The graphs indicate percentage weight loss with increasing temperature, which results from degradation of the amylose-based P2 films. The degradation of P2 occurred between 230 to 480 8C. UV absorbance at 254 nm (left panel) of trans-stilbene oxide M2 after being passed over column C2 packed with P2-based CSPs, c) fabricated by a single CAP ROMP reaction, and d) after two additional reinitiation steps. Right panels in c) and d) show the ratio of area under the peak between the two enantiomers. Low-resolution UV traces of some fractions are due to high noise/peak area ratio at low sample concentrations [39].

Zhang *et al.*, prepared chitosan bearing different amounts of persentages of 4-chlorophenylcarbamate–Urea and evaluated their chiral recognition abilities as chiral CSPs for HPLC. The derivatives containing 3,6-dicarbamate and 2-urea showed higher chiral recognition for most of racemates compared to the other derivatives containin biuret or allophanate substituents (Fig. **16**). Furthermore, the previous derivative contained pure 3,6-dicarbamate and 2-urea, implying that the current condition can lead to the chitosan derivatives with more preferable structures. The resolution of some racemates could be remarkably improved on the coated chitosan-based CSPs by using solvents, such as chloroform and methanol [40].

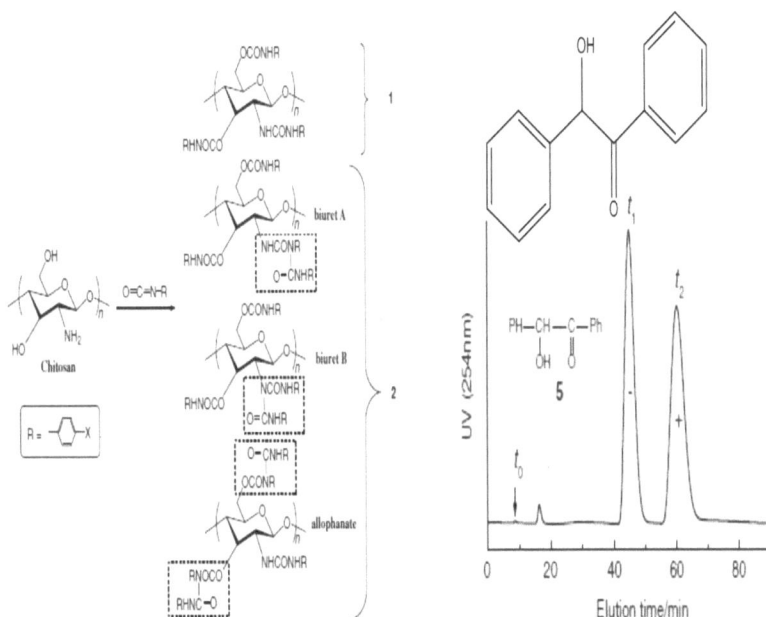

Fig. (16). Possible structures of chitosan derivatives and Chromatogram for the resolution of racemate 5 on CSP-a [40].

Inspired by the results of chitosan phenyl carbamate urea derivatives and to deliver a systematic study about the effect of the position, number, and nature of the substituent on the phenyl group, Zhang *et al.* synthesized fourteen chitosan 3,6-diphenylcarbamate-2-urea derivatives starting with deacetylated chitosan and the corresponding phenyl isocyanates followed by coating on silica gel. The derivatives were evaluated for their chiral recognition abilities in HPLC. These coated-type CSPs exhibited different chiral recognition abilities depending on the position, nature, and the number of the substituents introduced on the phenyl group. The introduction of either an electron-withdrawing or an electron-donating substituent improved the chiral recognition of the CSPs. Among the CSPs, the 2-substituted CSPs showed low chiral recognition abilities, while those with 3,5-dimethyl and 3,5-dichloro substituents showed relatively higher chiral recognition abilities, which enabled the baseline separation of some racemates. The CSPs could be used with some non-standard eluents containing chloroform demonstrating solvent versatility when compared to other coated polysaccharide-based CSPs. Some racemates were more efficiently resolved when non-standard eluents were used (Fig. **17**). Remarkably, the resolution of some racemates have been improved when using eluents containing chloroform and ethyl acetate, which can't be used with the cellulose and amylose-based coated-type CSPs [41].

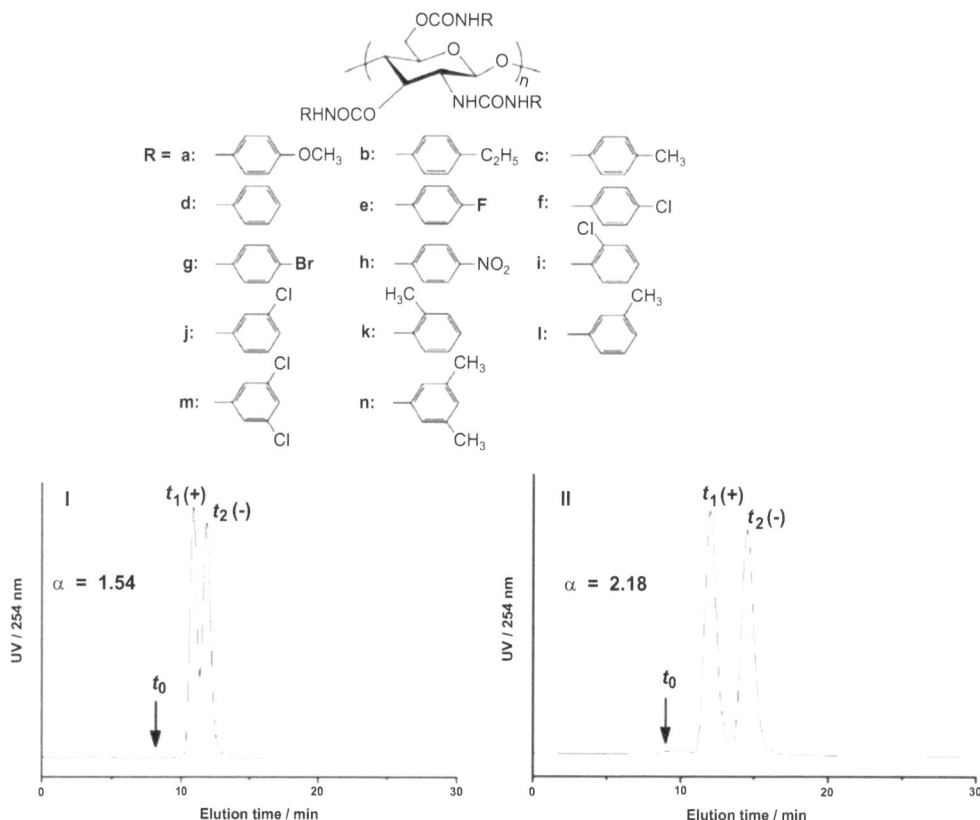

Fig. (17). Structure and chromatograms for the resolution of racemate 2 in hexane/2-propanol (90/10, v/v) (I) and hexane/chloroform/2-propanol (90/10%, v/v/v) (II) on derivative [41].

In a further study, Zhang *et al.* developed new chiral stationary phases (CSPs) with high chiral recognition capability and high compatibility with the non-standard organic solvents. Seven new chitosan bis-(phenylcarbamate)--N-cyclobutylformamide) derivatives were developed and used as chiral stationary phases (CSPs) in HPLC, (Fig. **18**). Compared to the well-identified CSPs of cellulose tris (3,5-dimethylphenylcarbamate) (CDMPC) and amylose tris(3,5-dimethylphenyl carbamate) (ADMPC), the newly derivatized CSPs showed better separations for some analytes [42].

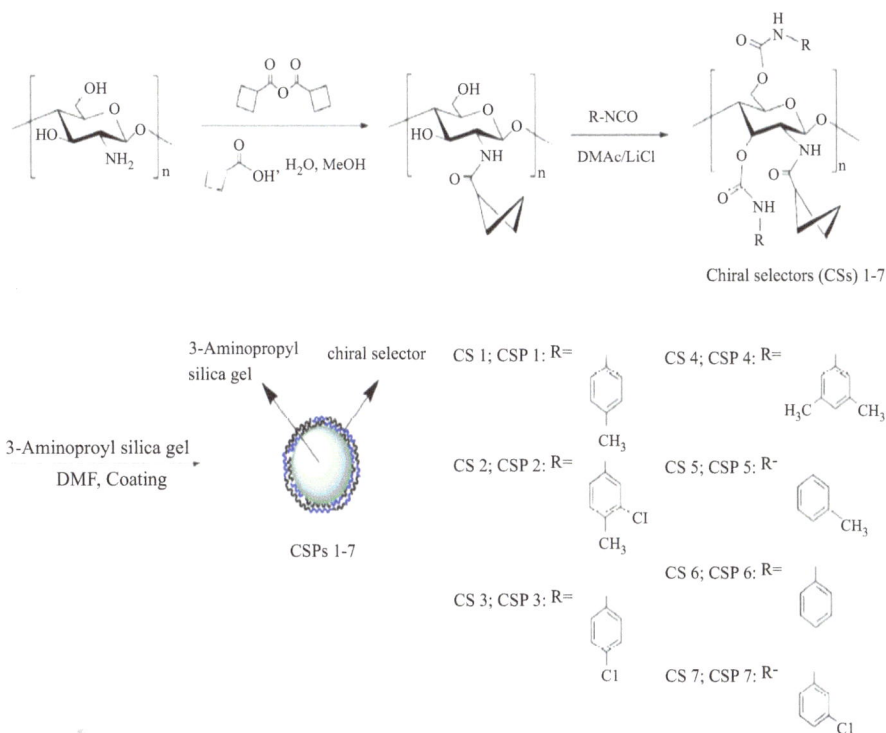

Fig. (18). Preparation scheme of CSs 1-7 and CSPs 1-7 [4?]

A further trial to prepare non-standard solvents compatible CSPs based on chitosan has been conducted by Feng *et al*. Chitosan was firstly acylated by various alkyl chloroformates yielding chitosan alkoxyformamides, then derivatized with 4-methylphenyl isocyanate to form chitosan bis(4-methylphenyl-carbamate)-(alkoxyformamide). A series of chiral stationary phases were prepared by coating these derivatives on 3-aminopropyl silica gel. The content of the derivatives on the chiral stationary phases was nearly 20% by weight. The chiral stationary phases prepared from chitosan bis(4-methylphenylcarbamate)-(ethoxy-formamide) and chitosan bis(4-methylphenylcarbamate)-(isopropoxyformamide) comparati- vely showed better enantioseparation capability than those prepared from chitosan bis(4-methylphenyl carbamate)-(n-pentoxyformamide) and chitosan bis(4-methyl- phenyl- carbamate)-(benzoxyformamide). Chitosan bis(4-methylphenylcarba- mate)-(ethoxyformamide) was stable in 100% ethyl acetate and 100% chloroform were used as mobile phases. The corresponding chiral stationary phases can be utilized in a wider range of mobile phases in comparison with conventionally coated type chiral stationary phases of cellulose and amylose derivatives (Fig. **19**) [43].

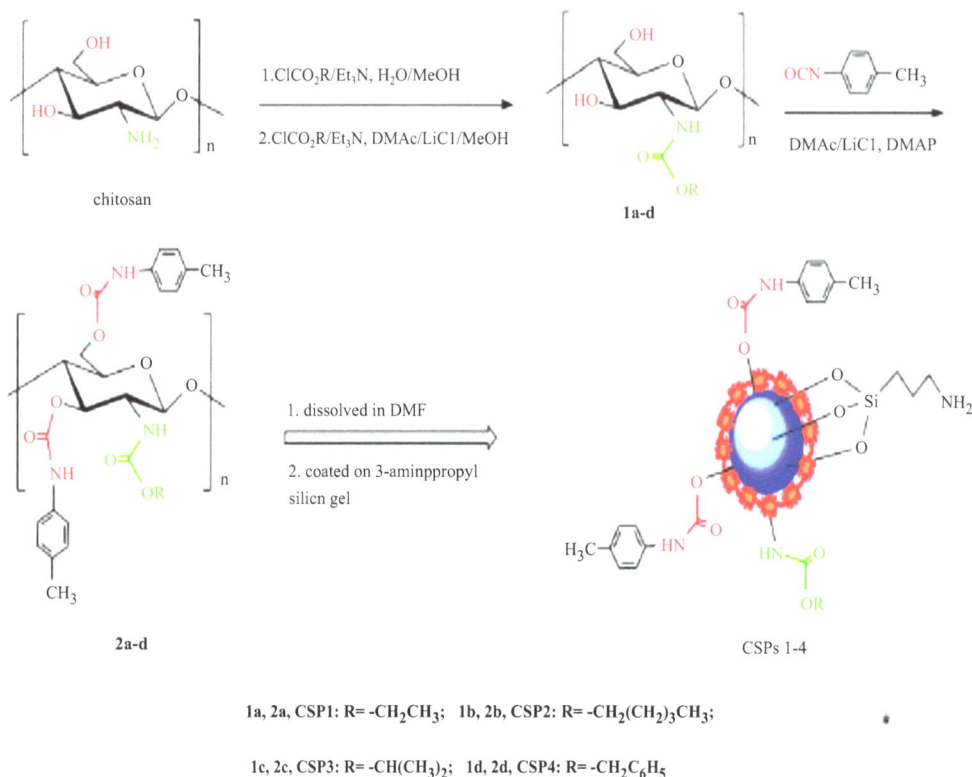

Fig. (19). Preparation scheme of chitosan derivatives and CSPs 1–4[43].

Tang *et al.* prepared chitosan derivatives coated onto 3-aminopropyl silica particles. The silica-support based columns were synthesized by carbamoylating chitosan isobutyrylamide with different methylphenyl isocyanates. This resulted in a series of new CSPs for HPLC. The observed chiral recognition abilities of these coated-type CSPs depended on the substituents on the phenyl moieties of the chitosan derivatives, the eluent composition, as well as the structure of racemates. The chitosan isobutyrylamide derivative CSPs demonstrated excellent chiral recognition ability, and meanwhile possessed satisfactory eluent tolerance in a wider range of solvents (Fig. **20**) [44].

In 2017, Wang *et al.* developed coated type CSPs with excellent enantioseparation capability and high tolerance against mobile phases. The CSPs consisted of synthesized chitosan bis(3,5-dimethylphenylcarbamate)-(alkyl urea)s coated on 3-aminopropyl silica gel. The newly prepared CSPs exhibited promising prospects for enantioseparation of several chiral compounds (Fig. **21**) [45].

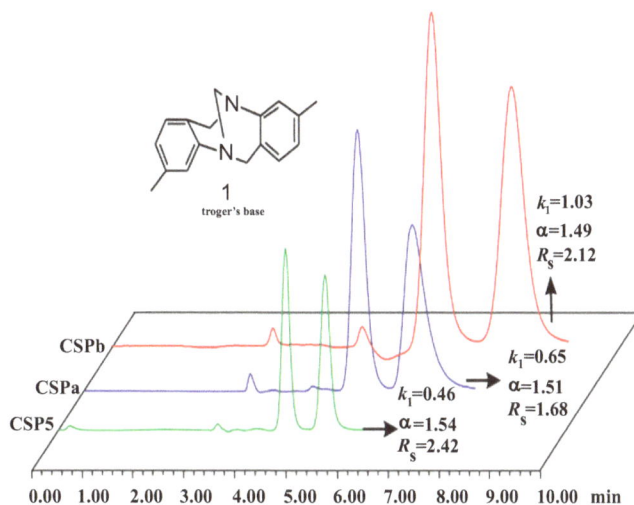

Fig. (20). Chromatograms of the resolution of troger's base in n-hexane/isopropanol on Chitosan bis(3-methylphenylcarbamate)-(isobutyrylamide) (CSP5), amylose tris(3,5-dimethylphenylcarbamate -based CSP (CSPa) and ellulose tris(3,5-dimethylphenylcarbamate -based CSP(CSPb) [44].

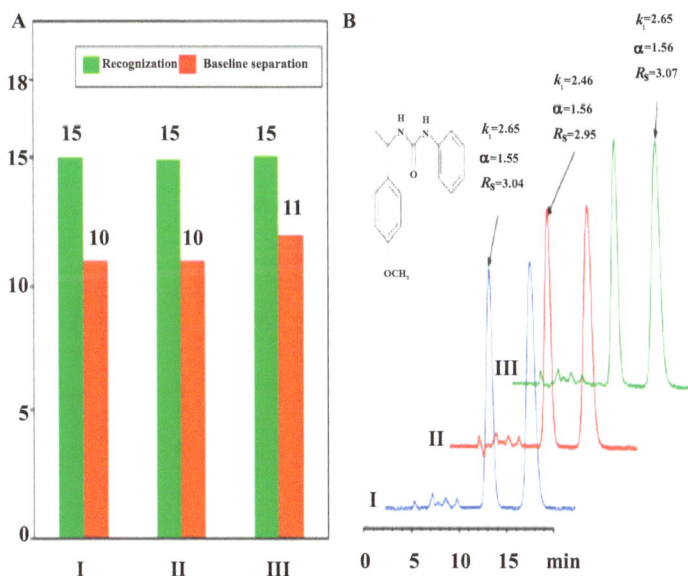

Fig. (21). Enantioseparation comparison of hitosan bis(4-chlorophenyl-carbamate)-(isobutyrylamid (CSP 1) under different conditions. A: total numbers of the compounds chirally recognized and baseline separated by CSP 1; B: chromatograms of compound 10 resolved by CSP 1. Eluent: n-hexane/ethanol (90/10, v/v). Conditions: I: the initial separation results; II: the separation results after CSP 1 was flushed with pure ethyl acetate; III: the separation results after CSP 1 was flushed with pure ethyl acetate and acetone [45].

Alternative derivatives of chitosan showed better analytical ability, while having good eluent tolerance in a wider range of solvents; especially when the eluents containing chloroform and ethyl acetate were used and the resolution of some racemates is improved. Furthermore, chitosan was used in conjunction with albumin to reduce the hydrophobic interaction between enantiomers and albumin and reduce the retention time of enantiomers [23 - 45].

Chitin

Chitin is a long-chain polymer of an N-acetylglucosamine found throughout the natural world. The structure of chitin is comparable to cellulose; chitin has been used for medicinal, industrial and biotechnological purposes.

Chitin bis(aryl carbamate)s were coated onto APS-Hypersil silica, using THF and N,N-dimethylacetamide (10:1 v/v) as the solvent. The chiral stationary phases prepared (CSP-1 and CSP-2) (Fig. **22**) were tested for the enantioseparation of a series of structurally different chiral compounds. It is known that chitin isolated from the same animal varies in the length of its molecular chain, number of acetyl groups, and in its crystallinity. Therefore, non-commercial chitin was investigated and the elemental analysis indicated a higher degree of derivatisation of hydroxyl groups. As with CSP-1 and CSP-2, the CSP-3 derivative was coated onto the support in 15% w/w ratio. The resulting column was evaluated with the previously selected racemic compounds and showed an increase in resolution when compared with CSP-1 and CSP-2. Because of its higher solubility and reactivity, chitin bis(aryl carbamate) has a good resolving ability [46].

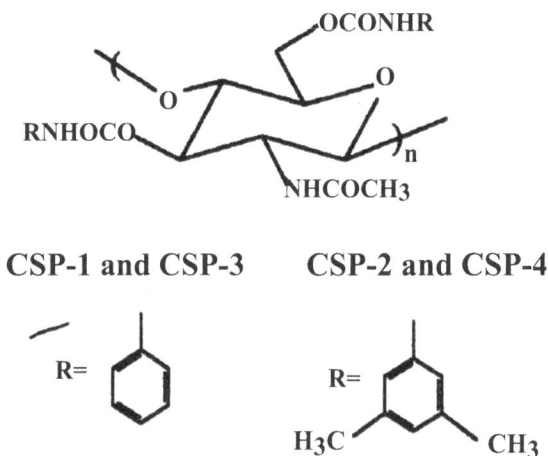

Fig. (22). Structure of chiral phases CSP-1 to CSP-4. CSP-1 and CSP-2: Sigma chitin. CSP-3 and CSP-4: CETEDRE chitin [46].

In 2000, Yamamoto *et al.* used chitin carbamate derivatives namely 4-substituted and 3,5-disubstituted phenylcarbamates, 1-phenylethylcarbamates, and cycloalkylcarbamates to coat macroporous silica gel. This was followed by a study to demonstrate their chiral recognition abilities as CSPs in HPLC. The efficiencies of the new chitin carbamate derivatives were significantly influenced by the substituents on their phenyl groups, and the derivatives 3,5-dimethylphenyl, 4-chlorophenyl, and 4-trifluoromethylphenylcarbamates demonstrated high-chiral recognition in enantioseparations. Chitin phenylcarbamates have a higher chiral recognition ability compared to cellulose and amylose due to the phenylcarbamate's acetamide residue which plays the same important role in chiral recognition as phenylcarbamate residues (Fig. **23**) [47].

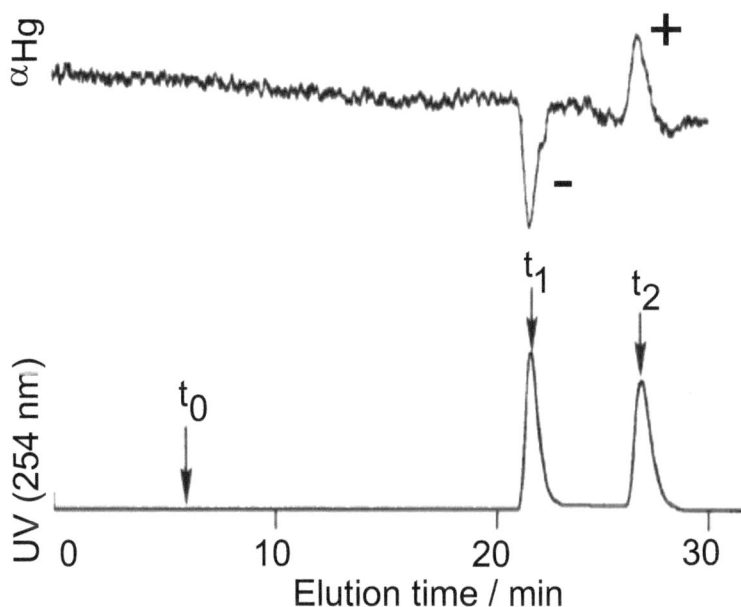

Fig. (23). Resolution of benzoin on chitin bis(3, 5-dichlorophenylcarbamate) [47].

In 2003, Yamamoto *et al.* produced three chitin derivatives, 3,6-bis(phenylcarbamate), 3,6- bis(3,5-dimethylphenylcarbamate), and 3,6-bis(3,5-dichloro-phenyl carbamate) to study their ability as chiral stationary phases for HPLC. The 3,5-dimethyl- and 3,5-dichlorophenylcarbamates phases demonstrated higher chiral recognition ability than 3,6-bis(phenylcarbamate) (Fig. **24**). Ibuprofen and ketoprofen were efficiently resolved using 3,5-dichlorophenyl-carbamate phase. Some chiral 2-arylpropionic acids were also efficiently resolved on the 3,5-dichlorophenylcarbamate [48].

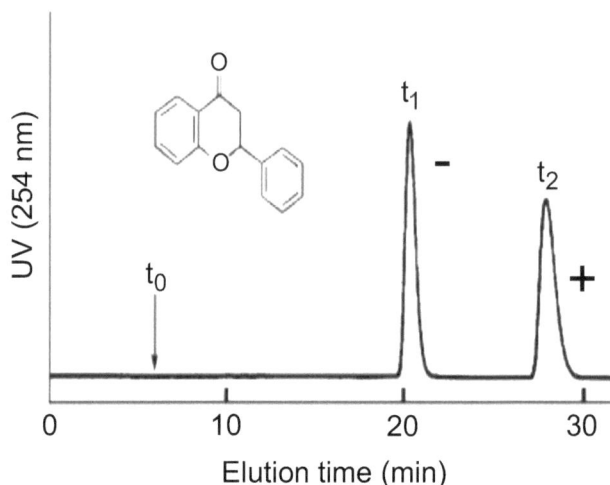

Fig. (24). Resolution of flavanone on chitin bis(4-chlorophenylcarbamate) [48].

In 2015, Wang *et al.* prepared new chiral stationary phases by blending chitin bis(3,5-dimethylphenylcarbamate) (Chi-DMPC) with high solvent tolerance and cellulose tris(4-methylbenzoate) (CMB) and cellulose tris(3,5-dimethylphenylcarbamate) (CDMPC) respectively to coat 3-aminopropyl silica gel, respectively. The motivation was to solve the problems of cellulose-derivative solvation or swelling in some organic solvents such as chloroform. The results revealed that the biselector ChiDMPC/CMB and ChiDMPC/ CDMPC possesses powerful chiral recognition capabilities compared to the corresponding single selector CSPs. These CSPs work in a wider range of mobile phases. Moreover, the supra molecular structure of polysaccharide derivatives in different mobile phases was variable, resulting in changeable enantioseparation ability [49]. For Chitin-based CSPs, alternative derivatives showed better analytical ability, and the biselector ChiDMPC / CMB and ChiDMPC / CDMPC have powerful chiral recognition.

Glycogen

Glycogen is a multibranched polysaccharide of glucose that serves as a form of energy storage in humans and animals. It represents the main storage form of glucose in the body. In 2010, the branched polysaccharide glycogen was firstly used by Chen *et al.* as a chiral selector to separate enantiomers of basic and acidic compounds using CE (Fig. **25**) [50]. Using 3.0% w/v glycogen with 90mM Tris-H_3PO_4 buffer (pH 7.0), enatomers of ibuprofen were base-line separated. By altering the conditions to use a 50mM Tris-H_3PO_4 buffer (pH 3.0) containing 3.0% glycogen, the enantiomers of citalopram, cetirizine and nefopam were also baseline-resolved. Such separations motivated Chen *et al.*, to pursue their study on glycogen, thus, the enantioselectivity of three glycogen-based dual selector systems with either chondroitin sulphate A (CSA, ionic polysaccharide), β-CD or hydroxypropyl-β-cyclodextrin (HP-β-CD)) were investigated for the chiral separation of racemic basic drugs. The dual systems showed good separations where glycogen/CSA obtained was the best seprators. Enhanced enantioseparations have been observed with the glycogen/CSA system when compared to single-selector systems for the tested analytes (Fig. **26**). This might be the result of interactions between glycogen and CSA [51].

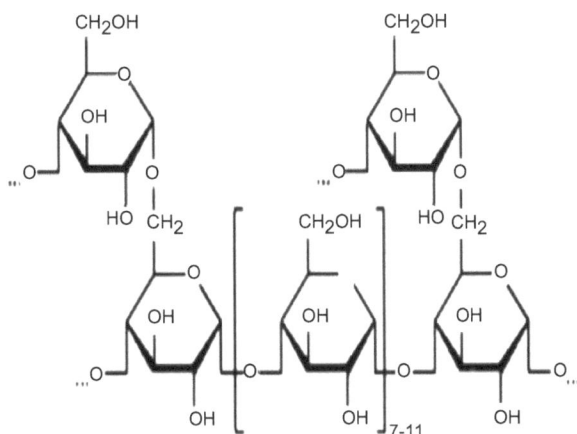

Fig. (25). Chemical structure of glycogen [50].

In the same vein, the synergistic effects of amino acid-based chiral ionic liquids (ILs) namely: tetramethylammonium-l-arginine (TMA-L-Arg) and tetramethyl ammonium-l-aspartic acid (TMA-L-Asp) with glycogen as chiral selector has been studied using CE. Improved separations of citalopram hydrobromide (CIT), Nefopam hydrochloride (NEF), and duloxetine hydrochloride (DUL) enantiomers were observed using chiral ILs/glycogen compared to glycogen alone based on three parameters (AAILs (amino acid ILs) concentration, glycogen concentration and buffer pH) (Fig. **27**) [52].

Pectin

Pectin is a structural heteropolysaccharide present in the cell walls of terrestrial plants. In 1994, Hicks *et al.* developed novel methods to prepare inositols and pectin oligosaccharides by TFA-catalyzed hydrolysis for use as CSPs in analytical and preparative HPLC [53].

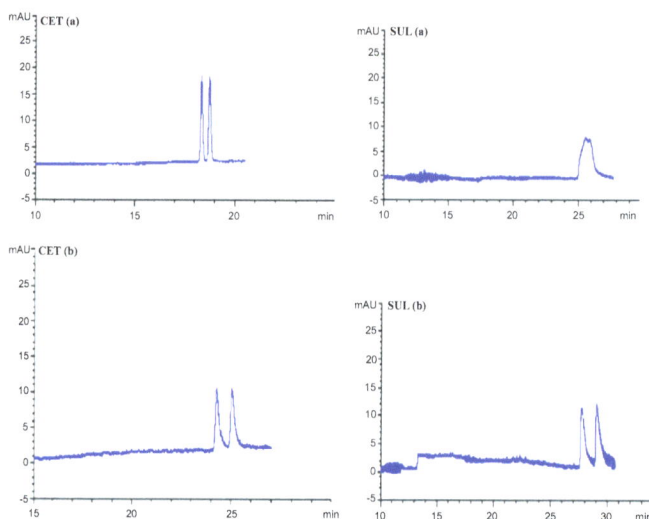

Fig. (26). Electropherograms of the chiral separations of CET and SUL in glycogen/CSA system. Conditions: fused-silica capillary, 50cm (41.5cm effective length)×50μm ID; BGE, 40mM phosphate buffer at pH 3.0; selector concentration, 3%glycogen alone for CET(a), 2%CSA alone for SUL(a), 3%glycogen + 2%CSA for CET(b) and SUL(b); applied voltage, 25 kV (CET) and 15 kV (SUL); capillary temperature, 20°C [51].

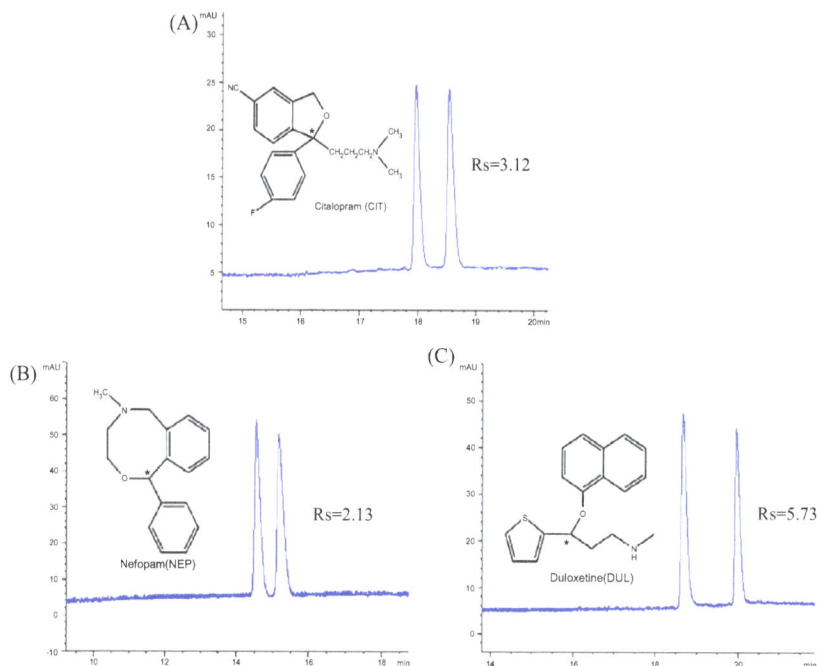

Fig. (27). Chiral separations of all tested drug enantiomers (A: CIT, B: NEF, C: DUL) in the final optimized TMA-L-Arg/glycogen synergistic system. Conditions: fused-silica capillary, 50 cm (41.5 cm effective length) ×50 μm id; applied voltage, 18.9 kV; capillary temperature, 19.2°C; BGE, 36.67 mM Tris/H3PO4 buffer (pH 3.0) containing2.5% glycogen and 60 mM TMA-L-Arg [52].

Pectin packed with aminopropyl silica gel and used in weak anion exchange mode, has high sample capacity for acidic oligosaccharides; however, for the separation of carbohydrates, it is better to use the cation-exchange HPLC columns because of their durability, easy operation and simple requirements such as flow rate, mobile phase and back-pressure [53, 54].

Amylopectin

Amylopectin is a highly branched polymer of glucose found in plants. The structure of the amylopectin tris(phenylcarbamate) CSP is different from cellulose or amylose CSPs, but very similar to β-cyclodextrin CSPs. In 1993, Felix and Zhang studied amylopectin tris(phenylcarbamate) as a chiral stationary phase for use in HPLC to investigate the effect of mobile phase composition and modifier on chiral discrimination. In their research, they separated twelve arylalcohol racemates (Fig. **28**) [55].

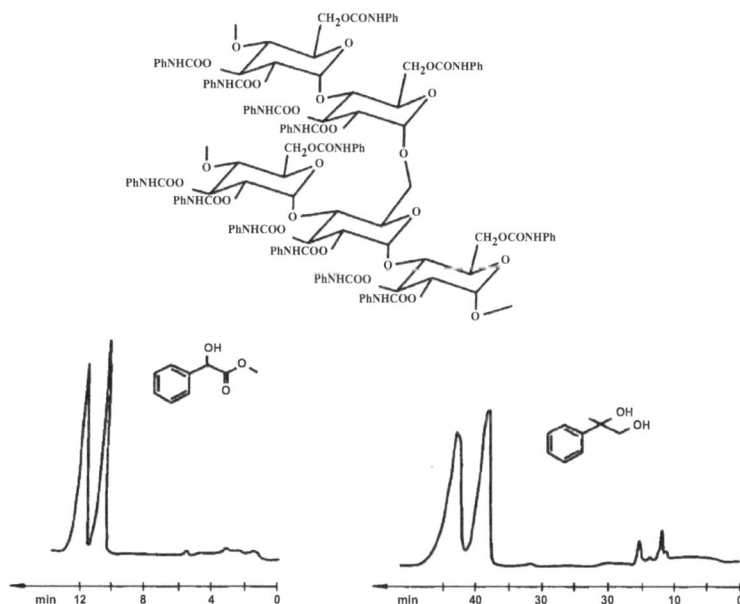

Fig. (28). Structure of amylopectin tris(phenylcarbamate) and separation of the racemates of compounds 2-phenyl-1,2-propanediol and methyl mandelate on amylopectin tris(phenylcarbamate): mobile phase, 0.65 M n-butanol in hexane; flow rate, 0.5 ml/min; detection, UV at 220 nm [55].

In 1995, Chassaing *et al*. tested the separation of five compounds under the same chromatographic conditions (hexane-isopropanol, 9:1 v/v; flow rate 0.5 ml/min) using tris(phenylcarbamate)-substituted amylopectins and compared the separations observed on the corresponding amylose derivatives. It was found that the enantioselectivity of polysaccharide stationary phases was highly dependent on the substituent on the phenyl group of the carbamate moiety, because the size of the substituted phenyl moiety has an influence on the discriminating power. They did not however, explain the enantioselective behavior of the amylopectin (APEC) CSPs [56].

In 2007, Wang *et al.* prepared a chiral stationary phase by synthesizing amylopectin-tris(phenylcarbamate) which was coated on to aminopropyl silica to separate fungicide enantiomers using HPLC. The influence of the percentage of isopropanol in the mobile phase on selectivity for twelve chiral fungicides was studied. The CSP showed stereoselectivity for seven chiral fungicides. The enantiomers of metalaxyl and benalaxyl were near baseline separated and myclobutanil, hexconazole, tebuconazole, uniconazole, and paclobutrazol enantiomers were completely separated (Fig. **29**) [57].

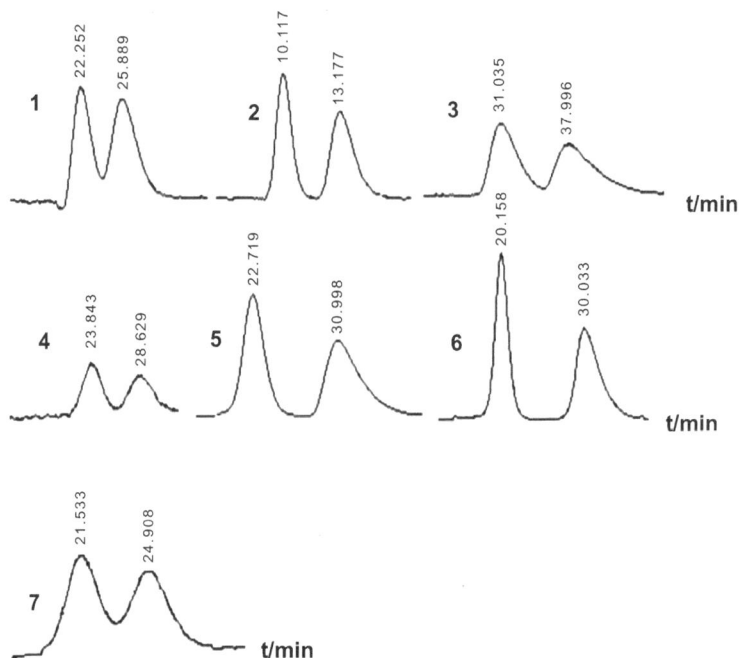

Fig. (29). The chromatograms of the chiral separations of the fungicide enantiomers on the ATPC chiral stationary phase. 1: metalaxyl, 10% isopropanol, 230 nm; 2: hexaconazole, 10% isopropanol; 3: myclobutanil, 10% isopropanol; 4: tebuconazole, 5% isopropanol; 5: uniconazole, 5% isopropanol; 6: paclobutrazol, 5% isopropanol and 7: benalaxyl, 3% isopropanol [57].

By reducing the percentage of isopropanol in the mobile hexane-isopropanol phase better separation and longer retention times were observed. Moreover, the separation efficiency of amylopectin stationary phase is related to the size of the substituent on the phenyl group of the carbamate moiety.

Miscellaneous

Other polysaccharides such as xylan, galactosamine and curdlan have been tested for their chiral separations efficiencies. The phenylcarbamate derivatives of cellulose, amylose and xylan were used for the chiral resolution of six dihydropyridines, nimodipine, nilvadipine, benidipine, nitrendipine, isradipine and nicardipine. The results revealed that the cellulose tris(3,5-dimethylphenylcarbamate) showed high optical resolving ability for many racemates. The cellulose tris(4-tert.-butylphenylcarbamate) and xylan bis(3,5-dichlorophenylcarbamate) almost completely resolved nicardipine in hexane- 2-propanol (90:10% v/v). Xylan bis(3,5-dimethylphenylcarbamate), however, provided a more expeditious separation in the same eluent [58].

Li *et al.* (2015) developed ten new xylan bisphenylcarbamate derivatives bearing meta- and parasubstituents on their phenyl groups, and tested their efficiencies as chiral stationary phases in HPLC after coating on macroporous silica. The result revealed that xylan bis(3,5 dimethylphenylcarbamate) based CSP possessed the highest resolving power for many racemates, and the meta-substituted CSPs also exhibited better chiral recognition than the para-substituted ones (Fig. **30**). The xylan bis(3,5-dimethylphenylcarbamate) derivative showed higher enantio-selectivity for some racemates due to the presence of an electron-donating substituent which is more preferable than an electron withdrawing group to improve the chiral recognition ability of the xylan phenylcarbamate derivatives [59].

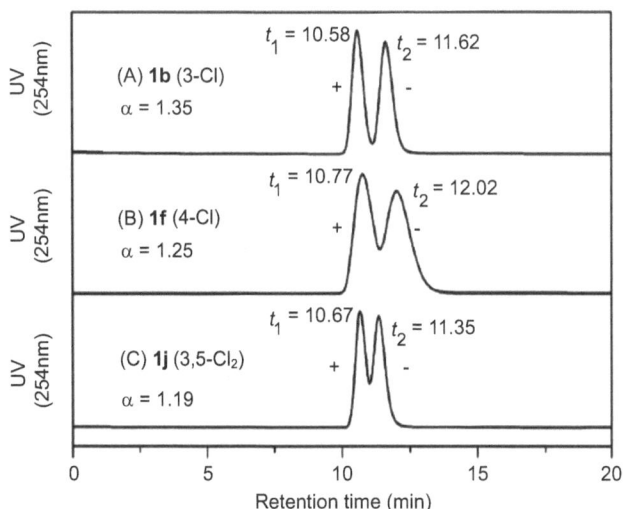

Fig. (30). Chromatograms for resolution of racemate 3 on xylan phenylcarbamates bearing 3-Cl (1b), 4-Cl (1f) and 3,5-Cl$_2$ (1j) substituents. Column: 25 × 0.20 cm (i.d.); eluent: hexane/2-propanol (90/10, v/v); flow rate: 0.1 mL/min [59].

Okamoto, Noguchi and Yashima (1998) prepared polysaccharide derivatives as chiral stationary phases, and found that the 3,5-dimethylphenylcarbamates of chitosan and xylan and the 3,5-dichlorophenylcarbamate of galactosamine demonstrated high chiral recognition ability. The 3,5-dimethylphenylcarbamates of cellulose and amylose often showed better chiral resolving power than 3,5-dimethylphenylcarbamates of chitosan and xylan and the 3,5-dichlorophenylcarbamate of galactosamine. Improved resolution; however, was better achieved using phenylcarbamates than the corresponding cellulose and amylose derivatives for some racemates and also some chiral drugs were better resolved on xylan bis(3,5-dimethyl- or 3,5dichlorophenylcarbamate) (Fig. **31**) [60].

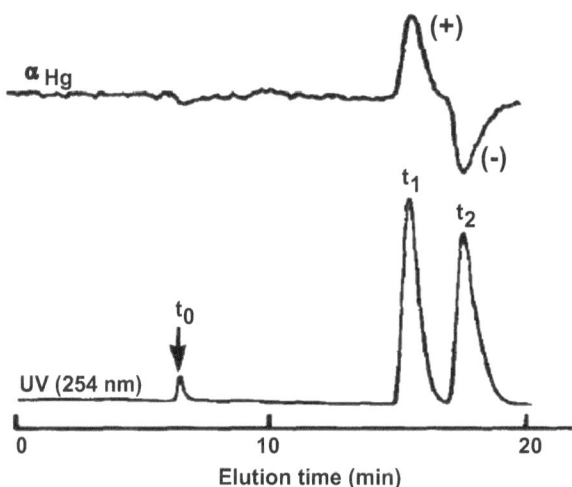

Fig. (31). Separation of enantiomers of (±)-1,2,2,2-tetraphenylethanol on bis(3,5-dimethylphenylcarbamate) of xylan (5a) [60].

In 1984, Ichida *et al.*, synthesized many polysaccharide derivatives, especially cellulose derivatives, which were used as chiral stationary phases for optical resolution of racemates by HPLC after being adsorbed on macroporous silica gel. The results revealed that cellulose triacetate (CTA-II) showed chiral recognition ability for many racemates similar to the chiral recognition with other cellulose derivatives such as cellulose tribenzoate (OB), cellulose trisphenylcarbamate (OC), cellulose tribenzyl ether (OE), and cellulose tricinnamate (OK). Curdfan triacetate was also shown to be an effective chiral selector [61]. The alternative derivatives exhibit better analytical ability than para-substituted derivatives [58 - 61].

CONCLUSIONS

Herein, non-commercially used polysaccharides as chiral selectors for enantioselective chromatography have been reviewed. Despite their unavailability in the market, glycogen, chitin, chitosan, pectin and amylopectin and others have been proven to be powerful chiral selectors in many chromatographic applications. The alternative derivatives of polysaccharides show better analytical seperations ability than the para-substituted derivatives of polysaccharides. In the n-hexane-2-propanol mobile phase system, decreasing the percentage of 2-propanol can achieve better separations and longer analysis time when using amylopectin stationary phases. Pectin, glycogen and chitin have been used as a biselector CSP; for examples, pectin packed with aminopropyl silica gel, glycogen combined with CSA, Chitin-DMPC / CMB and Chitin-DMPC / CDMPC have powerful chiral recognition and a better analytical ability. Chitosan has also been used in conjunction with albumin to reduce the hydrophobic interaction between enantiomers and albumin and to reduce the retention time of enantiomers. Thus, further investigations are advised to explore their recognition mechanisms and capabilities.

FUTURE STUDIES

The development of novel polysaccharide-based chiral stationary phases will continue to play important roles in separation science and technology, with strong focus on the synthesis of new derivatives of polysaccharide chiral selectors with multiple interaction sites. Mixed polysaccharide chiral selectors for increased chiral recognition ability might be beneficial in chiral separation. In the coming future, researchers should focus on fast enantioseparations and the development of novel CSs based columns should continue.

CONSENT FOR PUBLICATION

Not applicable.

CONFLICT OF INTEREST

The author confirms that he has no conflict of interest to declare for this publication.

ACKNOWLEDGEMENTS

Declared none.

REFERENCES

[1]   Aboul-Enein, HY; Wainer, I.W. *The Impact of Stereochemistry on Drug Development and Use*; John

Wiley & Sons, **1997**.

[2]   Ali, I; A.-E., HY. *Introduction. Chiral Pollutants: Distribution, Toxicity and Analysis by Chromatography and Capillary Electrophoresis.* ; John Wiley & Sons, Ltd, **2004**, pp. 1-35.

[3]   Younes, A.A.; Mangelings, D.; Vander Heyden, Y. Chiral separations in reversed-phase liquid chromatography: evaluation of several polysaccharide-based chiral stationary phases for a separation strategy update. *J. Chromatogr. A,* **2012**, *1269*, 154-167.
        [http://dx.doi.org/10.1016/j.chroma.2012.07.070] [PMID: 22921362]

[4]   Francotte, E.R. Enantioselective chromatography as a powerful alternative for the preparation of drug enantiomers. *J. Chromatogr. A,* **2001**, *906*(1-2), 379-397.
        [http://dx.doi.org/10.1016/S0021-9673(00)00951-1] [PMID: 11215898]

[5]   Gil-Av, E.; Feibush, B. Resolution of enantiomers by gas liquid chromatography with optically active stationary phases. Separation on packed columns. *Tetrahedron Lett.,* **1967**, *8*(35), 3345-3347.
        [http://dx.doi.org/10.1016/S0040-4039(01)89841-5]

[6]   Han, S.M. Direct enantiomeric separations by high performance liquid chromatography using cyclodextrins. *Biomed. Chromatogr.,* **1997**, *11*(5), 259-271.
        [http://dx.doi.org/10.1002/(SICI)1099-0801(199709)11:5<259::AID-BMC701>3.0.CO;2-U] [PMID: 9376706]

[7]   Ou, J.; Lin, H.; Tang, S.; Zhang, Z.; Dong, J.; Zou, H. Hybrid monolithic columns coated with cellulose tris(3,5-dimethylphenyl-carbamate) for enantioseparations in capillary electrochromatography and capillary liquid chromatography. *J. Chromatogr. A,* **2012**, *1269*, 372-378.
        [http://dx.doi.org/10.1016/j.chroma.2012.09.022] [PMID: 23022241]

[8]   Chankvetadze, B.; Kubota, T.; Ikai, T.; Yamamoto, C.; Kamigaito, M.; Tanaka, N.; Nakanishi, K.; Okamoto, Y. High-performance liquid chromatographic enantioseparations on capillary columns containing crosslinked polysaccharide phenylcarbamate derivatives attached to monolithic silica. *J. Sep. Sci.,* **2006**, *29*(13), 1988-1995.
        [http://dx.doi.org/10.1002/jssc.200500388] [PMID: 17017011]

[9]   Ahmed, M.; Gwairgi, M.; Ghanem, A. Conventional Chiralpak ID vs. capillary Chiralpak ID-3 amylose tris-(3-chlorophenylcarbamate)-based chiral stationary phase columns for the enantioselective HPLC separation of pharmaceutical racemates. *Chirality,* **2014**, *26*(11), 677-682.
        [http://dx.doi.org/10.1002/chir.22390] [PMID: 25271972]

[10]   Fanali, C.; Fanali, S.; Chankvetadze, B. HPLC Separation of Enantiomers of Some Flavanone Derivatives Using Polysaccharide-Based Chiral Selectors Covalently Immobilized on Silica. *Chromatographia,* **2016**, *79*(3-4), 119-124.
        [http://dx.doi.org/10.1007/s10337-015-3014-8]

[11]   Ghanem, A.; Ahmed, M.; Ishii, H.; Ikegami, T. Immobilized β-cyclodextrin-based silica vs polymer monoliths for chiral nano liquid chromatographic separation of racemates. *Talanta,* **2015**, *132*, 301-314.
        [http://dx.doi.org/10.1016/j.talanta.2014.09.006] [PMID: 25476312]

[12]   Sun, B. Preparation of regioselectively modified amylose derivatives and their applications in chiral HPLC. *Chromatographia,* **2012**, *75*(23-24), 1347-1354.
        [http://dx.doi.org/10.1007/s10337-012-2338-x]

[13]   Rocchi, S.; Fanali, S.; Farkas, T.; Chankvetadze, B. Effect of content of chiral selector and pore size of core-shell type silica support on the performance of amylose tris(3,5-dimethylphenylcarbamate)-based chiral stationary phases in nano-liquid chromatography and capillary electrochromatography. *J. Chromatogr. A,* **2014**, *1363*, 363-371.
        [http://dx.doi.org/10.1016/j.chroma.2014.05.029] [PMID: 24908153]

[14]   Cirilli, R.; Alcaro, S.; Fioravanti, R.; Secci, D.; Fiore, S.; La Torre, F.; Ortuso, F. Unusually high enantioselectivity in high-performance liquid chromatography using cellulose tris(4-methylbenzoate) as a chiral stationary phase. *J. Chromatogr. A,* **2009**, *1216*(22), 4673-4678.

[http://dx.doi.org/10.1016/j.chroma.2009.04.013] [PMID: 19394945]

[15]   Ikai, T.; Okamoto, Y. Structure control of polysaccharide derivatives for efficient separation of enantiomers by chromatography. *Chem. Rev.,* **2009**, *109*(11), 6077-6101.
[http://dx.doi.org/10.1021/cr8005558] [PMID: 19645486]

[16]   Ikai, T.; Yamamoto, C.; Kamigaito, M.; Okamoto, Y. Immobilization of polysaccharide derivatives onto silica gel Facile synthesis of chiral packing materials by means of intermolecular polycondensation of triethoxysilyl groups. *J. Chromatogr. A,* **2007**, *1157*(1-2), 151-158.
[http://dx.doi.org/10.1016/j.chroma.2007.04.054] [PMID: 17482625]

[17]   Ikai, T.; Yamamoto, C.; Kamigaito, M.; Okamoto, Y. Immobilized-type chiral packing materials for HPLC based on polysaccharide derivatives. *J. Chromatogr. B Analyt. Technol. Biomed. Life Sci.,* **2008**, *875*(1), 2-11.
[http://dx.doi.org/10.1016/j.jchromb.2008.04.047] [PMID: 18502192]

[18]   Okamoto, Y.; Ikai, T.; Shen, J. Controlled Immobilization of Polysaccharide Derivatives for Efficient Chiral Separation. *Isr. J. Chem.,* **2011**, *51*(10), 1096-1106.
[http://dx.doi.org/10.1002/ijch.201100025]

[19]   Shen, J.; Ikai, T.; Okamoto, Y. Synthesis and chiral recognition of novel amylose derivatives containing regioselectively benzoate and phenylcarbamate groups. *J. Chromatogr. A,* **2010**, *1217*(7), 1041-1047.
[http://dx.doi.org/10.1016/j.chroma.2009.07.027] [PMID: 19647833]

[20]   Shen, J. Synthesis and Immobilization of Amylose Derivatives Bearing a 4-tert-Butylbenzoate Group at the 2-Position and 3,5-Dichlorophenylcarbamate/3-(Triethoxysilyl)propylcarbamate Groups at 3- and 6-Positions as Chiral Packing Material for HPLC. *Chem. Lett.,* **2010**, *39*(5), 442-444.
[http://dx.doi.org/10.1246/cl.2010.442]

[21]   Tang, S. Immobilization of 3,5-dimethylphenylcarbamates of cellulose and amylose onto silica gel using (3-glycidoxypropyl)triethoxysilane as linker. *Journal of Separation Science,* **2010**, NA-NA.

[22]   Tang, S.; Ikai, T.; Tsuji, M.; Okamoto, Y. Immobilization and chiral recognition of 3,5-dimethylphenylcarbamates of cellulose and amylose bearing 4-(trimethoxysilyl)phenylcarbamate groups. *Chirality,* **2010**, *22*(1), 165-172.
[http://dx.doi.org/10.1002/chir.20722] [PMID: 19455617]

[23]   Preparation of β-Cyclodextrin-Modified N-Carboxymethylchitosan and Its Chromatographic Behavior as a Chiral HPLC Stationary Phase. *Anal. Sci.,* **1997**, *13*, 47-52.
[http://dx.doi.org/10.2116/analsci.13.47]

[24]   Okamoto, Y.; Noguchi, J.; Yashima, E. Enantioseparation on 3,5-dichloro- and 3,5-dimethylphenylcarbamates of polysaccharides as chiral stationary phases for high-performance liquid chromatography. *React. Funct. Polym.,* **1998**, *37*, 183-188.
[http://dx.doi.org/10.1016/S1381-5148(97)00135-1]

[25]   Franco, P. 3,5-Dimethylphenylcarbamates of amylose, chitosan and cellulose bonded on silica gel: Comparison of their chiral recognition abilities as high-performance liquid chromatography chiral stationary phases. *J. Chromatogr. A,* **1998**, *796*(2), 265-272.
[http://dx.doi.org/10.1016/S0021-9673(97)01004-2]

[26]   Senso, A.; Oliveros, L.; Minguillon, C. Chitosan derivatives as chiral selectors bonded on allyl silica gel: preparation, characterisation and study of the resulting high-performance liquid chromatography chiral stationary phases. *J. Chromatogr. A,* **1999**, *839*, 15-21.
[http://dx.doi.org/10.1016/S0021-9673(99)00072-2]

[27]   Shuang, L.; Fengbao, Z.; Guoliang, Z. Preparation of Bovine Serum Albumin Chiral Stationary Phase with a Hydrophilic Film. *Chin. J. Anal. Chem.,* **2006**, *34*(3), 385.

[28]   Son, S-H.; Jegal, J. Synthesis and characterization of the chiral stationary phase based on chitosan. *J. Appl. Polym. Sci.,* **2007**, *106*(5), 2989-2996.

[http://dx.doi.org/10.1002/app.26908]

[29]   Potekhina, E.V. Retentivity and separability of silica gel modified by chiral selector N-(3-sulfo-3-carboxy)-propionylchitosan. *Moscow Univ. Chem. Bull.,* **2007**, *62*(1), 37-41.
       [http://dx.doi.org/10.3103/S0027131407010099]

[30]   Yamamoto, C.; Fujisawa, M.; Kamigaito, M.; Okamoto, Y. Enantioseparation using urea- and imide-bearing chitosan phenylcarbamate derivatives as chiral stationary phases for high-performance liquid chromatography. *Chirality,* **2008**, *20*(3-4), 288-294.
       [http://dx.doi.org/10.1002/chir.20430] [PMID: 17597117]

[31]   Lü, Z.; Zhang, P.; Jia, L. Preparation of chitosan functionalized monolithic silica column for hydrophilic interaction liquid chromatography. *J. Chromatogr. A,* **2010**, *1217*(30), 4958-4964.
       [http://dx.doi.org/10.1016/j.chroma.2010.05.051] [PMID: 20566198]

[32]   Chen, J.L.; Hsieh, K.H. Nanochitosan crosslinked with polyacrylamide as the chiral stationary phase for open-tubular capillary electrochromatography. *Electrophoresis,* **2011**, *32*(3-4), 398-407.
       [http://dx.doi.org/10.1002/elps.201000410] [PMID: 21298667]

[33]   Chen, J.L. Molecularly bonded chitosan prepared as chiral stationary phases in open-tubular capillary electrochromatography: comparison with chitosan nanoparticles bonded to the polyacrylamide phase. *Talanta,* **2011**, *85*(5), 2330-2338.
       [http://dx.doi.org/10.1016/j.talanta.2011.07.091] [PMID: 21962650]

[34]   Lin, C. Progress in Recognition and Resolution of Enantiomers. *Journal of South China Normal University,* **2014**, *46*, 1-9.

[35]   Han, X. Enantioseparation Using Chitosan Tris(3-chlorophenylcarbamate) as a Chiral Stationary Phase for HPLC. *Chromatographia,* **2011**, *73*(11-12), 1043-1047.
       [http://dx.doi.org/10.1007/s10337-011-2014-6]

[36]   Hu, J.; Yin, L.; Jia, L. Chitosan-silica hybrid-coated open tubular column for hydrophilic interaction capillary electrochromatography. *J. Sep. Sci.,* **2011**, *34*(5), 565-573.
       [http://dx.doi.org/10.1002/jssc.201000688] [PMID: 21265018]

[37]   Yao, C. Preparation of a Monolith with Covalently Bound Bovine Serum Albumin for Capillary Electrochromatography. *Anal. Lett.,* **2012**, *45*(16), 2377-2388.
       [http://dx.doi.org/10.1080/00032719.2012.689793]

[38]   Chen, J-L.; Syu, H-J. Immobilization of chitosan in sol-gel phases for chiral open-tubular capillary electrochromatography. *Anal. Chim. Acta,* **2012**, *718*, 130-137.
       [http://dx.doi.org/10.1016/j.aca.2012.01.003] [PMID: 22305908]

[39]   Guntari, S.N. Fabrication of Chiral Stationary Phases *via* Continuous Assembly of Polymers for Resolution of Enantiomers by Liquid Chromatography. *Macromol. Mater. Eng.,* **2014**, *299*(11), 1285-1291.
       [http://dx.doi.org/10.1002/mame.201400103]

[40]   Zhang, L. Synthesis and Chiral Recognition Ability of Chitosan Derivatives with Different 4-Chlorophenylcarbamate–Urea Structures. *Chem. Lett.,* **2014**, *43*(1), 92-94.
       [http://dx.doi.org/10.1246/cl.130834]

[41]   Zhang, L.; Shen, J.; Zuo, W.; Okamoto, Y. Synthesis of chitosan 3,6-diphenylcarbamate-2-urea derivatives and their applications as chiral stationary phases for high-performance liquid chromatography. *J. Chromatogr. A,* **2014**, *1365*, 86-93.
       [http://dx.doi.org/10.1016/j.chroma.2014.09.002] [PMID: 25262030]

[42]   Francotte, E.; Zhang, T. Preparation and evaluation of immobilized 4-methylbenzoylcellulose stationary phases for enantioselective separations. *J. Chromatogr. A,* **2016**, *1467*, 214-220.
       [http://dx.doi.org/10.1016/j.chroma.2016.08.006] [PMID: 27503767]

[43]   Feng, Z.W.; Chen, W.; Bai, Z.W. Chiral stationary phases based on chitosan bis(4-methylphenylcarbamate)-(alkoxyformamide). *J. Sep. Sci.,* **2016**, *39*(19), 3728-3735.

[http://dx.doi.org/10.1002/jssc.201600680] [PMID: 27514503]

[44]   Tang, S.; Bin, Q.; Chen, W.; Bai, Z.W.; Huang, S.H. Chiral stationary phases based on chitosan
       bis(methylphenylcarbamate)-(isobutyrylamide) for high-performance liquid chromatography. *J.
       Chromatogr. A,* **2016**, *1440*, 112-122.
       [http://dx.doi.org/10.1016/j.chroma.2016.02.053] [PMID: 26931425]

[45]   Wang, J.; Xi, J.B.; Chen, W.; Huang, S.H.; Bai, Z.W. High performance chiral separation materials
       based on chitosan bis(3,5-dimethylphenylcarbamate)-(alkyl urea)s. *Carbohydr. Polym.,* **2017**, *156*,
       481-489.
       [http://dx.doi.org/10.1016/j.carbpol.2016.09.047] [PMID: 27842849]

[46]   Cass, Q.B.; Bassi, A.L.; Matlin, S.A. Chiral discrimination by HPLC on aryl carbamate derivatives of
       chitin coated onto microporous aminopropyl silica. *Chirality,* **1996**, *8*(1), 131-135.
       [http://dx.doi.org/10.1002/(SICI)1520-636X(1996)8:1<131::AID-CHIR19>3.0.CO;2-O]

[47]   Yamamoto, C. Enantioseparation by Using Chitin Phenylcarbamates as Chiral Stationary Phases for
       High-Performance Liquid Chromatography. *Chem. Lett.,* **2000**, (1), 12-13.
       [http://dx.doi.org/10.1246/cl.2000.12]

[48]   Yamamoto, C.; Hayashi, T.; Okamoto, Y. High-performance liquid chromatographic
       enantioseparation using chitin carbamate derivatives as chiral stationary phases. *J. Chromatogr. A,*
       2003, *1021*(1-2), 83-91.
       [http://dx.doi.org/10.1016/j.chroma.2003.09.017] [PMID: 14735977]

[49]   Wang, X. Enantioseparation Characteristics of Chiral Stationary Phases Based on Derivatives of
       Cellulose and Chitin. *Anal. Methods,* **2015**, (7), 2786-2793.
       [http://dx.doi.org/10.1039/C4AY02989E]

[50]   Chen, J.; Du, Y.; Zhu, F.; Chen, B. Glycogen: a novel branched polysaccharide chiral selector in CE.
       Electrophoresis, **2010**, *31*(6), 1044-1050.
       [http://dx.doi.org/10.1002/elps.200900534] [PMID: 20309915]

[51]   Chen, J.; Du, Y.; Zhu, F.; Chen, B. Evaluation of the enantioselectivity of glycogen-based dual chiral
       selector systems towards basic drugs in capillary electrophoresis. *J. Chromatogr. A,* **2010**, *1217*(45),
       7158-7163.
       [http://dx.doi.org/10.1016/j.chroma.2010.09.017] [PMID: 20926087]

[52]   Zhang, Q.; Du, Y. Evaluation of the enantioselectivity of glycogen-based synergistic system with
       amino acid chiral ionic liquids as additives in capillary electrophoresis. *J. Chromatogr. A,* **2013**, *1306*,
       97-103.
       [http://dx.doi.org/10.1016/j.chroma.2013.07.053] [PMID: 23910600]

[53]   Hicks, K.B. Analytical and preparative HPLC of carbohydrates: inositols and oligosaccharides derived
       from cellulose and pectin. *Carbohydr. Polym.,* **1994**, *25*(4), 305-313.
       [http://dx.doi.org/10.1016/0144-8617(94)90056-6]

[54]   Phinney, K.W.; Jinadu, L.A.; Sander, L.C. Chiral selectors from fruit: application of citrus pectins to
       enantiomer separations in capillary electrophoresis. *J. Chromatogr. A,* **1999**, *857*(1-2), 285-293.
       [http://dx.doi.org/10.1016/S0021-9673(99)00777-3] [PMID: 10536847]

[55]   G. and T. ZHANG, An amylopectin tris (phenylcarbamate) chiral stationary phase for high
       performance liquid chromatography. HRC. *J. High Resolut. Chromatogr.,* **1993**, *16*(6), 364-367.
       [http://dx.doi.org/10.1002/jhrc.1240160606]

[56]   Chassaing, C.; Thienpont, M.H.S.A.; Félix, G. Tris (phenylcarbamate) derivatives of amylopectin as
       chiral stationcoes in HPLC. *J. High Resolut. Chromatogr.,* **1995**, *18*(6), 389-391.
       [http://dx.doi.org/10.1002/jhrc.1240180615]

[57]   Wang, P.; Liu, D.; Jiang, S.; Gu, X.; Zhou, Z. The direct chiral separations of fungicide enantiomers
       on amylopectin based chiral stationary phase by HPLC. *Chirality,* **2007**, *19*(2), 114-119.
       [http://dx.doi.org/10.1002/chir.20353] [PMID: 17096377]

[58]   Okamoto, Y. Optical resolution of dihydropyridine enantiomers by high-performance liquid chromatography using phenylcarbamates of polysaccharides as a chiral stationary phase. *J. Chromatogr. A,* **1990**, *513*, 375-378.
[http://dx.doi.org/10.1016/S0021-9673(01)89459-0]

[59]   Li, G.; Shen, J.; Li, Q.; Okamoto, Y. Synthesis and Enantioseparation Ability of Xylan Bisphenylcarbamate Derivatives as Chiral Stationary Phases in HPLC. *Chirality,* **2015**, *27*(8), 518-522.
[http://dx.doi.org/10.1002/chir.22472] [PMID: 26039871]

[60]   Okamoto, Y.; Noguchi, J.; Yashima, E. Enantioseparation on 3,5-dichloro- and 3,5-dimethylphenylcarbamates of polysaccharides as chiral stationary phases for high-performance liquid chromatography. *React. Funct. Polym.,* **1998**, *37*(1), 183-188.
[http://dx.doi.org/10.1016/S1381-5148(97)00135-1]

[61]   Ichida, A. Resolution of enantiomers by HPLC on cellulose derivatives. *Chromatographia,* **1984**, *19*(1), 280-284.
[http://dx.doi.org/10.1007/BF02687754]

CHAPTER 7

Ru(II)-polypyridyl Complexes as Potential Sensing Agents for Cations and Anions

Goutam Kumar Patra*, **Anupam Ghorai** and **Amit Kumar Manna**

Department of Chemistry, Guru Ghasidas Vishwavidyalaya, Bilaspur (C.G), India

Abstract: Design, synthesis and applications of cation and anion sensing selective Ru(II)- Polypyridyl complexes have attracted a considerable attention because of their multipurpose and promising biological insinuation. Ruthenium(II)-polypyridyl ligands such as 2,2-bipyridine (bpy), 1,10-phenanthroline (phen), and/ or ortho-phenanthroline *etc.* on reaction with Ru(II) forms Ru(II)-polypyridyl complexes which have various significant benefits like metal-to-ligand charge transfer based on excited visible light and emission intensity, high chemical and photochemical stabilities, high shifts in Stokes (particularly more than 150 nm), greater response efficiency, low cytotoxicity and good water solubility. Owing to the MLCT emission, Ru(II) polypyridyl complexes were mostly used as essential materials in electro-generated chemi-luminescence analysis offers unique properties like high rigidity, selectivity and less sensitive to environment. The multifaceted nature of Ru(II) polypyridyl-complexes in sensing is found as electrochemical sensors, solid state sensors, amperometric sensors, and straightforward chemosensors. A methodical report of different Ru(II) polypyridyl-complex chromophores emphasizing fluorophore frame work has been thoroughly discussed in this chapter. The binding mechanism has been proposed precisely. Unlike other metal complexes, the versatility and uniqueness of Ru(II) polypyridyl-complexes as chemosensor is observed in sensing of less common analytes which are rarely detected according to the literature. Polypyridyl Ru(II)-complexes being covalently bonded with photo conducting polymer may also be used as photosensitizers.

Keywords: Anions, Cations, Chemosensor, Colorimetric sensor, Fluorescence sensor, Hazardous metal ions, Molecular probe, Phosphorescence, Sensitive and selective sensor, MLCT, Ru(II)-polypyridyl complexes.

INTRODUCTION

The design and synthesis of molecular probes which are able to detect cations and anions has received extensive importance because of their significance in industrial, biological and environmental field. In connection with this, lots of

---
* Corresponding Author Goutam Kumar Patra: Department of Chemistry, Guru GhasidasVishwavidyalaya, Bilaspur (C.G), India; Tel: 91 7587312992; Fax: 07752-260148; E-mail: patra29in@yahoo.co.in

research work has been carried out on colorimetric and/or fluorescent systems able to the detection of various types of analytes [1 - 4] in comparison to those of the organic emitters because of their following characteristics: (a) long emissive lifetimes (b) remarkable Stoke shifts which allow simple discrimination of excitation and emission (c) environmental outsized shifts in emission and (c) large emission shifts from the local environment. Contrary to the conventional organic fluorophores, which are singlet emitters, transition metal complexes exhibit triplet emission due to spin-orbit coupling conferred by the heavy atom effect, which resulted in the efficient singlet-triplet state mixing and enrichment of phosphorescence quantum efficiency. The phosphorescence behaviour of metal complexes has found potential for the construction of organic light-emission diodes (OLEDs) or lighting applications [5 - 8]. Polypyridyl- Ru(II) complexes proved to be one of the most explored probe amongst the widely studied luminescent transition metal complexes, because they are chemically stable and provide interesting redox and photo-physical features. Based on these favourable properties, optical and electrochemical sensors of ruthenium poly-pyridylcomplexes have been rapidly developed for the detection of heavy metal cations, biologically important anions and other different analytes.

After excitation ruthenium system shows complex behaviour in comparison to organic probes. They generally exhibit metal-to-ligand charge transfer (MLCT), ligand- to-ligand charge transfer (LLCT), intra-ligand charge transfer (ILCT), ligand-to-metal charge transfer (LMCT), (MMLCT), ligand-to-metal-metal charge transfer (LMMCT) and (MLLCT) states [9, 10].These properties depend upon the sensitivity of central metal ion, ligand types, and the environmental factors facilitates the designing of metal complexes for certain applications by controlling the photo physical property of metal complexes *i.e* emission wavelength, lifespan and luminescence intensity *etc*. Certainly, the charge separation in the MLCT excited state would result in the complexes being usually sensitive to their environment and with appropriate incorporation of receptors, they may act as reporters of the environment. Thus, various sensing applications by making use of polypyridyl- ruthenium(II) complexes could be achieved through proper design of suitable pyridyl ligands. This chapter is a brief survey of the Ru(II) polypyridyl-complexes which cater the need of detection of various cations and anions with excellent selectivity and sensitivity.

SENSING OF CATIONS

Metal cations are very important in biological and environmental systems. A number of d- block transition metal ions are useful in normal human physiology and pathology [11, 12]. Nevertheless, certain heavy metal ions like Hg^{2+} and Cd^{2+} are toxic to living organisms. With these metals, other transition metals also have great control over environment and living being. Aluminium is the most abundant metallic element in the earth's crust and the abundance of aluminium in the biosphere is around 8% of the total mineral components [13 - 15]. Compounds of aluminium are widely distributed in the environment in various from, *i.e.* paper industry, food additives, water treatment plants, textile industry, medicines, deodorants, aluminium cookware, cans, production of light alloys and bleached flour *etc*. The amount of free Al^{3+} in surface water increases due to leaching from soil during acid rain and it is poisonous to fish [16]. Although, aluminium is an important building block of living systems but the ionic radius (0.051 nm) and charge of Al^{3+} makes it a competitive inhibitor of Mg^{2+} (0.066 nm), Ca^{2+} (0.099 nm) and Fe^{3+} (0.064 nm) elements of similar characteristics [17 - 19]. High concentration of Al^{3+} causes damage not only plant growth but also the human nervous system which further leads to lots of diseases such as microcytic hypochromic anemia, encephalopathy, myopathy, Alzheimer's disease, Parkinson's disease, amyotrophic lateral sclerosis *etc* [20 - 22]. It reduces liver and kidney function, memory loss, speech problems, softening of the bones and aching muscles [23 - 26]. Furthermore, aluminium toxicity is responsible for about 40% of the world's acidic soil. Phytotoxic Al^{3+} predominates in acidic soils mainly at pH < 5.0 which leads to increase in toxicity of aluminium in plants, is a subject of significant concern [27, 28].The electrochemical gradient in animal cells are regulated by Na^+/K^+ pump [29].

Therefore, it is most common that continuous effort has also been devoted into the development of chemosensors for the detection of cations. Ru(II)-polypyridyl complexes offer the requirement for this purpose in sensing the cations.

There are several review articles on the use of transition metal complexes as cation mapping chemosensors [1, 11, 12]. But these articles lack the detail description of recent luminescent metal complexes particularly on Ru(II)-polypyridyl complexes. This chapter deals with the sensing properties of Ru(II)-polypyridyl complexes for the detection of ions performing host and guest coordination between the ligand and metal ion. The optical behavior is also being emphasized by analyzing their asorbance as well as emission property upon the accumulation of analytes to the complex with particular care to the changes in their luminescence behaviours. Specific examples related to response of of Ru(II)-polypyridyl complexes towards cation, anion and neutral molecules have been cited.

Specific Examples

Peter Comba *et al*. [30] reported three Ruthenium(II) complexes carrying heteroditopic 1,10-phenanthroline ligands $[Ru(bpy)_2(H_2L^1)](PF_6)_2$, H_2L^1 =5,6-

bis(2-pyridyl carboxamido)-1,10-phenanthroline (1); [Ru(bpy)$_2$(H$_2$L^2)](PF$_6$)$_2$, H$_2$L^2=5,6-bis[(4-methoxy-2- pyridyl)carboxamido]-1,10-phenanthroline (2); and [Ru(H$_2$L^1)$_3$](PF$_6$)$_2$ (3); which were used as effective luminescent probe for Cu^{2+} (Fig. 1). These complexes displayed characteristic MLCT absorption and emission bands at 450 and 620 nm, respectively. The emission intensity gradually reduced on the addition of Cu^{2+} ion resulting in the complete luminescence quenching.

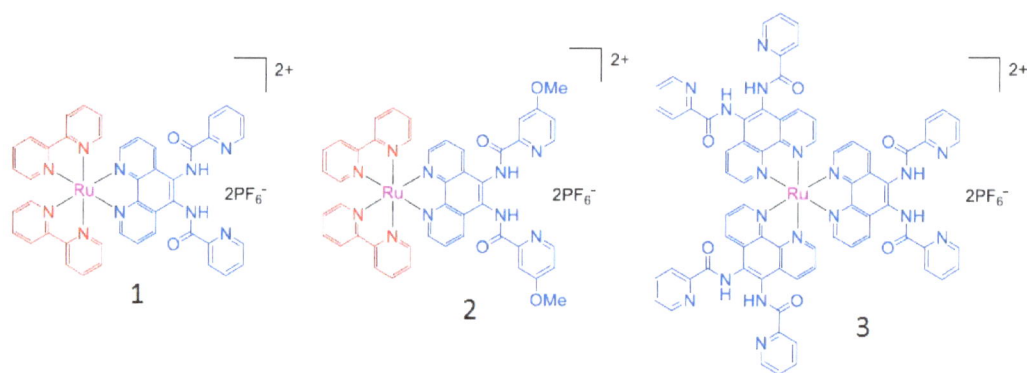

Fig. (1). Molecular representation of the complexes **1**, **2** and **3**.

In another report, Paul*et al.* [31] employed a macrocyclic ionophore for the detection of Cu^{2+} ion. They prepared [Ru(bpy)$_2$(L)][PF$_6$]$_2$), where, L is, 4,5,6,7,8,9,10,11,12,13-decahydro- [1, 4, 7, 10, 13, 16] dioxatetraaza-cyclooctadecino[2,3-f] [1, 10]phenanthroline-3,14(2H,15H)-dione) (**4**). In acetonitrile the absorption spectra of **4** displayed a ligand centered charge transfer (LCCT) at 286 nm and MLCT transition for bpy at 452 nm. At the excited state, Ru(bpy)$_2$(L)][PF$_6$]$_2$) displayed highest MLCT emission at 603 nm with a quantum yield (ϕ) 0.0593. Upon the gradual addition of Cu^{2+} ion to **4**, strong quenching in the emission intensity was noticed (Fig. 2). Though compound **4** was basically designed for anions, the cavity size and the donor site of **4** are favourable for Cu^{2+} ion sensing.

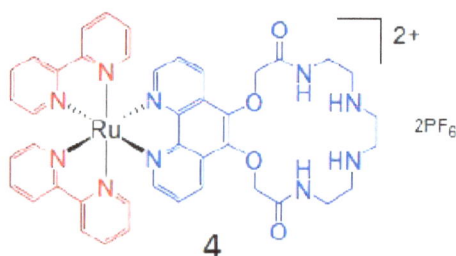

Fig. (2). Ru(II)-bpy derivative for sensing of Cu^{2+}.

Further, Zhang *et al.* reported a ruthenium(II) complex [Ru(phen)$_2$HPIP]$^{2+}$, HPIP = 2-(2- hydroxyphenyl) imidazo [4,5-f] [1, 10]phenanthroline)] (**5**) for fluorescence imaging of Cu^{2+} in live pea aphids [32]. When Cu^{2+} conc. is increased, the absorbance of complex **5** was increased at 357 nm and 495 nm and the intensities of the original absorption peak at 263 nm and 456 nm were decreased (Fig. **3**). Moreover, the addition of opper salts like copper nitrate, copper sulphate, copper acetate and copper chloride exhibits a similar trend in fluorescence quenching, indicated that counter anions does not change the emission property and binding capacity of complex **5**. Complex **5** binds with Cu^{2+} ion in a 1:1 ratio to give the association constant value of 1.83x10^{5}M^{-1} with the detection limit of 7.23x 10^{-8} M.

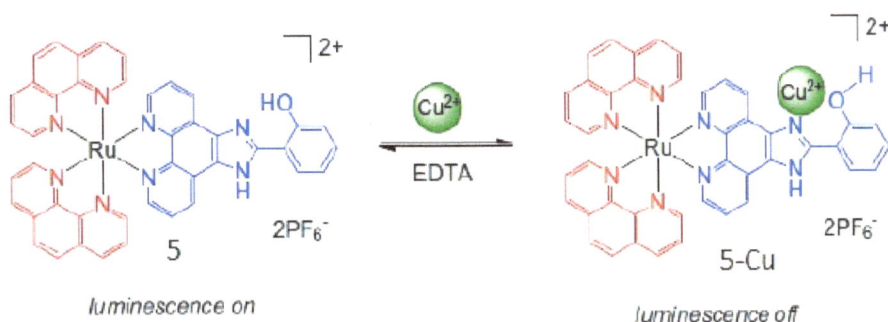

luminescence on   *luminescence off*

Fig. (3). Reversible sensing action of **5**.

Cu^{2+} recognition property of two dinuclear Ru(II) polypyridyl complexes was demonstrated by Cheng and co-workers [33] (Fig. **4**). Having vacant coordination sites, [{Ru(bpy)$_2$}$_2$(μ$_2$-L$_1$)]$^{4+}$, L$_1$ = 2,5-di(1H-imidazo[4,5-f] [1, 10]phenanthroline-2-yl)furan (**6**) and [{Ru(bpy)$_2$}$_2$(μ$_2$-L$_2$)]$^{4+}$, L$_2$ = 2,5-di(1Himidazo[4,5-f] [1, 10]phenanthroline-2-yl)thiophene) (**7**) undergo photo-induced electron transfer (PET) processes from the lone pair electrons of the sulphur/oxygen to the chromophore units in **6** and **7**. The detection limit reached up to 5.78x10^{-7} M.

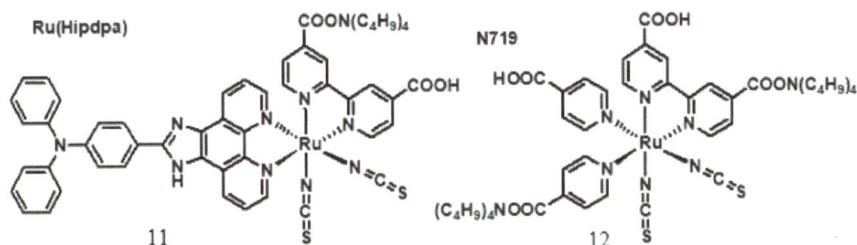

Fig. (4). Ru(II) poly-pyridyl complexes for Cu^{2+} recognition.

Zheng *et al*. [34] reported a Ru(II) complex of $[Ru(bpy)_2(Htppip)](ClO_4)_2$ ·H_2O·DMF (RuL) {bpy =2,2′-bipyridine and Htppip=2-(4-(2,6-di(pyrid-n-2-yl)pyridin-4-yl)phenyl)-1Himidazo[4,5-f] [1, 10]phenanthroline}(Fig. **5**) and tested its cation-sensing properties in both pure CH_3CN and HEPES buffer (pH = 7.2)/CH_3CN (71/1, v/v) solutions with the help of UV–visible absorption, emission, and ^1H NMR spectra.

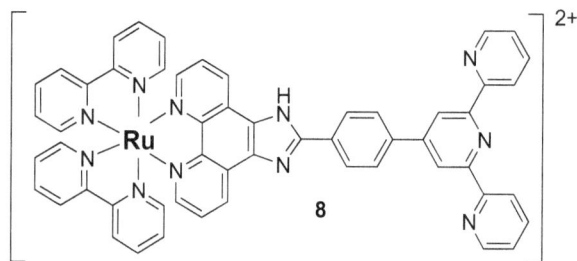

Fig. (5). Chemical structure of $[Ru(bpy)_2(Htppip)]^{2+}$.

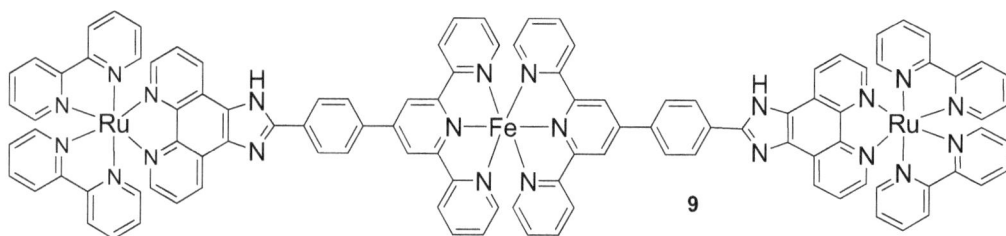

Fig. (6). Mechanism of response of Ru-polypyridyl complex to Fe^{2+} ions.

RuL treated as an effective colorimetric sensor for Fe^{2+} accompanied with clear colour change from pale yellow to light red-purple (Fig. **6**). These chemosensors exhibit high selectivity towards Fe^{2+} upon the addition of different cations such as Na^+, Mg^{2+}, Ba^{2+}, Mn^{2+}, Fe^{3+}, Co^{2+}, Ni^{2+}, Cu^{2+}, Zn^{2+}, Cd^{2+}, Hg^{2+}, and Ag^+. The detection limit of Fe^{2+} was found to be ~4.58×10^{-8} M in CH_3CN:H_2O where as ~4.46×10^{-8} M in pure CH_3CN. This result proved that RuL can be used as a good colorimetric chemosensor towards Fe^{2+} in CH_3CN:H_2O solution

Thus, **RuL** has an advantage over the pure organic ligand of Htppip, which could tolerate a water content of less than 40% by volume, where as **RuL** acts as an efficient sensor with a water content of 96.4% in CH_3CN:H_2O solvent. The binding stoichiometry between Fe^{2+}-RuL complex was approximately 1:2 (as depicted in Fig. **6**), which has been confirmed by Job plot analysis. Hg metal is carcinogenic in nature, which is accumulated through the food chain even at lower concentrations, leading to a threat to aquatic, animal, plant and human lives [35 - 37]. Therefore, for the environmental and toxicological monitoring, regulating and specific determination of Hg concentration in water is a matter of concern. A

Ru(II) complex,(cis- ruthenium-bis [2,2-bipyridine]-bis[4-aminothiophenol]-bis [hexafluorophosphate])(Ru-4-ASP) was found to have the potentiality to detect Hg^{2+} in aqueous solution. It has been synthesized and characterized by Hamid *et al.* [38].

In this probe, Ru(II)-bis(bpy) worked as a potential transporter for the sulfur-containing 4- aminothiophenol moiety which performed as the host part for the preparation of a new chemosensor to detect mercuric ions as shown in (Fig. **7**). Based on the chromophore-spacer-receptor concept, the 4-aminothiophenol ligand assimilating Hg(II) cations could probably control the binding properties of the Ru(II)-bipyridine center and explored a new effective chemosensor.

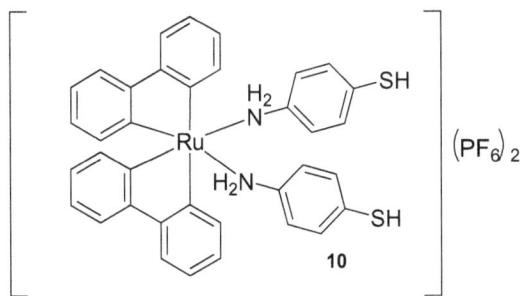

Fig. (7). Structure of the *cis*-Ruthenium-bis[2,2'-bipyridine]-bis[4-aminothiophenol]-bis[hexafluorophosphate],(Ru-4-ASP).

On treatment with Hg(II), the colour of the Ru-4-ASP solution was changed at once from light brown-reddish to orange-reddish. The limit of detection (LOD) for Hg(II) ions was likely to be ~ 0.4 ppm. The Investigation revealed that a highly efficient and active chemosensor was designed for the recognition of Hg^{2+}ions in water-ethanol (1:1) solution even in the presence of other highly concentrated interferring metal cations.

Both chromogenic and fluorogenic chemosensor for selective recognition of mercury ion was reported by Fan *et al.* [39]. The main core of this chemosensor was a heteroleptic ruthenium dye $[Ru(Hipdpa)(Hdcbpy)(NCS)_2]^-$. $0.5H^+$. $0.5[N(C_4H_9)_4]^+$ Ru(Hipdpa) {where Hdcbpy = monodeprotonted-4,4'-dicarbo-y-2,2'-bipyridine and Hipdpa = 4-(1H-imidazo[4,5-f] [1, 10]phenanthrolin-2-y-)-N,N-diphenylaniline}. The binding mechanism was based on the increase of energy gap between the highest occupied molecular orbital (HOMO) and the lowest unoccupied molecular orbital (LUMO) of the optical sensor upon the interaction of electron-deficient Hg^{2+}ion with the electron-rich sulfur atom of the thiocyanate (NCS) ligand in the Ru(Hipdpa). Ru(Hipdpa) was found to have lower Hg^{2+} detection limit and enhanced linear region as compared to di(tetrabutylammonium) cis- bis (isothiocyanato) bis(2,2'-bipyridine-4-carboxylic acid-4′-carboxylate) ruthenium(II) marked as N719 (Fig. **8**).

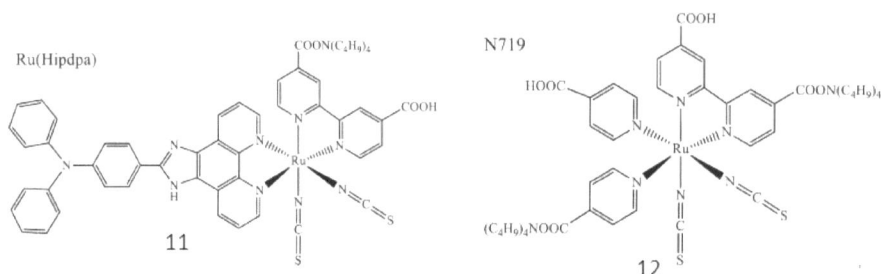

Fig. (8). The chemical structures of Ru(Hipdpa) (11) and N719 (12).

Baitalik *et al.* [40] also published a monometallic Ru(II) complex of composition [(bipy)$_2$Ru(tpy-Hbzim-dipy)](ClO$_4$)$_2$, (Fig. **9**)(where tpy-Hbzim-dipy = 4′-[4-(4-5-di-pyridin-2-yl-1H-imidazol-2-yl)-phenyl]-[2,2′;6′,2″]terpyridine), which can act as an optical sensor for Fe^{2+} ion by showing colour change from yellow orange to deep red-violet in CH$_3$CN solution with a low LOD value of 6.68×10^{-9} M. The demerits of the chemosensors are less water solubility and tedious synthetic protocol.

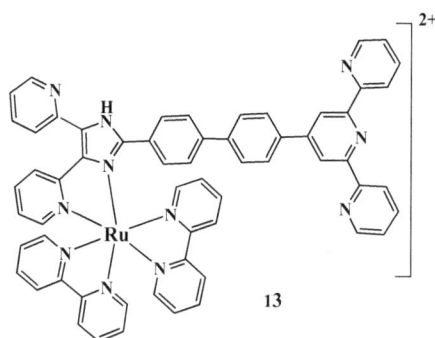

Fig. (9). Chemical structure of [(bipy)$_2$Ru(tpy-Hbzim-dipy)]$^{2-}$.

Another novel colorimetric probe based on a mesoporous nano crystalline TiO$_2$ film sensitized with a ruthenium dye (**14**, Fig. **10a**) has been synthesized by J. R. Durrant and co-workers. They efficiently exploited it for detection of mercury ion with easily discernible colour change. Moreover, an user friendly system for the immerse sensing of mercury (II) in aqueous solution has been developed by adsorbing the sensor molecules onto the high surface-area mesoporous metal oxide films. In order to make these films reusable they are treated with an aqueous solution of KI which practically discarded the absorbed mercury from the surface by forming a stable iodide complex of it [41].

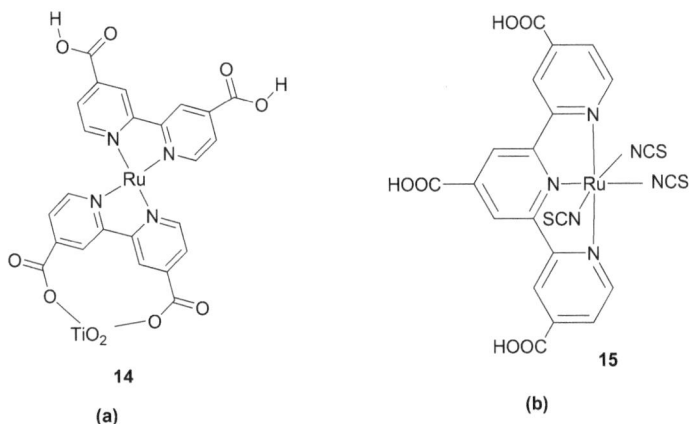

14

(a)

15

(b)

Fig. (10. a). TiO$_2$ film sensitized with a ruthenium dye, **10.b**. Ru-polypyridyl complex for detection of Hg^{2+}ions.

The same researchers later designed and synthesized another novel Ru-polypyridyl complex (15, Fig. **10b**) for detection of mercury ion. The reusing of an entirely different approach, Hg^{2+} has been sensed by means of a coordination complex of Ru with substituted bipyridine and thiocyanate ions. Hg^{2+} coordinates to the sulfur atom of the NCS groups present on the dye reversibly and showing a colour change even at sub-micromolar level concentration of mercury [42]. The *bis*(2,2′-bipyridine)(4-methyl-2,2′-bipyridine-4′-carboxylic   acid)ruthenium(II) ·2PF$_6^-$ complex, as a fluorescent probe to identify Cu(II) in ethanol/H$_2$O (1:1, v/v) solution was first employed by C.-L. He [43]. There the ruthenium (II) ion complexed chemosensor having three bipyridyl units with having carboxyl group in one of the bipyridyl units (Fig. **11**).

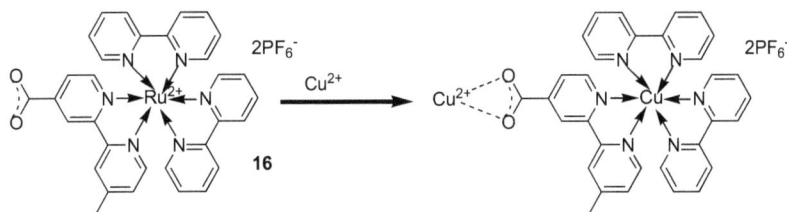

16

Fig. (11). Proposed binding mode of the Ru(II)-bipyridyl complex with Cu(II).

This carboxyl group interacts with the Cu(II), which leads to quenching in emission of complex. A linear range covering from 5.0×10^{-8} to 1.0×10^{-4} M and LOD value of 4.2×10^{-8} M of Cu(II) was attainable by this chemosensor. It is to be noted that the response behaviour of the complex to Cu(II) is pH independent in medium condition (pH 4.0 - 8.0), however, at higher pH range Cu(II) ion gets precipitated to $Cu(OH)_2$, which reduces the complexation process with sensor, as a result fluorescence intensity increases. In case of lower pH (pH<4.0), the emission intensity of the probe decreases clearly. The proton-induced quenching was the possible mechanisms proposed for the pH-dependent emission of these complexes. Looking into the sensitivity and sensing speed of the probe, all the experiments were carried out at pH of 6. The advantages of this sensor are high selectivity, neutral medium and large shift in Stokes *etc*.

Luminescent Metal Complex Chemodosimeters for Cations

For sensing trace amounts of metal ions, fluorescent chemodosimeters have been recognized as potential device because they are easy to handle, highly sensitive and short responsive time. Presently utilization of upconversion luminescence (UCL) has become an attractive tool in the field of both sensing and bioimaging for the signal detection [44 - 50]. This distinctive UCL mechanism offers several advantages in bioimaging as photoluminescent probes, such as a large anti-Stokes shift upto several hundred nanometers, lack of auto-fluorescence from biological samples [51], notable deep penetration of light in tissue and no photo- bleaching [51, 52]. For these advantages Liu *et al.* synthesized a novel chromogenic assembled nanophosphor- ruthenium complex- (N719-UCNPs) (Scheme **1**) and exploited it for bio imaging of intracellular mercury ions as an extremely efficient water-soluble probe based on upconversion luminescence sensing [53].

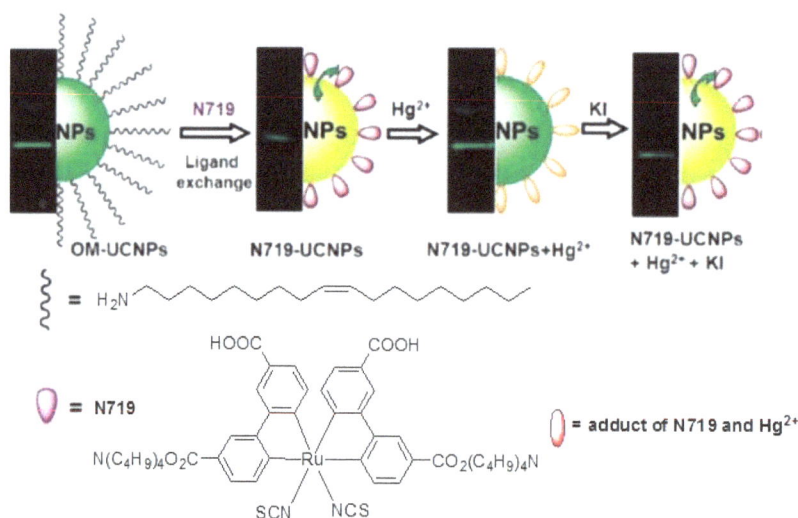

Scheme 1. Synthesis of the ruthenium complex N719-modified upconversionnanophosphor (N719-UCNPs) and their upconversion luminescence (UCL) response to Hg^{2+}.

The different features of synthesized nanophosphors were investigated using X-ray powder diffraction studies (XRD), transmission electron microscopy (TEM), energy- dispersive X-ray analysis (EDXA), Fourier transform infrared spectroscopy (FTIR), and X-ray photoelectron spectroscopy (XPS). The different application of N719-UCNPs towards sensing of Hg^{2+} was established by optical titration experiment and upconversion luminescence live cell imaging. Using the ratiometric upconversion luminescence as a detection signal, the nanoprobe showed a low value of LOD for Hg^{2+} up to 1.95 ppb in aqueous solution which is less than the maximum level (2 ppb) of Hg^{2+} in drinking water set by the United States EPA. The nanoprobe N719-UCNPs has been proved to be efficient for observing changes in the distribution of Hg^{2+} in living cells by UCL bioimaging.

Yunfei Zhang and co- workers constructed a novel water-soluble Ru(II) complex as a turn-on Cu^{2+} chemosensor based on o-(phenylazo)aniline moeity (Fig. **12**) [54].

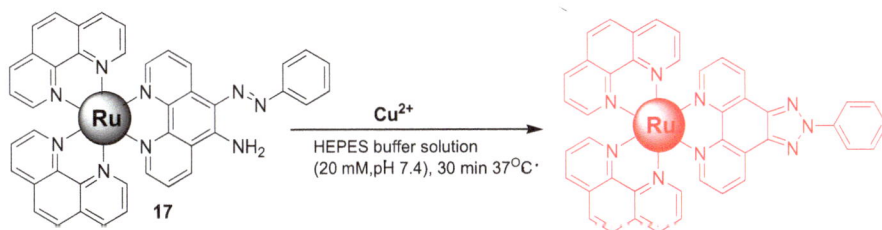

Fig. (12). Mechanism of response of Ru-polypyridyl complex to Cu^{2+} ions.

The azo group present in receptor 17 after the oxidative cyclization reaction with copper(II) ions changed into a highly luminescent benzotriazole, which showed considerable increments in luminescence property within the linearity range of copper(II) ions (Fig. **12**). This chemodosimeter was different interms of its selectivity, high sensitivity (almost 80-fold enhancement in fluorescence) and small LOD value of 4.42 nM in aqueous solution. The presence of other metal ions or amino acids has negligible effect on the luminescence response of the probe. On light irradiation, the copper (II) luminescent sensor showed good photo-stability. Moreover, this chemosensor can be successfully utilised in biological samples analyses and copper(II) ion imaging in living pea aphids.

Khatua *et al*reported a *bis*-heteroleptic Ru(II) complex of benzimidazole-substituted 1,2,3-triazole pyridine ligand (**18**) [55]. It can detect cationic analyte $Ag^+$and phosphate ions in CH_3 CN solvent upto ppb level. A novel cation sensing mechanism through the unique cyclometalated Ag^+-triazolide complex formation was established where the Ag^+ion is detected through photoluminescence spectroscopy (Fig. **13**).

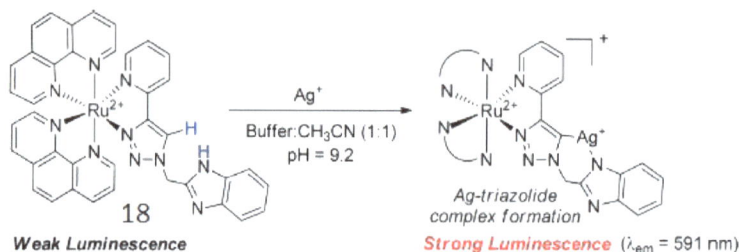

Fig. (13). Sensing of Ag⁺ by Ru(II)-poly pyridyl complex.

Sensing Mechanisms

Similar to the transition metal chemosensors, polypyridyl - Ru(II) complexes which were widely investigated as chemosensors for the detection of cations commonly incorporate a receptor motif on their ancillary ligands for the recognition of cations. Another way of cation sensing is chemodosimeter approach where a probe irreversibly reacts with cations showing some visible colour change and significant variation in its luminescence property. Change in luminiscence intensity, fluorescence lifetime or a shift of emission wavelength are the key factors for detecting optical signal of fluoroscent sensing material. General mechanism for fluorescence signal transduction was attributed to intermolecular charge transfer (ICT), inhibition of photo induced electrontransfer (PET), fluorescence resonance energy transfer (FRET), chelation enhanced fluorescence (CHEF), planar intra-molecular charge transfer (PICT) and twisted intra-molecular/ intermediate charge transfer (TICT). Interaction between the coordination site and the guest moiety utilized hydrogen bonding, electrostatic force, metal-ligand coordination, hydrophobic and van der Waals forces for the development of wide range of sensing materials [56 - 60].

RUTHENIUM COMPLEXES AN ANION SENSOR

Anions are omnipresent throughout biological systems. Among the anions, the detection of fluoride ion is of great interest because of its both beneficial and detrimental effects on health. Excess of fluoride can cause dental and skeletal fluorosis and deficiency causes osteoporosis. Acetate ion is very important for various metabolic processes. Production and oxidation rate of acetate ion have been used as a marker of organic disintegration in oceanic sediments. Phosphate anions play vital role in living organisms. Phosphate-binding proteins (PBP) that occur naturally, combine with hydrogen phosphate strongly and selectively. The uses of nitrite (NO_2^-) ions as intestinal relaxant or laxative medicine, vasodilator, bronchodilator *etc*. are well known [61]. Yet, highly concentrated NO_2^- ions are harmful to living beings. The invention of new phosphorescent metal complex sensors for the detection of anions is an area of growing interest among the supramolecular chemists owing to their significance in biological and environmental processes [62]. Transition metal complexes possess unique geometries as compared to organic molecules for the correct positioning of the ligands in the right conformations in such a manner that receptor "host" groups can efficiently bind the external anion "guest". Among all metal complexes, polypyridyl -Ru(II) complexes are extensively tested for sensing of anion [63]. There are three main approaches in designing of various Ru(II)-polypyridyl complexes for anions. Firstly, to bring in an acidic hydrogen group into the ligand moiety of the Ru(II)-polypyridyl complexes. Hydrogen bonds are formed by weakly basic anions but strongly basic anions can bring about de-protonation. In the second approach, there is specific Lewis acid-base reaction occurs between the Lewis acids and anions. Then chemodosimeter approach is the final one that exploits a precise chemical interaction between a labile group on the metal complex and anion. A sequence of artificial receptors, based on Ru(II)-polypyridyl complex system, have been developed for detection of anions. In this section the sensing of anions by Ru(II)-polypyridyl complexes has been illustrated on the basis of these three main approaches.

Anion Sensing Through Acidic Hydrogen

Gunnlaugsson *et al*. investigated 1,8-naphthalimide derivative of $[Ru(bpy)_3]^{2+}$ complex for anion sensing (19, Fig. **14a**) [64]. In the absence of anions, the complex produced MLCT emission at 615 nm along with a weak shoulder at 505 nm when excited at 432 nm, attributed to the naphthalimide emission, signifying that efficient energy transfer takes place between the naphthalimide singlet excited state and the MLCT. After the addition of anions showed inhibition of emission which affect strongly F^- ion, followed by the other two anions *i.e* AcO^- and Cl^-. An important characteristic of this design is that the anion detection event at the naphthalimide site interferes with the energy transfer mechanism from the naphthalimide moiety to the Ru(II) MLCT excited state, resulting the effective quenching of both the naphthalimide and Ru-centred emissions. Emission titration spectroscopy indicated 1:1 stoichiometry between anion and metal complex where as NMR spectroscopic studies revealed that the anions mainly binds to the 4-amino moiety of the naphthalimide unit through hydrogen bonding. Afterwards, the same group introduce an aryl urea to one of the ligands in $[Ru(bpy)_3]^{2+}$ to synthesize complex (b) for anion detection (20, Fig. **14b**). Surprisingly, opposite behavior was observed. Acetate, phosphate and pyrophosphate ions showed positive response to the MLCT emission but fluoride ion showed absolutely no response [65]. The luminescence intensity of the Ru(II) complex increased after the addition of Phosphate ion, whereas in case of both pyrophosphate and acetate ions emission intensity decreases. Almost complete quenching (>90%) of luminescence was observed for pyrophosphate ion. Acidic hydrogen atoms present in its urea functionality was mainly responsible for the negative behavior of (b) to F^- ions. NMR observations confirmed that there was direct interaction between the urea protons of complex and anions

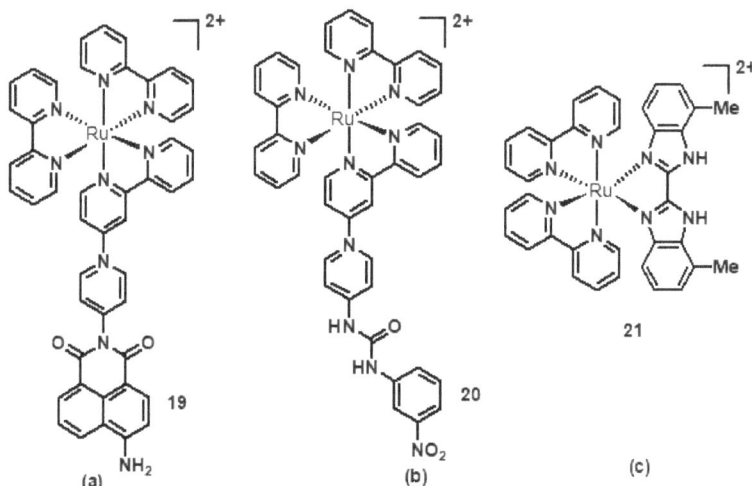

Fig. (14). Ru(II)-polypyridyl complexes as anion sensor.

$[Ru(bpy)_2(DMBbimH_2)](PF_6)_2$ (DMBbimH$_2$ = 7, 7′-dimethyl- 2,2′-bibenzimidazole) has been tested as a luminescent sensor for anions (**21**, **Fig. 14c**) [66]. The imidazole scaffold contains bi-functional unit in which electron lone pairs coordinated to the central metal ion and an -NH group available in the second coordination sphere for hydrogen bonding to anions. $[Ru(bpy)_2(DMBbimH_2)]^{2+}$ exhibited an intermediate pK_a value of 6.2, where as its analogues 2,2′-biimidazole and 2,2′-bibenzimidazole units showed 7.2 and 5.7 respectively reported earlier. The sensing ability of weakly basic anions such as Cl^-, Br^-, I^-, or NO_3^- is due to the moderate acidity of the ligand and hydrogen bond formation avoiding unnecessary deprotonation. After binding with weakly basic anions, there was about 3 to 4 fold enhancement in emission intensity originally at 617 nm with a small positive shift . The increase of emission intensity was attributed to the rigid nature of hydrogen bond present in the -NH group, resulted in the reduction of non-radiative decay of the MLCT state through the -NH bond. On the other hand, the metal complex was de-protonated upon interaction with strongly basic anions such as F^- and AcO^- by decreasing the luminescence intensity. This mainly happens through intramolecular quenching of the emissive MLCT state by electron-transfer from the anionic de-protonated amide moiety to MLCT state. This study revealed that the hydrogen bonding tendency of these complexes can be finely tuned by small changes to the peripheral groups of the anion binding site to attain the desired acidity for anion recognition and separation.

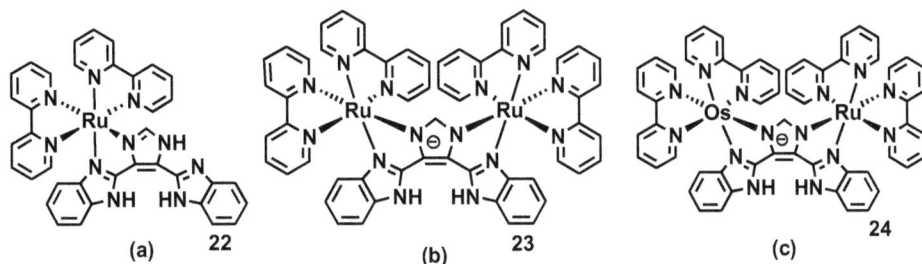

Fig. (15). (**a**) Mono-metallic, (**b**) Homo di-metallic and (**c**) Hetero di-metallic complexes for anion sensing.

Baitalik and co-workers investigated 4, 5-bis(benzimidazole-2-yl)imidazole (H3Imbzim) moiety for constructing monometallic Ru^{II} (22, Fig. **15a**), or homo bimetallic RuII-RuII (23, Fig. **15b**) and hetero metallic Ru(II)-Os(II) complexes (24, Fig. **15c**) for anion detection [67]. The central anionic bridges the two metal centres in the bimetallic complexes, benzimidazole subunits provides both formation of coordinative bond to the metal and -NH groups for anion sensing in the second coordination sphere. The imidazole moiety in the bi-metallic complex works in two different pathways: at the first co ordination sphere the imines scaffold binds to the Ru(II)(bpy)$_2$ unit and at the same time the proximal imidazole NH protons may provide hydrogen bond to the anions in the second coordination sphere which was revealed by the colour change. The chelating structure of the 4, 5-bis(benzimidazole-2-yl)imidazole bidentate ligand enforces a syn conformation for the two -NH groups of the benzimidazole subunits. Again Photo-induced energy transfer takes place from the ruthenium unit to the osmium unit, in case of hetero metallic Ru(II)-Os(II) complex. Luminescence intensity of these complexes were quenched by the addition of F$^-$ and AcO$^-$ ions where as negligible changes was observed in case of other ions .When F$^-$ and AcO$^-$ ions concentration is low in solution, a 1:1 hydrogen bonded complex was formed but at higher concentration of anions, two benzimidazole-NH groups undergoes stepwise de-protonation for anion recognition event.

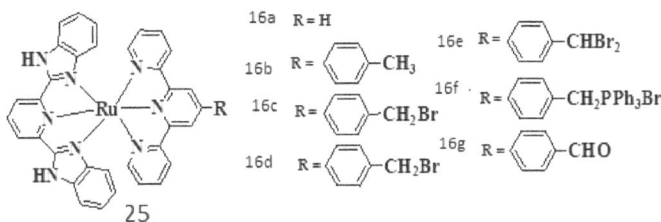

Fig. (16). Ru(II) terpyridine complexes as anion sensor.

The same research group also reported a series of luminescent heteroleptic bis-tridentate [Ru(tpy)(H$_2$pbbzim)]$^{2+}$ complexes derived from tridentate 2,6-bis(benzimidazole-2-yl)pyridine (H$_2$pbbzim) and terpyridine (tpy) motif (Fig. **16**). Electron-withdrawing groups were mainly placed in the tpymoiety. In order to stabilize MLCT state and regulate the electronic properties of the complex. At room temperature [Ru(tpy)$_2$]$^{2+}$ or [Ru(H$_2$pbbzim)$_2$] showed non-luminescent property with very short excited state but after introduction of electron withdrawing groups in the tpy moiety, the heteroleptic complexes showed enhancement in the life time period of excited state. However, weakly basic anions like Cl$^-$, Br$^-$, I$^-$, NO$_3^-$ and ClO$_4^-$ had negligible effect on the luminescence behavior of complexes whereas quenching in emission and positive red shift was observed incase of fluoride and acetate ion. Colorimetric, electrochemical and NMR titration results proved that the complex was prone to F$^-$ and AcO$^-$ ion sensing. The mechanism was explained on the basis of deprotonated -NH proton of the benzimidazole moiety as well as electron-withdrawing character of the metal centre. Subsequently, a series of homoleptic and heteroleptic tridentate Ru (II) complexes containing the phenanthrene-imidazole (HImzphen) moiety were reported by other group for anion recognition (Fig. **17**) [68]. Like the previous results, the luminescence intensity of the complexes was quenched by the accumulation of basic anions.

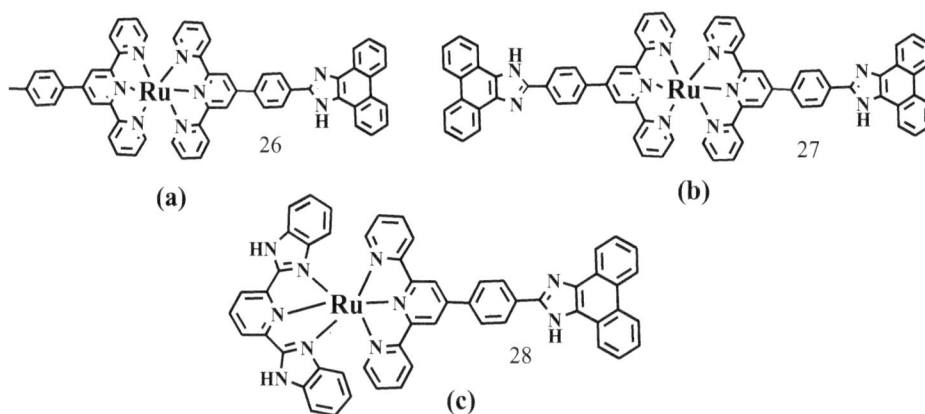

(a)   26   **(b)**   27   **(c)**   28

Fig. (17). Ru(II)-phenanthrene-imidazole complexes as anion sensor.

Interestingly, upon the addition of F$^-$ ion to [Ru (tpy)$_2$]$^{2+}$ derivatives, a remarkable blue -shift was obtained, while the corresponding MLCT absorption band showed a positive red-shift. Different degrees of destabilization of the π* orbital of the tpy-HImzphen and the Ru(II) metal -centred orbital both in the ground state and excited state of complexes are encountered for this shift. The absorption spectra, NMR spectroscopic studies and cyclic voltammetric results can thoroughly explain the anion recognition ability of the Ru(II)-phenanthrene-imidazole complexes.

Shang and co workers showed the potential of Ru -sulfonamido system for the construction of a series of sensors (Fig. **18**). UV-vis, ^{1}H-NMR titration and electrochemical experiments were used to determine the action of these receptors with biologically significant anions [69]. These chromophores exhibited high binding affinity for fluoride (F$^-$) or acetate (AcO$^-$), moderate response for dihydrogen phosphate (H$_2$PO$_4^-$) or hydroxyl (OH$^-$) and almost no effect for chloride (Cl$^-$), bromide (Br$^-$) and iodide (I$^-$) .The hydrogen bond was formed during the coordination between the ligand and anions. The two -NH group of the sulfonamido scaffold in these complexes are in ortho-configuration. DFT studies revealed that there is enough space available in these chromophores for the better fitting of outsized fluoride and acetate anion. Consequently, they have a strong attraction for fluoride and acetate anion. The advantages of using receptors (Fig. **18**) for the recognition of a variety of anions are like this:(1) the probes show strong recognition affinity for F$^-$ and AcO$^-$ ion; (2) The sensors show colour changes in visible light upon binding with F$^-$ and AcO$^-$. Hence, the „sulfonamido complexes **29** can be tested for naked-eye detection of F$^-$ in toothpaste.

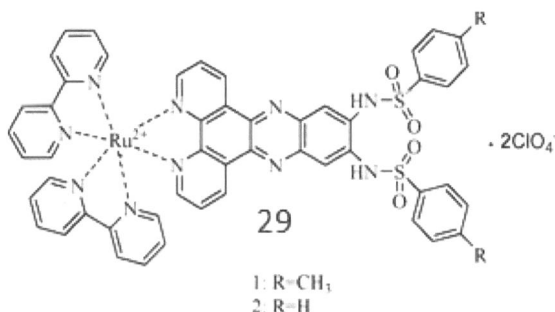

Fig. (18). Ru(II) polypyridylsulfonamido complexes as anion sensor.

The selective phosphate-sensing property of a bisheteroleptic Ru-II complex, 30[PF$_6$]$_2$, having a halogen bonding iodotriazole unit, was demonstrated by Ghosh and co-workers [70], and shown to be superior to its hydrogen-bonding analogue, 31[PF$_6$]$_2$. Complex 30[PF$_6$]$_2$, exploiting halogen-bonding interactions, shows enhanced phosphate recognition in both acetonitrile and aqueous acetonitrile compared with its hydrogen-bonding analogue, owing to considerable amplification of the Ru-IIcenter- based metal-to-ligand charge transfer (MLCT) emission response and luminescence period (Fig. **19**). Detailed solution state studies reveal a higher association constant, lower limit of detection, and greater change in lifetime for complex 30[PF$_6$]$_2$in the presence of phosphates compared with its hydrogen bonding analogue, complex 31[PF$_6$]$_2$.

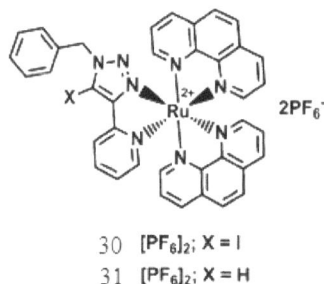

30   [PF₆]₂; X = I
31   [PF₆]₂; X = H

Fig. (19). Ru(II) complexes for phosphate sensing.

Anion Sensing through Lewis Acidic Sites

Lewis acidic groups present in the ligand have the capacity to bind anions effectively. That is why they can be proposed as a metal luminescent sensor for anion recognition. For example, Boron compounds having triple coordination are able to combine F^- and CN^- ion. Such anion binding at the lewis acidic sites, leads to distortion in the π-π interaction around ligand center and also control the photo physical properties of the ligand showing change in emission.

Wang *et al* reported a group of $[Ru(bpy)_3]^{2+}$-based complexes having moeity for sensing of anion [71] (Fig. **20**). The original $[Ru(bpy)_3]^{2+}$ molecule shows emission at 252 nm while the triaryl boron functionalized complex exhibit positive red shift in MLCT emission. LUMO energy of π -interaction between the bpy moiety and electron withdrawing triaryl boron group was supressed which leads to change in MLCT emission. Luminescence energy of t -butyl-bipyridyl complexes was less than their parent bipyridyl analogues, because of excitation of t_{2g} energy level of HOMO by electron- donating t -bu groups. Almost 10- 40 nm blue-shift was observed in the emission after the addition of fluoride or cyanide ions to the complex which could be attributed to shifting of triaryl boron group from electron withdrawing to electron donating upon interaction with analyte. This was the first study in case of Ru(II) complexes which exhibits change of MLCT phosphorescence upon the addition of fluoride and cyanide anions.

N. Singh *et al*. [72] reported one Ru- complex of benzimidazole ligand (Fig. **21**), which can sense chloride ion upto 5 mM level in aqueous CH_3CN solution, and established that its PPh_3 derivatives are not suitable for anion recognition by binding site signalling approach. It has been tested for the determination of chloride concentration in a series of four commercially available oral rehydration salts.

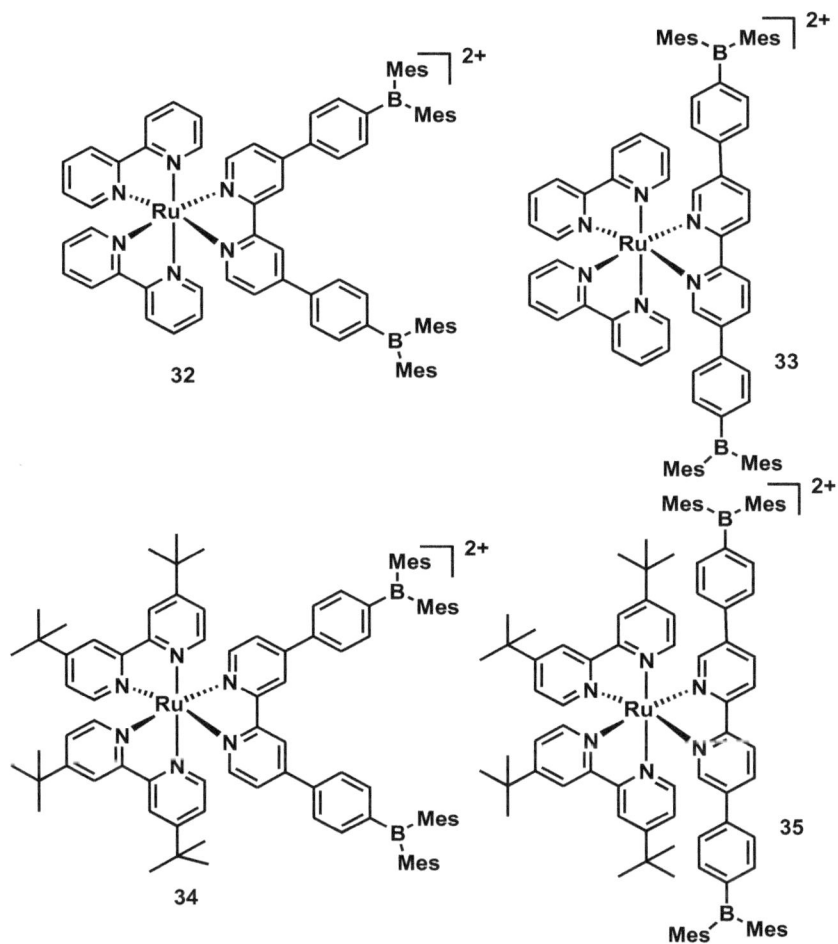

Fig. (20). Ru(II)-polypyridyltriarylboron complexes as anion sensor.

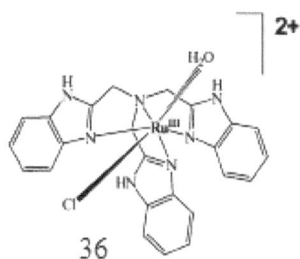

Fig. (21). Ru- complex of benzimidazole for Cl⁻ sensing.

Anion Sensing through Chemodosimeter Approach

Chemodosimeter approaches analyte in two different ways. First step is chemical transformations through breaking of bond and another is new bond formation [73]. Chemodosimeter generally consists of two parts: (i) an active functional moiety, which is performing a chemical transformation in the existence of a specific ion and (ii) functional units that recognise the chemical reaction with a transformed spectroscopic indication. The design of chemodosimeters involving chemical transformation process is highly specific in nature; hence they can be employed for better selectivity of analytes.

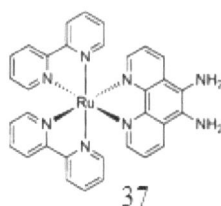

Fig. (22). Ru(II) polypyridyl complex as nitrite anion sensor.

A luminescent chemodosimeter of Ru (II) complex was explored by das and group for nitrite ion recognition (Fig. **22**) [74]. The basis of system was mainly focused on the interaction between N_2O_3 group generated from NO_2^- and a vicinal diamine group present on the phenanthroline moiety, which results in the formation of a new triazole derivative. A positive red shift coupled with doubling of emission intensity was observed in case of triazole derivative in comparision with diamine complexes. Remarkably the probe exhibited a highly selective response to NO_2^- among other tested anions such as NO_3^-, H_2O_2, O_2^- and ClO^- having active oxygen and nitrogen species.

A trinuclearheterobimetallicRu(II)-Cu(II) donor-acceptor complex, {RuII(*tbubpy*)(CN)$_4$-[Cu^{II}(dien)]$_2$}(ClO$_4$)$_2$ (*tbubpy*)4,4'-di-tert-butyl-2,2'-bipyridine; dien) diethylenetriamine) (Fig. **23**), has been successfully designed as a chemodosimeter for recognition of cyanide ion in DMF/water solvent [75]. The MLCT shift and emission properties indicated the solvatochromic behaviour of [RuII(tbubpy)(CN)$_4$]$^{2-}$ which acts as donor coordinated with two Cu(II) acceptors but restored in the presence of CN$^-$. The cleavage of the cyano bridge between Ru(II) and Cu(II) of the chemodosimetric assembly after the interaction of cyanide to the Cu(II) centres was confirmed by NMR studies and mass spectroscopy. The association constant was calculated to be $(7.39 \pm 0.23) \times 10^6$ M^{-1}. A limit of detection 1.2 μM (0.03 ppm) of CN$^-$ in DMF/H$_2$O (pH 7.4) was obtained. The outcome revealed that the anion selectivity of chemodosimeter is encountered with the relative stability of the donor-acceptor complex to that of adducts formed between the acceptor metal centre and the anions.

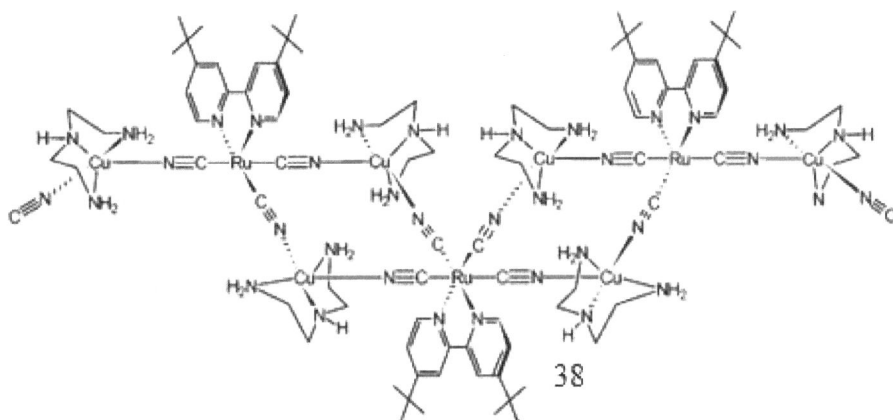

Fig. (23). HeterobimetallicRu(II)-Cu(II) donor-acceptor complex, {RuII(*tbubpy*)(CN)$_4$-[CuII(dien)]$_2$}(ClO$_4$)$_2$ (*tbubpy*)4,4'-di-tert-butyl-2,2'-bipyridine; dien) diethylenetriamine).

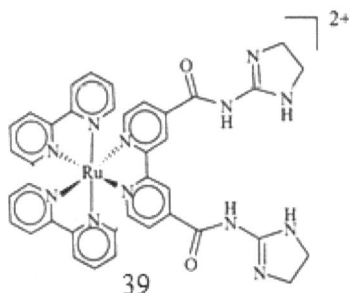

Fig. (24). [Ru(bpy)$_3$]$^{2+}$complex with amide functionalities.

Another [Ru(bpy)$_3$]$^{2+}$ based assembly with amide functionalities (Fig. **24**), reported by Watanabe *et al.*, binds anionic phosphor-diesters [76]. For instance, nearly 1.3 fold enhancement in the emission intensity was marked by addition of ten equivalents of anionic tetraethyl ammonium diphenyl phosphate to a 0.02 mM acetone solution of the complex. The authors proposed that the increase in luminescence intensity is a combining effect of binding-induced rigidification and quenching in vibrational modes responsible for non-radiative decay. Again, another possibility is that anion binding electrostatically destabilizes the coordinated bpy ligand created in the excited MLCT state. Destabilization, sequentially increase the energy gap between excited state and ground state, thereby decreasing the non-radiative decay rate and facilitates the emission quantum yield.

Schmittel *et al*. [77] reported two new ruthenium complexes $[Ru(bipy)_2(PDA)]^{2+}$ (**40**) and $[Ru(phen)_2(PDA)]^{2+}$ (**41**) (PDA = 1,10-phenanthroline-4,7-dicarboxaldehyde) (Fig. **25**) for the detection of cyanide ion in pure CH_3CN and aqueous CH_3CN resulted in construction of cyanohydrins both by absorption and emission spectroscopy. The limit of detection reached upto sub-micromolar level. The mechanism of cyanide sensing has been given in Scheme **2**.

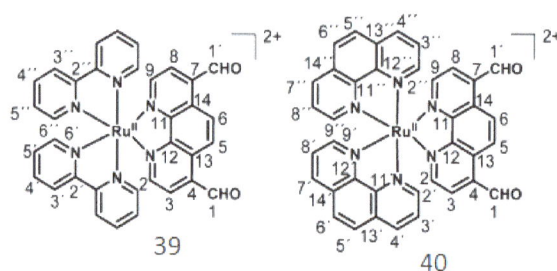

Fig. (25). Ru(II) complexes for cyanide sensing.

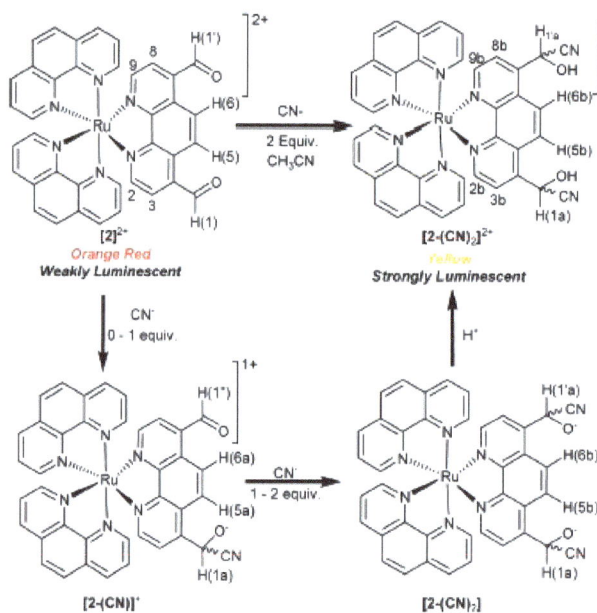

Scheme 2. Sensing Mechanism of cyanide sensing.

CONCLUSION & PERSPECTIVE

Ru(II) polypyridyl complexes provides a wide range of attractive benefits that make them superior to organic molecules as effective chemosensors. At first, they have large emission quantum yields and long phosphorescence lifetime that can permit them to behave efficiently in the existence of a high fluorophore unit by manipulating time-resolved emission spectroscopy. Another advantage is that their photophysical properties are extremely responsive to changes in structure or in the environment, such as upon coordination or reaction of the analyte with a recognition motif conjugated to a supporting ligand. Lastly, the fine-tuning of their chemical and/or photophysical properties are modulated through efficient synthesis of a variety of analogues bearing different functional groups. In this chapter, recent reported examples of Ru(II) polypyridylcomplex chemosensors for both cations and anions in the last couple of years are mostly highlighted. Marching ahead, one possible criteria for the betterment of developed fluorescent chemosensors with higher selectivity and sensitivity was the potentiality of real sample analysis. Although, the number of Ru(II) polypyridyl complex sensors are comparatively greater than that of other metal complexes utilized for detection of ions, still there is a scarcity of those Ru(II) polypyridyl complex sensors which can recognize metal cations and anions in biological and environmental samples both *in vivo* and *in vitro*. They can also provide fluorescence image of metal cation-contaminated living cells, stained with the complex solution. Moreover, highly advantageous probes excited at visible region for intracellular detection *in vivo* and trace level determination of metal cations are still deficient. Hence, there is a massive opportunity for the researchers to investigate in this broad arena that remains still unexplored. At last but not least, naked eye detection of metal cations and anions at trace level is highly desirable.

CONSENT FOR PUBLICATION

Not applicable.

CONFLICT OF INTEREST

The author confirms that he has no conflict of interest to declare for this publication.

ACKNOWLEDGEMENTS

Authors are grateful to Guru GhasidasVishwavidyalaya, Bilaspur for various help.

REFERENCES

[1]   Keefe, M.H.; Benkstein, K.D.; Hupp, J.T. Luminescent sensor molecules based on coordinated metals: a review of recent developments. *Coord. Chem. Rev.,* **2000**, *205*, 201-228.

[http://dx.doi.org/10.1016/S0010-8545(00)00240-X]

[2] Sun, S-S.; Lees, A.J. Transition metal based supramolecular systems: synthesis, photophysics, photochemistry and their potential applications as luminescent anion chemosensors. *Coord. Chem. Rev.,* **2002**, *230*, 171171-171192.
[http://dx.doi.org/10.1016/S0010-8545(02)00043-7]

[3] Rogers, C.W.; Wolf, M.O. Luminescent molecular sensors based on analyte coordination to transition-metal complexes. *Coord. Chem. Rev.,* **2002**, *341*, 233-234.

[4] Fletcher, N.C.; Lagunas, M.C.; Bozecin, H.L. , **2010**.

[5] Zhou, G.; Wong, W-Y.; Yang, X. New design tactics in OLEDs using functionalized 2-phenylpyridine-type cyclometalates of iridium(III) and platinum(II). *Chem. Asian J.,* **2011**, *6*(7), 1706-1727.
[http://dx.doi.org/10.1002/asia.201000928] [PMID: 21557486]

[6] Happ, B.; Winter, A.; Hager, M.D.; Schubert, U.S. Photogenerated avenues in macromolecules containing Re(I), Ru(II), Os(II), and Ir(III) metal complexes of pyridine-based ligands. *Chem. Soc. Rev.,* **2012**, *41*(6), 2222-2255.
[http://dx.doi.org/10.1039/C1CS15154A] [PMID: 22080248]

[7] Kalinowski, J.; Fattori, V.; Cocchi, M.; Williams, J.A.G. Light-emitting devices based on organometallic platinum complexes as emitters. *Coord. Chem. Rev.,* **2011**, *255*, 2401-2425.
[http://dx.doi.org/10.1016/j.ccr.2011.01.049]

[8] You, Y.; Park, S.Y. Phosphorescent iridium(III) complexes: toward high phosphorescence quantum efficiency through ligand control. *Dalton Trans.,* **2009**, (8), 1267-1282.
[http://dx.doi.org/10.1039/B812281D] [PMID: 19462644]

[9] Yam, V.W-W.; Wong, K.M-C. Luminescent metal complexes of d^6, d^8 and d^{10} transition metal centres. *Chem. Commun. (Camb.),* **2011**, *47*(42), 11579-11592.
[http://dx.doi.org/10.1039/c1cc13767k] [PMID: 21956216]

[10] Zhao, Q.; Huang, C.; Li, F. Phosphorescent heavy-metal complexes for bioimaging. *Chem. Soc. Rev.,* **2011**, *40*(5), 2508-2524.
[http://dx.doi.org/10.1039/c0cs00114g] [PMID: 21253643]

[11] Boal, A.K.; Rosenzweig, A.C. Structural biology of copper trafficking. *Chem. Rev.,* **2009**, *109*(10), 4760-4779.
[http://dx.doi.org/10.1021/cr900104z] [PMID: 19824702]

[12] Li, Y.; Zamble, D.B. Nickel homeostasis and nickel regulation: an overview. *Chem. Rev.,* **2009**, *109*(10), 4617-4643.
[http://dx.doi.org/10.1021/cr900010n] [PMID: 19711977]

[13] Yokel, R.A. Toxicity of aluminum exposure to the neonatal and immature rabbit. *Fundam. Appl. Toxicol.,* **1987**, *9*(4), 795-806.
[http://dx.doi.org/10.1016/0272-0590(87)90186-2] [PMID: 3692032]

[14] Muller, G.; Bernuzzi, V.; Desor, D.; Hutin, M.F.; Burnel, D.; Lehr, P.R. Developmental alterations in offspring of female rats orally intoxicated by aluminum lactate at different gestation periods. *Teratology,* **1990**, *42*(3), 253-261.
[http://dx.doi.org/10.1002/tera.1420420309] [PMID: 2274891]

[15] Archer, T.; Hiltunen, A.J.; Järbe, T.U.; Kamkar, M.R.; Luthman, J.; Sundström, E.; Teiling, A. Hyperactivity and instrumental learning deficits in methylazoxymethanol-treated rat offspring. *Neurotoxicol. Teratol.,* **1988**, *10*(4), 341-347.
[http://dx.doi.org/10.1016/0892-0362(88)90037-2] [PMID: 3226377]

[16] Alstad, N.E.W.; Kjelsberg, B.M.; Vøllestad, L.A.; Lydersen, E.; Poléo, A.B.S. The significance of water ionic strength on aluminium toxicity in brown trout (Salmo trutta L.). *Environ. Pollut.,* **2005**, *133*(2), 333-342.

[http://dx.doi.org/10.1016/j.envpol.2004.05.030] [PMID: 15519464]

[17]   Banerjee, A.; Sahana, A.; Das, S.; Lohar, S.; Guha, S.; Sarkar, B.; Mukhopadhyay, S.K.; Mukherjee, A.K.; Das, D. A naphthalene exciplex based Al^{3+} selective on-type fluorescent probe for living cells at the physiological pH range: experimental and computational studies. *Analyst (Lond.)*, **2012**, *137*(9), 2166-2175.
[http://dx.doi.org/10.1039/c2an16233d] [PMID: 22416274]

[18]   Das, S.; Karak, D.; Lohar, S.; Banerjee, A.; Sahana, A.; Das, D. Interaction of a naphthalene based fluorescent probe with Al^{3+}: experimental and computational studies. *Anal. Methods*, **2012**, *4*, 3620-3624.
[http://dx.doi.org/10.1039/c2ay25825k]

[19]   Williams, R.J.P. Recent aspects of aluminium chemistry and biology: a survey. *Coord. Chem. Rev.*, **2002**, *228*, 93-97.
[http://dx.doi.org/10.1016/S0010-8545(02)00072-3]

[20]   Baral, M.; Sahoo, S.K.; Kanungo, B.K. Tripodal amine catechol ligands: a fascinating class of chelators for aluminium(III). *J. Inorg. Biochem.*, **2008**, *102*(8), 1581-1588.
[http://dx.doi.org/10.1016/j.jinorgbio.2008.02.006] [PMID: 18472165]

[21]   Kawahara, M.; Muramoto, K.; Kobayashi, K.; Mori, H.; Kuroda, Y. Aluminum promotes the aggregation of Alzheimer's amyloid beta-protein in vitro. *Biochem. Biophys. Res. Commun.*, **1994**, *198*(2), 531-535.
[http://dx.doi.org/10.1006/bbrc.1994.1078] [PMID: 7507666]

[22]   Paik, S.R.; Lee, J.H.; Kim, D.H.; Chang, C.S.; Kim, J. Aluminum-induced structural alterations of the precursor of the non-A beta component of Alzheimer's disease amyloid. *Arch. Biochem. Biophys.*, **1997**, *344*(2), 325-334.
[http://dx.doi.org/10.1006/abbi.1997.0207] [PMID: 9264546]

[23]   Lin, J.L.; Kou, M.T.; Leu, M.L. Effect of long term low-dose aluminum-containing agents on hemoglobin synthesis in patients with chronic renal insufficiency. *Nephron*, **1996**, *74*(1), 33-38.
[http://dx.doi.org/10.1159/000189278] [PMID: 8883017]

[24]   Good, P.F.; Olanow, C.W.; Perl, D.P. Neuromelanin-containing neurons of the substantia nigra accumulate iron and aluminum in Parkinson's disease: a LAMMA study. *Brain Res.*, **1992**, *593*(2), 343-346.
[http://dx.doi.org/10.1016/0006-8993(92)91334-B] [PMID: 1450944]

[25]   Yokel, R.A. Ruthenium d-orbital delocalization in bis(bipyridine)ruthenium derivatives of redox active quinonoid ligands. *Coord. Chem. Rev.*, **2002**, *228*, 97-113.
[http://dx.doi.org/10.1016/S0010-8545(02)00078-4]

[26]   Darbre, P.D. Aluminium, antiperspirants and breast cancer. *J. Inorg. Biochem.*, **2005**, *99*(9), 1912-1919.
[http://dx.doi.org/10.1016/j.jinorgbio.2005.06.001] [PMID: 16045991]

[27]   Rout, G.R.; Roy, S.S.; Das, P. Aluminium toxicity in plants: a review. *Agronomie*, **2001**, *21*, 3-21.
[http://dx.doi.org/10.1051/agro:2001105]

[28]   Brown, D.R.; Kozlowski, H. Biological inorganic and bioinorganic chemistry of neurodegeneration based on prion and Alzheimer diseases. *Dalton Trans.*, **2004**, (13), 1907-1917.
[http://dx.doi.org/10.1039/b401985g] [PMID: 15252577]

[29]   Li, Y.; Zamble, D.B. Nickel homeostasis and nickel regulation: an overview. *Chem. Rev.*, **2009**, *109*(10), 4617-4643.
[http://dx.doi.org/10.1021/cr900010n] [PMID: 19711977]

[30]   Comba, P.; Krämer, R.; Mokhir, A.; Naing, K.; Schatz, E. Synthesis of New Phenanthroline-Based Heteroditopic Ligands - Highly Efficient and Selective Fluorescence Sensors for Copper(II) Ions. *Eur. J. Inorg. Chem.*, **2006**, 4442-4448.

[http://dx.doi.org/10.1002/ejic.200600469]

[31]   Patra, S.; Boricha, V.P.; Sreenidhi, K.R.; Suresh, E.; Paul, P. Luminescent metalloreceptors with pendant macrocyclic ionophore: Synthesis, characterization, electrochemistry and ion-binding study. *Inorg. Chim. Acta,* **2010**, *363*, 1639-1648.
[http://dx.doi.org/10.1016/j.ica.2010.01.003]

[32]   Zhang, Y.; Liu, Z.; Zhang, Y.; Xu, Y.; Li, H.; Wang, C.; Lu, A. Sun, A reversible and selective luminescent probe for Cu^{2+} detection based on a ruthenium(II) complex in aqueous solution. S. *Sens. Actuators B Chem.,* **2015**, *211*, 449-455.
[http://dx.doi.org/10.1016/j.snb.2015.01.116]

[33]   Cheng, F.; He, C.; Ren, M.; Wang, F.; Yang, Y. Two dinuclear Ru(II) polypyridyl complexes with different photophysical and cation recognition properties. *Spectrochim. Acta A Mol. Biomol. Spectrosc.,* **2015**, *136*(Pt B), 845-851.
[http://dx.doi.org/10.1016/j.saa.2014.09.103] [PMID: 25459607]

[34]   Zheng, Z.B.; Duan, Z.M.; Ma, Y.Y.; Wang, K.Z. Highly sensitive and selective difunctional ruthenium(II) complex-based chemosensor for dihydrogen phosphate anion and ferrous cation. *Inorg. Chem.,* **2013**, *52*(5), 2306-2316.
[http://dx.doi.org/10.1021/ic301555r] [PMID: 23410247]

[35]   Clarkson, T.W.; Magos, L.; Myers, G.J. The toxicology of mercury--current exposures and clinical manifestations. *N. Engl. J. Med.,* **2003**, *349*(18), 1731-1737.
[http://dx.doi.org/10.1056/NEJMra022471] [PMID: 14585942]

[36]   Wang, Y.; Yang, F.; Yang, X. Colorimetric biosensing of mercury(II) ion using unmodified gold nanoparticle probes and thrombin-binding aptamer. *Biosens. Bioelectron.,* **2010**, *25*(8), 1994-1998.
[http://dx.doi.org/10.1016/j.bios.2010.01.014] [PMID: 20138750]

[37]   Yang, Q.; Tan, Q.; Zhou, K.; Xu, K.; Hou, X.J. Direct detection of mercury in vapor andaerosol from chemical atomization and nebulization at ambient temperature: exploiting the flame atomic absorption spectrometer. *J. Anal. At. Spectrom.,* **2005**, *20*, 760-762.
[http://dx.doi.org/10.1039/b505298j]

[38]   Hamid, A.A.; Amer, A.G.; Mohammad, A.K.; Ziyad, A.; Mahmoud, Q.; Safwan, O.; Rawashdeh, M.; Monem, A.A. Selective chemosensor for mercuric ions based on 4-aminothiophenol-ruthenium(II) bis(bipyridine) complex. *Int. J. Inorg. Chem.,* **2011**, 843051.
[http://dx.doi.org/10.1155/2011/843051]

[39]   Fan, S-H.; Shen, J.; Wu, H.; Wang, K-Z.; Zhang, A-G. A highly selective turn-on colorimetric and luminescence sensor based on triphenylamine-appended ruthenium(II) dye for detecting mercury ion. *Chin. Chem. Lett.,* **2015**, *26*, 580-584.
[http://dx.doi.org/10.1016/j.cclet.2014.11.031]

[40]   Das, S.; Karmakar, S.; Mardanya, S.; Baitalik, S. Synthesis, structural characterization, and multichannel anion and cation sensing studies of a bifunctional Ru(II) polypyridyl-imidazole based receptor. *Dalton Trans.,* **2014**, *43*(9), 3767-3782.
[http://dx.doi.org/10.1039/c3dt52424h] [PMID: 24441933]

[41]   Palomares, E.; Vilar, R.; Durrant, J.R. *Heterogenious colorimetric sensor for mercuric salts, Chem*; Comm, **2004**, pp. 362-363.

[42]   Coronado, E.; Galán-Mascarós, J.R.; Martí-Gastaldo, C.; Palomares, E.; Durrant, J.R.; Vilar, R.; Gratzel, M.; Nazeeruddin, M.K. Reversible colorimetric probes for mercury sensing. *J. Am. Chem. Soc.,* **2005**, *127*(35), 12351-12356.
[http://dx.doi.org/10.1021/ja0517724] [PMID: 16131215]

[43]   He, C.L.; Ren, F.L.; Zhang, X.B.; Dong, Y.Y.; Zhao, Y. A fluorescent chemosensor for copper(II) based on a carboxylic acid-functionalized tris(2,2'-bipyridine)-ruthenium(II) complex. *Anal. Sci.,* **2006**, *22*(12), 1547-1551.
[http://dx.doi.org/10.2116/analsci.22.1547] [PMID: 17159313]

[44]   Kumar, M.; Nyk, R.; Ohulchanskyy, T.Y.; Flask, C.A.; Prasad, P.N. Combined Optical and MR Bioimaging Using Rare Earth Ion Doped NaYF4 Nanocrystals. *Adv. Funct. Mater.,* **2009**, *19*, 853-859. [http://dx.doi.org/10.1002/adfm.200800765]

[45]   Nyk, M.; Kumar, R.; Ohulchanskyy, T.Y.; Bergey, E.J.; Prasad, P.N. High contrast in vitro and in vivo photoluminescence bioimaging using near infrared to near infrared up-conversion in Tm3+ and Yb3+ doped fluoride nanophosphors. *Nano Lett.,* **2008**, *8*(11), 3834-3838. [http://dx.doi.org/10.1021/nl802223f] [PMID: 18928324]

[46]   Xiong, L.; Chen, Z.; Tian, Q.; Cao, T.; Xu, C.; Li, F. High contrast upconversion luminescence targeted imaging *in vivo* using peptide-labeled nanophosphors. *Anal. Chem.,* **2009**, *81*(21), 8687-8694. [http://dx.doi.org/10.1021/ac901960d] [PMID: 19817386]

[47]   Idris, N.M.; Li, Z.; Ye, L.; Sim, E.K.W.; Mahendran, R.; Ho, P.C.L.; Zhang, Y. Tracking transplanted cells in live animal using upconversion fluorescent nanoparticles. *Biomaterials,* **2009**, *30*(28), 5104-5113. [http://dx.doi.org/10.1016/j.biomaterials.2009.05.062] [PMID: 19539368]

[48]   Wang, C.; Cheng, L.; Liu, Z. Drug delivery with upconversion nanoparticles for multi-functional targeted cancer cell imaging and therapy. *Biomaterials,* **2011**, *32*(4), 1110-1120. [http://dx.doi.org/10.1016/j.biomaterials.2010.09.069] [PMID: 20965564]

[49]   Zhou, J.; Sun, Y.; Du, X.; Xiong, L.; Hu, H.; Li, F. Dual-modality in vivo imaging using rare-earth nanocrystals with near-infrared to near-infrared (NIR-to-NIR) upconversion luminescence and magnetic resonance properties. *Biomaterials,* **2010**, *31*(12), 3287-3295. [http://dx.doi.org/10.1016/j.biomaterials.2010.01.040] [PMID: 20132982]

[50]   Cao, T.; Yang, Y.; Gao, Y.; Zhou, J.; Li, Z.; Li, F. High-quality water-soluble and surface-functionalized upconversion nanocrystals as luminescent probes for bioimaging. *Biomaterials,* **2011**, *32*(11), 2959-2968. [http://dx.doi.org/10.1016/j.biomaterials.2010.12.050] [PMID: 21262531]

[51]   Yu, M.; Li, F.; Chen, Z.; Hu, H.; Zhan, C.; Yang, H.; Huang, C. Laser scanning up-conversion luminescence microscopy for imaging cells labeled with rare-earth nanophosphors. *Anal. Chem.,* **2009**, *81*(3), 930-935. [http://dx.doi.org/10.1021/ac802072d] [PMID: 19125565]

[52]   Park, Y.; Kim, J.H.; Lee, K.T.; Jeon, K.S.; Na, H.B.; Yu, J.H.; Kim, H.M.; Lee, N.; Choi, S.H.; Aik, S.B. Nonblinking and NonbleachingUpconverting Nanoparticles as an Optical Imaging Nanoprobe and T1 Magnetic Resonance Imaging Contrast Agent. *Adv. Mater.,* **2009**, *21*, 4467-4471. [http://dx.doi.org/10.1002/adma.200901356]

[53]   Liu, Q.; Peng, J.; Sun, L.; Li, F. High-efficiency upconversion luminescent sensing and bioimaging of Hg(II) by chromophoric ruthenium complex-assembled nanophosphors. *ACS Nano,* **2011**, *5*(10), 8040-8048. [http://dx.doi.org/10.1021/nn202620u] [PMID: 21899309]

[54]   Zhang, Y.; Liu, Z.; Yang, K.; Zhang, Y.; Xu, Y.; Li, C.; Wang, Y.; Lu, A.; Sun, S. A ruthenium(II) complex as turn-on Cu(II) luminescent sensor based on oxidation cyclization mechanism and its application *in vivo*. *Sci. Rep.,* **2015**, *5*, 8172. [http://dx.doi.org/10.1038/srep08172]

[55]   Sheet, S.K.; Sen, B.; Thounaojam, R.; Aguan, K.; Khatua, S. Ruthenium(II) complex-based luminescent bifunctional probe for Ag$^+$ and phosphate ions: Ag$^+$-assisted detection and imaging of rRNA. *Inorg. Chem.,* **2017**, *56*(3), 1249-1263. [http://dx.doi.org/10.1021/acs.inorgchem.6b02343] [PMID: 28098980]

[56]   Kubik, S.; Goddard, R.; Otto, S.; Pohl, S.; Reyheller, C.; Stüwe, S. Optimization of the binding properties of a synthetic anion receptor using rational and combinatorial strategies. *Biosens. Bioelectron.,* **2005**, *20*(11), 2364-2375. [http://dx.doi.org/10.1016/j.bios.2004.01.035] [PMID: 15797340]

[57]   Namasivayam, C.; Sangeetha, D.; Gunasekaran, R. Removal of Anions, Heavy Metals, Organics and Dyes from Water by Adsorption onto a New Activated Carbon from Jatropha Husk, an Agro-Industrial Solid Waste. *Process Saf. Environ.,* **2007**, *85*, 181-184.
[http://dx.doi.org/10.1205/psep05002]

[58]   *Supramolecular Chemistry of Anions*; Wiley-VCH: Weinheim, **1997**.

[59]   Beer, P.D.; Gale, P.A. Anion Recognition and Sensing: The State of the Art and Future Perspectives. *Angew. Chem. Int. Ed. Engl.,* **2001**, *40*(3), 486-516.
[http://dx.doi.org/10.1002/1521-3773(20010202)40:3<486::AID-ANIE486>3.0.CO;2-P]   [PMID: 11180358]

[60]   Steed, J.W. Coordination and organometallic compounds as anion receptors and sensors. *Chem. Soc. Rev.,* **2009**, *38*(2), 506-519.
[http://dx.doi.org/10.1039/B810364J] [PMID: 19169464]

[61]   Wenzel, M.; Hiscock, J.R.; Gale, P.A. Anion receptor chemistry: highlights from 2010. *Chem. Soc. Rev.,* **2012**, *41*(1), 480-520.
[http://dx.doi.org/10.1039/C1CS15257B] [PMID: 22080279]

[62]   Moragues, M.E.; Martínez-Máñez, R.; Sancenón, F. Chromogenic and fluorogenic chemosensors and reagents for anions. A comprehensive review of the year 2009. *Chem. Soc. Rev.,* **2011**, *40*(5), 2593-2643.
[http://dx.doi.org/10.1039/c0cs00015a] [PMID: 21279197]

[63]   Beer, P.D.; Hayes, E.J. Transition metal and organometallic anion complexation agents. *Coord. Chem. Rev.,* **2003**, *240*, 167-189.
[http://dx.doi.org/10.1016/S0010-8545(02)00303-X]

[64]   Elmes, R.B.P.; Gunnlaugsson, T. Luminescence anion sensing *via* modulation of MLCT emission from a naphthalimide-Ru(II)-polypyridyl complex. *Tetrahedron Lett.,* **2010**, *51*, 4082-4087.
[http://dx.doi.org/10.1016/j.tetlet.2010.05.127]

[65]   Kitchen, J.A.; Boyle, E.M.; Gunnlaugsson, T. Synthesis, structural characterisation and luminescent anion sensing studies of a Ru(II)polypyridyl complex featuring an aryl urea derivatised 2,2'-bpy auxiliary ligand. *Inorg. Chim. Acta,* **2012**, *381*, 236-242.
[http://dx.doi.org/10.1016/j.ica.2011.10.026]

[66]   Mo, H.J.; Niu, Y.L.; Zhang, M.; Qiao, Z.P.; Ye, B.H. Photophysical, electrochemical and anion sensing properties of Ru(II) bipyridine complexes with 2,2'-biimidazole-like ligand. *Dalton Trans.,* **2011**, *40*(32), 8218-8225.
[http://dx.doi.org/10.1039/c0dt01446j] [PMID: 21731958]

[67]   Saha, D.; Das, S.; Bhaumik, C.; Dutta, S.; Baitalik, S. Monometallic and bimetallic ruthenium(II) complexes derived from 4,5-bis(benzimidazol-2-yl)imidazole (H3Imbzim) and 2,2'-bipyridine as colorimetric sensors for anions: synthesis, characterization, and binding studies. *Inorg. Chem.,* **2010**, *49*(5), 2334-2348.
[http://dx.doi.org/10.1021/ic9022272] [PMID: 20170201]

[68]   Bhaumik, C.; Saha, D.; Das, S.; Baitalik, S. Synthesis, structural characterization, photophysical, electrochemical, and anion-sensing studies of luminescent homo- and heteroleptic ruthenium(II) and osmium(II) complexes based on terpyridyl-imidazole ligand. *Inorg. Chem.,* **2011**, *50*(24), 12586-12600.
[http://dx.doi.org/10.1021/ic201610w] [PMID: 22098484]

[69]   Shang, X.F.; Li, J.; Lin, H.; Jiang, P.; Cai, Z.S.; Lin, H.K. Anion recognition and sensing of ruthenium(II) and cobalt(II) sulfonamido complexes. *Dalton Trans.,* **2009**, (12), 2096-2102.
[http://dx.doi.org/10.1039/b804445g] [PMID: 19274287]

[70]   Chowdhury, B.; Khatua, S.; Dutta, R.; Chakraborty, S.; Ghosh, P. Bis-heteroleptic ruthenium(II) complex of a triazole ligand as a selective probe for phosphates. *Inorg. Chem.,* **2014**, *53*(15), 8061-

8070.
[http://dx.doi.org/10.1021/ic5010598] [PMID: 25052811]

[71] Sun, Y.; Hudson, Z.M.; Rao, Y.; Wang, S. Tuning and switching MLCT phosphorescence of [Ru(bpy)$_3$]$^{2+}$ complexes with triarylboranes and anions. *Inorg. Chem.,* **2011**, *50*(8), 3373-3378.
[http://dx.doi.org/10.1021/ic1021966] [PMID: 21413739]

[72] Sharma, H.; Guadalupe, H.J.; Narayanan, J.; Hofeld, H.; Pandiyan, T.; Singh, N. Pyridyl- and benzimidazole-based ruthenium(III) complex for selective chloride recognition through fluorescence spectroscopy. *Anal. Methods,* **2013**, *5*, 3880-3887.
[http://dx.doi.org/10.1039/c3ay40434j]

[73] Yang, Y.; Zhao, Q.; Feng, W.; Li, F. Luminescent chemodosimeters for bioimaging. *Chem. Rev.,* **2013**, *113*(1), 192-270.
[http://dx.doi.org/10.1021/cr2004103] [PMID: 22702347]

[74] Ghosh, A.; Das, P.; Saha, S.; Banerjee, T.; Bhatt, H.B.; Das, A. Diamine derivative of Ru(II)-polypyridyl complex for chemodosimetric detection of nitrate ion in aqueous solution. *Inorg. Chim. Acta,* **2011**, *372*, 115-119.
[http://dx.doi.org/10.1016/j.ica.2011.01.066]

[75] Chow, C.F.; Lam, M.H.; Wong, W.Y. A chemodosimetric probe based on a conjugated oxidized bis-indolyl system for selective naked-eye sensing of cyanide ions in water. *Inorg. Chem.,* **2004**, *43*, 8387-8393.
[http://dx.doi.org/10.1021/ic0492587] [PMID: 15606187]

[76] Watanabe, S.; Onogawa, O.; Komatsu, Y.; Yoshida, K. Luminescent metaloreceptor with a neutral bis(acylaminoimidazoline) binding site: optical sensing of anionic and neural phosphodiesters. *J. Am. Chem. Soc.,* **1998**, *120*, 229-230.
[http://dx.doi.org/10.1021/ja973263a]

[77] Khatua, S.; Samanta, D.; Bats, J.W.; Schmittel, M. Rapid and highly sensitive dual-channel detection of cyanide by bis-heteroleptic ruthenium(II) complexes. *Inorg. Chem.,* **2012**, *51*(13), 7075-7086.
[http://dx.doi.org/10.1021/ic2022853] [PMID: 22416978]

SUBJECT INDEX

A

Achiral stationary phase 14, 15
Acids 18 19, 96, 42, 43, 44, 45, 49, 50, 51, 81,
    82, 97, 136, 142, 147, 151, 153, 155
    acetic 42, 43, 44, 45, 49, 50, 51, 136, 142,
        147, 155, 237
    lactic 96, 151, 153, 155
    phenolic 81, 82, 97
    phenylacetic 18, 19
    succinic 237
Acrylamide 80, 81, 91, 103
Acrylonitrile 142, 147, 149, 151, 153, 155,
    159, 161
Activated carbon 152, 154
    conventional 154
Activation analysis 179, 180, 186, 194, 196,
    203, 208, 215
    radiochemical neutron 196, 208
Activity 52, 187, 189, 202, 208
    esterase 52
    induced 187, 189, 202
    strong 208
Affinity membrane 134, 159, 160
Air 152, 153, 154, 215
    filtration applications 152, 153, 154
    pollution 152, 215
Albumin 233, 248, 257
Alcohols 42, 142, 147, 148, 153, 155, 159
    furfuryl 42
    vinyl 142, 147, 148, 153, 155, 159
Amphetamine 3, 7, 9, 15, 17, 18, 19, 20, 21,
    22, 23, 24, 26, 29, 31, 32
    acetylated 21
    enantiomers 24, 26, 31
    illicit 18, 21
Amplifier gain 199, 200
Amylopectin 228, 230, 253, 254, 257
    tris 253
Amylose 23, 229, 230, 244, 247, 249, 255,
    256

Analytical 1, 18, 62, 74, 80, 143 181, 208,
    212, 216
    chemistry, modern 62, 74
    techniques, powerful 212, 216
Anion 33, 274, 275, 276, 277, 278, 279, 280,
    281, 282, 283
    acetate 279
    polymeric 33
    recognition 276, 278, 280
    sensor 274, 276, 277, 278, 279, 281
Aqueous 63, 73, 75, 76, 77, 78, 79, 82, 85, 88,
    89, 90, 91, 92, 98, 269, 270, 273
    extract 77, 78, 79, 88, 90, 91
    sample matrix 63
Aromatic amines 69, 70, 77, 87, 89, 98, 100,
    101, 102, 108, 148
    water HPLC-UV 69, 70
    hydrocarbons, polycyclic 77, 87, 100, 108
Association of racing commissioners
    international (ARCI) 24
Asymmetric 32, 96, 228, 229
    synthesis 228, 229
    aerosol sample 96
    pressure chemical ionization (APCI) 32

B

Beam intensity 206, 207
Benzimidazole subunits 277
Biodiesel industry 42, 44
Biological 140, 144, 192, 208, 215, 273
    samples analyses 273
Biomass, lignocellulosic 41, 42
Biosensors, enzymatic 144, 146, 147

C

Capillary 1, 12, 26, 27, 28, 29, 107, 239, 240
    electro chromatography (CEC) 239, 240
    electrophoresis 1, 12, 26, 27, 28, 29, 107
Carbon nanotubes, multiwalled 85, 107, 147,
    148, 151

www.ingramcontent.com/pod-product-compliance
Lightning Source LLC
Chambersburg PA
CBHW050811220326
41598CB00006B/174